THE
COAST GUARDSMAN'S MANUAL

11th Edition

THE
COAST GUARDSMAN'S MANUAL

11th Edition

Jim Dolbow

NAVAL INSTITUTE PRESS
ANNAPOLIS, MARYLAND

Naval Institute Press
291 Wood Road
Annapolis, MD 21402

First edition published 1952. Tenth edition 2012.

ISBN: 978-1-68247-189-0 (paperback)
ISSN 0530-0045 (print)

♾ Print editions meet the requirements of ANSI/NISO z39.48-1992 (Permanence of
Paper).
Printed in the United States of America.

Unless otherwise noted, all photographs are official U.S. Coast Guard photographs.

25 24 23 22 21 20 19 18 17   9 8 7 6 5 4 3 2 1
First printing

SEMPER PARATUS

From Aztec Shore to Arctic Zone, to Europe and Far East,
The Flag is carried by our ships
In times of war and peace;
And never have we struck it yet,
In spite of foemen's might,
Who cheered our crews and cheered again
For showing how to fight.

CHORUS
So here's the Coast Guard marching song,
We sing on land or sea.
Through surf and storm and howling gale,
High shall our purpose be,
"Semper Paratus" is our guide,
Our fame, our glory, too.
To fight to save or fight and die!
Aye! Coast Guard, we are for you.

Surveyor and *Narcissus*,
The *Eagle* and *Dispatch*,
The *Hudson* and the *Tampa*,
These names are hard to match;
From Barrow's shores to Paraguay,
Great Lakes' or Ocean's wave,
The Coast Guard fights through storms and winds
To punish or to save.

Aye! We've been "Always Ready"
To do, to fight, or die!
Write glory to the shield we wear
In letters to the sky.
To sink the foe or save the maimed
Our mission and our pride.
We'll carry on 'til Kingdom Come
Ideals for which we've died.

—*Capt. F. S. Van Boskerck, USCG*

Contents

V. NAUTICAL AND MILITARY SKILLS

VI. SAILORS AT SEA AND ASHORE

VII. PERSONAL FITNESS AND SURVIVAL SKILLS

APPENDICES

Illustrations

The U.S. Coast Guard's Role and Values

The Coast Guard:

- Is one of America's five Armed Services and maintains joint interoperability;
- Is a maritime law enforcement, regulatory, environmental, and humanitarian agency;
- Is the maritime operating arm of the Department of Homeland Security;
- Is a 24/7 maritime first responder—Always Ready—first on scene when crises strike;
- Has provided a persistent presence, from inland/inshore waters through the littorals to offshore waters, for over 226 years;
- Is locally based, nationally deployed, and globally connected;
- Builds and maintains robust partnerships to ensure unity of effort at all levels;
- Influences global maritime organizations and industry to benefit U.S. economic, safety, security, and environmental interests; and
- Projects presence around the globe in support of national objectives.

We do all of this with the same people and assets, ready and agile to shift among missions as needed. We have the impact of many agencies in one. We are a unique, effective, and efficient instrument of national security.

No one else can do all that we do. This is what we do. We are the United States Coast Guard.

COAST GUARD CORE VALUES

Honor: Integrity is our standard. We demonstrate uncompromising ethical conduct and moral behavior in all of our personal actions. We are loyal and accountable to the public trust.

Respect: We value our diverse work force. We treat each other with fairness, dignity, and compassion. We encourage individual opportunity and growth. We encourage creativity through empowerment. We work as a team.

Devotion to Duty: We are professionals, military and civilian, who seek responsibility, accept accountability, and are committed to the successful achievement of our organizational goals. We exist to serve. We serve with pride.

These core values are more than just Coast Guard rules of behavior. They are deeply rooted in the heritage that has made our organization great. They demonstrate who we are and guide our performance, conduct, and decisions every minute of every day. Because we each represent the Coast Guard to the public, we must all embrace these values in our professional undertakings as well as in our personal lives.

United States Coast Guard Ethos

I am a Coast Guardsman.

I serve the people of the United States.

I will protect them.

I will defend them.

I will save them.

I am their Shield.

For them I am Semper Paratus.

I live the Coast Guard Core Values.

I am proud to be a Coast Guardsman

We are the United States Coast Guard.

Your U.S. Coast Guard's Average Day

On an average day, the United States Coast Guard:

- Saves 10 lives and over $1.2 million in property;
- Conducts 45 search-and-rescue cases;
- Seizes 874 pounds of cocaine and 214 pounds of marijuana;
- Escorts 5 high-capacity passenger vessels;
- Conducts 57 waterborne patrols of critical maritime infrastructure;
- Interdicts 17 illegal migrants trying to enter the United States;
- Conducts 24 security boardings in and around U.S. ports;
- Services 82 buoys and fixed aids to navigation;
- Screens 360 merchant vessels for potential security threats prior to arrival in U.S. ports;
- Conducts 105 marine inspections;
- Completes 26 safety examinations on foreign vessels entering U.S. ports;
- Conducts 14 fisheries conservation boardings to ensure compliance with fisheries laws;
- Investigates 14 marine casualties involving commercial vessels;
- Investigates 35 pollution incidents; and
- Facilitates movement of $8.7 billion worth of goods and commodities through the Nation's Maritime Transportation System.

About the Book

Regardless of rank or time in service, everyone wearing a U.S. Coast Guard uniform will find this manual to be essential to their professional development. Its value as a ready source of guidance is attested to by generations of men and women who have made it a part of their personal libraries since 1952, when the first edition was published. Today it remains the basic training manual for new Coast Guard recruits. It is also full of helpful information to civilian boaters.

This eleventh edition is designed to enhance your professionalism as a member of the U.S. Coast Guard. Updated information is offered on Coast Guard cutters, aircraft, organization, and history, among other subjects. Continued emphasis is placed on the Coast Guard's role as an armed service, safety of life, protection of national assets, and defending the homeland.

I
BACKGROUND

Figure 1-1.
SM1c Douglas Munro, USCG, is the only Coast Guardsman ever to be awarded
the Medal of Honor. He was killed in action on Guadalcanal
while on a dangerous mission to evacuate Marines from behind enemy lines.

CHAPTER 1

History

INTRODUCTION

The U.S. Coast Guard is one of this nation's five armed services and seven uniformed services. It is also the only branch of the U.S. military within the Department of Homeland Security. The legal basis for the Coast Guard is Title 14 of the United States Code, which states: "The Coast Guard as established January 28, 1915, shall be a military service and a branch of the armed forces of the United States at all times."

As members of a military service, Coast Guardsmen are subject to the Uniform Code of Military Justice and receive the same pay and allowances as members of the same pay grades in their four sister services. In peacetime, the Coast Guard operates under the Department of Homeland Security and the commandant of the Coast Guard reports to the secretary of Homeland Security. In time of war or when directed by the president, the Coast Guard operates as a service within the Department of the Navy.

By being aware of the Coast Guard's proud history and heritage, you will be able to understand where the Coast Guard is going and how you can best fit in. Remember, the operations and missions of today will be the history of tomorrow. You have an important part to play in helping preserve and promote our history and heritage for future generations.

Coast Guard history can be viewed as a series of important but unrelated events that can be selectively presented because they establish precedents, reinforce traditions, provide a neat comparison with another military organization, or have some special local interest. The Coast Guard is rich in this kind of history, and it encourages pride in our service.

The history of the service is very complicated because it is the amalgamation of five federal agencies. These agencies—the Revenue Cutter Service, the Lighthouse Service, the Steamboat Inspection Service, the Bureau of Navigation, and the Life-Saving Service—were originally independent but had overlapping authorities and were shuffled around the government. They sometimes received new names, and they were all finally united under the umbrella of the Coast Guard. The multiple missions and responsibilities of the modern service are directly tied to this diverse heritage and the magnificent achievements of all of these agencies.

One of the first orders of business our nation's founders had to contend with was maritime trade. In those days, almost all trade beyond the borders of the United States was done by ships. The following timeline reflects the many organizational changes of the predecessor agencies and when they became part of what is now the United States Coast Guard as well as changes in the organizational structure of the Coast Guard itself:

7 August 1789: The service, eventually to be known as the U.S. Lighthouse Service, was established under the control of the Treasury Department.

4 August 1790: This date is important to the Coast Guard because it is celebrated as the Coast Guard's birthday. Congress authorized Secretary of the Treasury Alexander Hamilton to create a maritime service to enforce customs laws. It didn't have an official name. It was simply referred to as "the cutters" or "the system of cutters." Hamilton established this fleet to enforce tariff laws, so these cutters were armed but manned by civilian crews under the Treasury Department. Since the Continental Navy was disbanded in 1785, there was no navy initially under the Constitution, and the cutters were the only maritime force available to the new government. Between 1790 and 1798 Hamilton's cutters were the only armed vessels protecting the coast, trade, and maritime interests of the new republic.

7 July 1838: A service to "provide better security of the lives of passengers on board of vessels propelled in whole or in part by steam" was established under the control of the Justice

Department. This service later became the Steamboat Inspection Service.

14 August 1848: Congress appropriated funds to pay for lifesaving equipment to be used by volunteer organizations.

30 August 1852: The Steamboat Act established the Steamboat Inspection Service under the control of the Treasury Department.

9 October 1852: The Lighthouse Board, which administered the nation's lighthouse system until 1 July 1910, was organized.

18 June 1878: The U.S. Life-Saving Service was established as a separate agency under the control of the Treasury Department.

18 June 1878: Congress enacted the Posse Comitatus Act. This act limited military involvement in civilian law enforcement, leaving the Revenue Cutter Service as the only service branch consistently charged with federal law enforcement on the high seas and U.S. waters.

5 July 1884: The Bureau of Navigation was established under the control of the Treasury Department.

14 February 1903: The Department of Commerce and Labor was created. The Bureau of Navigation and the Steamship Inspection Service were transferred to the new department.

28 January 1915: President Woodrow Wilson signed into law the Act to Create the Coast Guard, an act passed by Congress on 20 January 1915 and combining the Life-Saving Service and Revenue Cutter Service to form the U.S. Coast Guard.

6 April 1917: With the declaration of war against Germany, the Coast Guard was transferred by executive order to the control of the Navy Department.

28 August 1919: The Coast Guard reverted to the Treasury Department after President Wilson signed Executive Order 3160.

30 June 1932: The Steamboat Inspection Service and the Bureau of Navigation were combined to form the Bureau of Navigation and Steamboat Inspection. The new agency remained under Commerce Department control.

27 May 1936: Public Law 622 reorganized and changed the name of the Bureau of Navigation and Steamboat Inspection Service to the Bureau of Marine Inspection and Navigation. The bureau remained under Commerce Department control.

1 July 1939: The Lighthouse Service became part of the Coast Guard.

19 February 1941: The Coast Guard Reserve was established by the passage of the Coast Guard Reserve and Auxiliary Act on this date. This legislation also established the Coast Guard Auxiliary under its present name (the Coast Guard Auxiliary had originally been called the Coast Guard Reserve). The new Coast Guard Reserve was modeled after the Naval Reserve as a military component. It was composed of two broad classifications: regular reservists and temporary reservists.

1 November 1941: President Franklin Roosevelt's Executive Order 8929 transferred the Coast Guard to Navy Department control.

28 February 1942: Executive Order 9083 transferred the Bureau of Marine Inspection temporarily to the Coast Guard under Navy Department control.

23 November 1942: Congress enacted Public Law 77-773 establishing the Women's Reserve as a branch of the Coast Guard. Members of this branch became known as SPARs, an acronym drawn from the service's motto, "Semper Paratus, Always Ready."

1 January 1946: In compliance with Executive Order 9666, the Coast Guard was returned to Treasury Department control.

April 1946: The Coast Guard created the Eastern, Western, and Pacific Area commands to coordinate cases that required the assets of more than one district.

16 July 1946: Pursuant to Executive Order 9083 and Reorganization Plan No. 3, the Bureau of Marine Inspection was abolished and became a permanent part of the Coast Guard under Treasury Department control.

1 April 1967: The Coast Guard was transferred from the Treasury Department to the newly created Department of Transportation.

January 1973: The Coast Guard renamed the Eastern and Western Areas to the Atlantic and Pacific Areas, respectively.

28 December 2001: The Coast Guard became the newest member of the U.S. Intelligence Community (IC) with the signing into law of the Intelligence Authorization Act for Fiscal Year 2002, which mandated the inclusion of the Coast Guard's national intelligence element formally into the IC.

1 March 2003: The Coast Guard was formally transferred from the Department of Transportation to the newly created Department of Homeland Security (DHS).

2004: To create unity of command in America's ports, better align field command structures, and improve Coast Guard operational effectiveness, sector commands were created throughout the Coast Guard by integrating groups, marine safety offices (MSOs), vessel traffic services (VTSs), and, in some cases, air stations.

2013: The U.S. Coast Guard Cyber Command (CGCYBER-COM) was established to provide cyber capabilities that foster excellence in the execution of Coast Guard operations, support DHS cyber missions, and serve as the Service Component Command to U.S. Cyber Command.

While the Coast Guard has undergone many changes over the years, it never stopped the Coast Guard and its predecessor agencies from answering our nation's call to duty.

THE COAST GUARD AT WAR

Coast Guardsmen and their forefathers have fought in every war since the ratification of the U.S. Constitution. The U.S. Coast Guard standard proudly exhibits forty-three battle streamers, each one designed to remind us of our proud history and heritage. Many of the men and women who served our country in these wars sacrificed their lives for us, who now enjoy the freedoms they so jealously guarded. This record of military operations can be traced back to the Quasi-War with France and brought forward to today's war against terrorism.

Quasi-War of 1798–1800

Aggressive military and naval activity by France resulted in the seizure of several merchant vessels by armed French vessels on the high seas. The naval response of the United States was to employ the revenue cutters in the defense of our merchant fleet. In these operations, the cutters responded effectively, as they would in future conflicts in support of the nation's naval missions. Especially outstanding were the *Pickering*, COMO Edward Preble's first command, and the *Eagle*. The *Eagle* fought a memorable engagement with the French privateer *La Vengeance* (*Revenge*) in 1799, recapturing the American vessels *Nancy* and *Mehitable*.

War of 1812

From 1801 until the war with England in 1812, the cutters were busy enforcing the Embargo and Non-Intercourse Acts. After the War of 1812 began, the cutter *Jefferson* captured the first enemy prize in June. The cutter *Vigilant* captured the *Dart*. The most famous action was fought by the cutter *Eagle*. When she was caught by two British ships, her captain ran the ship ashore on Long Island. The crew dragged guns to a cliff top and beat off the enemy. That night they refloated their ship and headed for New Haven, Connecticut, but she could make little speed and the enemy finally captured her.

The end of the war brought no peace for the cutters. For the next quarter century, they fought pirates, slavers, smugglers, and Indians. The *Alabama* and *Louisiana* broke up the notorious Lafitte gang of pirates in the Gulf of Mexico, and the *Dallas* and *Jackson* fought the Seminole Indians in Florida.

Seminole Wars

During the Seminole Wars (1835–42) eight revenue cutters supported Army and Navy operations. Duties performed by these vessels included attacks on war parties, breaking up rendezvous points, picking up survivors of Seminole raids, carrying dispatches, transporting troops, blocking rivers to the passage of Seminole forces, and dispatching landing parties and artillery for the defense of settlements. These duties were performed along the entire coast of Florida.

Figure 1-2. The revenue cutter *Harriet Lane* fires a shot across the bow of the U.S. Mail steamer *Nashville* when that merchantman appeared with no colors flying near Charleston, South Carolina, 12 April 1861. *Harriet Lane* is credited with firing the first naval shot of the Civil War.
By Howard Koslow; courtesy of the U.S. Coast Guard Art Collection and the artist

Mexican War

The two principal naval operations carried out during the Mexico War (1846–48) were amphibious landings and blockading the enemy's coasts. The U.S. Navy was critically short of the shallow-draft vessels needed for the landings. Five cutters were engaged in amphibious operations and performed important services during a number of landings, particularly those at Alvarado and Tabasco. Cutters also served on blockade duty.

Civil War

The sympathies of the cutter force were divided between the North and the South during the American Civil War (1861–65). In a famous dispatch to Treasury Department agents, Secretary of the Treasury John A. Dix ordered, "If any one attempts to haul down the American flag, shoot him on the spot." Transmitted on the evening of 15 January 1861, this order was to ensure federal control of the cutter *Robert McClelland*, then in the port of New Orleans. Despite this message, many cutter men, including those on the *Robert McClelland*, chose to join the Confederacy. It was at this time that the service received its first official name, the Revenue Cutter Service.

The principal wartime duties of Union cutters were patrolling for commerce raiders and providing fire support for troops ashore. Meanwhile,

Figure 1-3. The *Hudson* prepares to tow the disabled torpedo boat *Winslow* beyond enemy fire at the battle of Cardenas, Cuba, 11 May 1898.
By Dean Ellis; courtesy of the U.S. Coast Guard Art Collection and the estate of the artist

Confederate cutters were principally used as commerce raiders. Cutters were also involved in notable individual actions. The first naval shot of the Civil War was fired by the cutter *Harriet Lane* when it challenged the steamer *Nashville* with a shot across its bow. The steamer was attempting to enter Charleston harbor without displaying the U.S. flag. The *Harriet Lane* also took part in the capture of Hatteras Inlet. Following this action, the cutter was transferred to the Navy. The cutter *Miami* carried President Abraham Lincoln and his party to Fort Monroe in May 1862, preparatory to the Peninsular Campaign. *Reliance*'s commanding officer was killed as the cutter engaged Confederate forces on the Great Wicomico River in 1864. On 21 April 1865 cutters were ordered to search all outbound ships for the assassins of President Lincoln. During the Civil War, 219 Revenue Cutter Service officers served, and 1 officer was killed in action.

Spanish-American War

The Revenue Cutter Service also rendered conspicuous service during the Spanish-American War (1898). Eight cutters carrying forty-three guns were in ADM William Sampson's fleet and on the Havana blockade. The *McCulloch*, with a complement of ten officers and ninety-five men and carrying six guns, was at the Battle of Manila Bay and was later employed by ADM George Dewey as his dispatch boat.

In the action off Cardenas on 11 May 1898, the cutter *Hudson*, LT Frank H. Newcomb commanding, sustained the fight against Spanish gunboats and shore batteries side by side with the torpedo boat USS *Winslow*. When half of *Winslow*'s crew had been killed and its commander wounded, *Hudson* rescued the torpedo boat from certain destruction. In recognition of this act of heroism, Congress authorized a gold medal for Lieutenant Newcomb, a silver medal for each of the officers, and a bronze medal for the enlisted members of the crew.

Also during the Spanish-American War, the Navy assigned the task of coast watching to the U.S. Life-Saving Service. As a result, approximately two-thirds of the Navy's coastal observation stations were Life-Saving stations. Although the Spanish fleet never approached the U.S. coast, this Coast Guard predecessor service dutifully maintained its vigilance throughout the war. During the Spanish-American War 660 Revenue Cutter Service personnel served with 1 fatality due to enemy action.

World War I

The outbreak of World War I in Europe in the summer of 1914 saw cutters become responsible for enforcing U.S. neutrality laws. It also marked the first time the Coast Guard was transferred to the Navy Department in wartime. The Coast Guard augmented the Navy with 47 vessels of all types and 279 stations scattered along the entire United States coastline. It was in this same period that Coast Guard aviation was born. LT E. F. Stone, USCG, one of several early Coast Guard flyers, was copilot on the Navy NC-4 flying boat, the first aircraft to cross the Atlantic.

During World War I, the Coast Guard continued to enforce rules and regulations that governed the anchorage and movements of vessels in American harbors. The Espionage Act, passed in June 1917, gave the Coast Guard further power to protect merchant shipping from sabotage. This act included the responsibility to safeguard waterfront property, supervise vessel movements, and establish anchorages and restricted areas, and the right to control and remove people on board ships.

A large number of Coast Guard officers held important commands during World War I. Twenty-four commanded naval warships in the war

Figure 1-4. "Man's best friend" was an important member of the Coast Guard during World War II. On a lonely outpost on the Atlantic, this alert, trained dog gave a warning signal to its Coast Guard handler, who then challenged any suspected enemy spies and saboteurs.

zone, five commanded warships attached to the American Patrol detachment in the Caribbean Sea, twenty-three commanded warships attached to naval districts, and five Coast Guard officers commanded large training camps. Six were assigned to aviation duty; two of those commanded important air stations including one in France. Shortly after the armistice in 1918, four Coast Guard officers were assigned to command large naval transports engaged to bring the troops home from France. Officers not assigned to command served in practically every phase of naval activity, on transports, cruisers, cutters, patrol vessels, in naval districts, as inspectors, and at training camps.

During World War I, 8,835 men served in the Coast Guard. Of those, 192 would pay the ultimate price due to combat, accident, or illness,

Figure 1-5. The U.S. Army's 1st Infantry Division disembarks from a Coast Guard–manned landing craft, vehicle, personnel (LCVP) from the USS *Samuel Chase* (APA-26) on the morning of 6 June 1944 at Omaha Beach, Normandy, France.

to include the 115 crew members and 16 passengers on board USCGC *Tampa*, which was sunk on September 26, 1918—the largest loss of life on any U.S. combat vessel during World War I.

World War II

In World War II Coast Guardsmen operated escort destroyers, frigates, corvettes, patrol boats, and sub chasers, while planes patrolled the coasts and assisted in antisubmarine warfare. Beach patrols and port security units ensured coastal safety.

Immediately after the Japanese attack on Pearl Harbor on 7 December 1941, cutters were ordered to escort duty. This included going into action against Adolf Hitler's vaunted submarine fleet, nicknamed "hearses" by the Coast Guardsmen who fought them to the death on the open seas. During the war the U.S. Navy credited Coast Guard forces with sinking or assisting in the sinking of thirteen of Hitler's U-boats, although the number was probably only eleven. In the Pacific the Navy credited Coast Guard warships with sinking one Japanese submarine, but they probably sank two. Coast Guardsmen also captured two Nazi surface vessels, and they can take pride in knowing that they were the only United States service to do so during World War II. Additionally, two U-boats surrendered to Coast Guard–manned warships at the end of hostilities, including one, *U-234*, that was bound for Japan transporting a cargo of uranium and the latest German rocket and jet technology.

In the Pacific war, amphibious operations employed thousands of Coast Guardsmen, where their small-boat experience proved invaluable. Coast Guardsmen served on combat ships and in the support forces, where they manned naval transports, attack transports, supply vessels, and shore units.

During World War II 241,093 men and women served in the Coast Guard. Of these, 1,917 died, a third losing their life in action; 4 members of the Coast Guard were held as prisoners of war. The personnel strength of the Coast Guard peaked in 1944. Headquarters reported that as of 30 June 1944, there were 9,874 commissioned officers, 3,291 warrant officers, and 164,560 enlisted personnel serving in the regular and reserve (not including the Temporary Reserve or Auxiliary). Almost 2,000 Coast Guardsmen were decorated. Douglas Munro, a signalman, was posthumously awarded the only Medal of Honor earned by a member of the Coast Guard. He heads a long list of Coast Guard personnel who were cited for courage during World War II. The Coast Guard returned to the Treasury Department on 1 January 1946.

Korean War

In 1946 a Coast Guard team was sent to Seoul, Korea, to organize, supervise, and train a Korean coast guard. This Korean group became the National Maritime Police. A shipyard was set up, a supply and communications system was arranged, training facilities were established, and nineteen ships were transferred to Korea from the Japanese and U.S. navies.

With the outbreak of the Korean War (1950–53), marked by the invasion by North Korean Communist forces southward across the thirty-eighth parallel into the Republic of South Korea on 25 June 1950, the U.S. Coast Guard performed a variety of tasks. The Coast Guard established air detachments throughout the Pacific. These detachments, located

Figure 1-6. The 5-inch gun on board the 311-foot USCGC *Half Moon* (WHEC-378) blasts at a Viet Cong stronghold in the Song Ong Doc area in South Vietnam.

at Sangley Point in the Philippines and in Guam, Wake, Midway, Adak, and Barbers Point in the Hawaiian Islands, conducted search and rescue to safeguard the tens of thousands of troops that were being airlifted across the Pacific.

The Coast Guard manned a number of Navy destroyer escorts, performing a variety of missions in the Western Pacific. A chain of mobile LORAN (long-range navigation) stations was established in the Far East to support the air and sea navigational needs of UN forces. One of those Coast Guard stations was situated at Pusan, South Korea. The Coast Guard also provided communications and meteorological services, plus ensured port security and proper ammunition handling.

The Coast Guard remained under the Department of the Treasury throughout the Korean War. Eight thousand five hundred Coast Guardsmen were eligible for the Korean Service Medal. There were no Coast Guard casualties in the Korean War.

Vietnam War

At the outset of the Vietnam War the Navy lacked shallow-water craft needed for inshore operations. To help fill this need, the Coast Guard sent twenty-six eight-two-foot cutters to Vietnam. The cutters spent some 70 percent of their time under way. They inspected junks for contraband, intercepted and destroyed North Vietnamese and Viet Cong craft, and provided fire support for friendly forces.

While the eighty-two-foot cutters helped patrol inshore, larger cutters were needed to form a deep-water barrier against infiltration. For this task, the Coast Guard deployed high endurance cutters. Thirty high endurance cutters served on this duty between 1967 and 1971.

The U.S. Army had the difficult task of setting up harbor security and getting cargo safely unloaded and moved into the country. Since almost all munitions entered South Vietnam by ship, the Army asked for the assistance of the Coast Guard. The men of the Coast Guard Port Security and Waterways Detail traveled throughout Vietnam inspecting ports and harbors for security against enemy attack and for the safe storage of hazardous materials. Coast Guard explosives loading detachments were established at major ports to supervise the offloading of ships.

The Coast Guard also set up and operated a LORAN-C system in Southeast Asia in order to assist the U.S. Air Force warplanes with precision navigation. It was a difficult task finding transmitting sites, bringing in equipment, and building the system. LORAN stations were established in Lampang, Sattahip, and Udorn, Thailand, and in Con Son, Vietnam. A fifth station was later added in Tan My, Vietnam.

The rapid development of deep-water ports in Vietnam brought an expanded need for navigational aids to prevent vessel accidents. South Vietnam's small aids-to-navigation force, with its one buoy tender, could not meet the demand. Coast Guard buoy tenders in the Pacific made periodic trips to Vietnam installing and maintaining buoys. A Coast Guard aids-to-navigation detail was set up in Saigon to coordinate workloads for these visits as well as to keep buoys and range markers lighted.

At the height of the war more than three hundred merchant ships were engaged in the sealift of matériel to Vietnam. The Coast Guard Merchant Marine Detail resolved merchant seaman problems and ensured that these ships moved in and out with as little delay as possible.

Coast Guard pilots flew combat search and rescue with the Air Force in Southeast Asia, under an interservice exchange program that was initiated in 1967. Most of the time the pilots were assigned to the 37th Aerospace Rescue and Recovery Squadron at Danang. They flew Sikorsky HH-3F "Jolly Green Giants" in some of the most dangerous operations undertaken during the war. LT Jack Rittichier (Coast Guard Aviator #997) became the first Coast Guardsman to die from enemy action in Vietnam when his helicopter was shot down on 9 June 1968 during an attempt to pull a downed Marine Corps pilot from enemy-held territory.

Some eight thousand Coast Guardsmen served in Vietnam. Seven lost their lives and sixty were wounded.

Operations Desert Shield and Desert Storm

On 2 August 1990 Iraq invaded the country of Kuwait. A coalition of countries from around the world, led by the United States, quickly responded to this invasion. The Coast Guard's unique operational capabilities and maritime expertise were tapped quickly and often by the Departments of Defense, State, and Transportation, and by governments of the Allied Coalition.

On 22 August 1990 President George H. W. Bush authorized the call-up of members of the selected reserve to active duty in support of Operation Desert Shield. Three port security units (PSUs) consisting of a total of 550 Coast Guard reservists were deployed to Saudi Arabia in support of Operation Desert Shield to provide for waterside security. (This was the first involuntary overseas mobilization of Coast Guard Reserve PSUs in the Coast Guard Reserve's fifty-year history.) Law enforcement detachments were dispatched to augment and most often lead the naval boarding teams enforcing the UN economic embargo. These teams conducted hundreds of boardings and uncovered tons of illicit cargo. Moreover, during Operation Desert Shield, the initial Coast Guard efforts included breaking out sixty-eight ships of the Ready Reserve and locating merchant seamen to safely sail the vessels. The loading of hazardous materials and ammunition was supervised in U.S. ports.

Coast Guard personnel also augmented the staff of the Commander, Middle East Forces to coordinate interdiction and search-and-rescue (SAR) efforts of the myriad naval forces deployed in the Persian Gulf and Red Sea. Coast Guard oil-spill experts quickly responded to a request from the Saudi Arabian government to assist them in dealing with the massive oil spill. These experts and Aireye-equipped aircraft provided invaluable technical advice and assistance in combating this act of environmental terrorism.

With the commencement of offensive action against the forces of Iraq under the auspices of Operation Desert Storm on 17 January 1991, Coast Guard Marine Safety Offices increased both the level and tempo of port safety and security activities in the port to protect critical commercial and military waterfront facilities from the threat of terrorism. The Coast Guard increased both shoreside and waterside security patrols in ports using regular and reserve augmentation forces and established and enforced waterside security zones around key facilities.

Operations Support Democracy and Uphold Democracy

From November 1993 to August 1995 (Operation Support Democracy) and October 1994 to March 1995 (Operation Uphold Democracy), Coast Guard cutters and port security units supported UN-led operations to restore democratic institutions in Haiti. Two port security units, a harbor defense unit, five law enforcement detachments, and thirteen cutters carried out operations that included conducting maritime surveillance and interdiction, providing search-and-rescue coverage for in-transit U.S. aircraft, and establishing and restoring aids to navigation.

Operation Allied Force

Operation Allied Force was the seventy-eight-day NATO air campaign from 24 March 1999 to 10 June 1999 against Serbian military targets in the former Federal Republic of Yugoslavia. USCGC *Bear* (WMEC-901) deployed to the Adriatic Sea in support of Operation Allied Force as part of USS *Theodore Roosevelt* (CVN-71) battle group. *Bear* provided surface surveillance and SAR response for the Sea Combat Commander, and force protection for the Amphibious Ready Group that operated near Albania. *Bear* provided combat escort for U.S. Army vessels transporting military cargo between Italy and Albania. This escort operation took *Bear* up to the Albanian coastline, well within enemy surface to surface missile range.

9/11

On 11 September 2001 terrorists from the Al Qaeda network hijacked four commercial aircraft, crashing two into the World Trade Center in New York and one into the Pentagon in Washington, D.C. (The fourth aircraft crashed around Shanksville, Pennsylvania, when passengers on board attempted to regain control of the aircraft from the terrorists.) USCG units from New York City were among the first military units to respond in order to provide security and render assistance to those in need.

In response to the terrorist threat and to protect our nation's coastline, ports, and waterways, six U.S. Navy *Cyclone*-class patrol coastal warships were assigned to Operation Noble Eagle on 5 November 2001. This was the first time that U.S. Navy ships were employed jointly under Coast Guard command.

In response to 9/11, President George W. Bush proposed the establishment of the Department of Homeland Security. The Coast Guard was slated to be transferred from the Department of Transportation to the new department.

On 25 November 2002 President Bush signed HR 5005, creating the Department of Homeland Security. Soon after, Tom Ridge, former governor of Pennsylvania, was confirmed as the department's first secretary. On 25 February 2003 Transportation Secretary Norman Mineta transferred leadership of the U.S. Coast Guard to Secretary Ridge, formally recognizing the change in civilian leadership over the Coast Guard and ending the Coast Guard's almost thirty-six-year term as a member of the Department of Transportation. As a result of the attacks, homeland security moved to the forefront of the service's primary missions.

Operation Iraqi Freedom

Coast Guard forces deployed to Southwest Asia in support of the U.S.-led coalition engaged in Operation Iraqi Freedom early in 2003. Coast Guard vessels and land-based personnel brought many vital capabilities to the theater of operations, including inshore patrol and port security. At the height of operations there were 1,250 Coast Guard personnel deployed, including about 500 reservists. The forces also included two large cutters, a buoy tender, eight patrol boats, four port security units, law enforcement detachments, and support staff to the U.S. Central Command and U.S. European Command.

On 25 April 2004 DC3 Nathan Bruckenthal, USCG, from Smithtown, New York, and two U.S. Navy sailors were killed in the line of duty while conducting maritime intercept operations in the North Arabian Gulf. Bruckenthal and six other coalition sailors had attempted to board a small boat near an Iraqi oil facility. As they boarded the boat, it exploded. Petty Officer Bruckenthal later died from injuries sustained in the explosion; he was the first Coast Guardsman killed in action since the Vietnam War. Bruckenthal had been assigned to Tactical Law Enforcement South in Miami, Florida, and had deployed with Coast Guard Patrol Forces Southwest Asia on board USS *Firebolt* (PC-10).

DISASTERS AND CONTINGENCY OPERATIONS

Closer to home, the Coast Guard provides our nation with tremendous value in service to the public during times of natural and man-made disasters and other contingencies. A sampling of key domestic events and missions for the Coast Guard includes the following:

Sinking of the *Titanic*, 1912

Following the sinking of the *Titanic*, Congress authorized the International Ice Patrol.

Camarioca Boatlift, 1965

The Camarioca Boatlift was the first large-scale exodus of Cuban immigrants attempting to enter the United States. Thereafter, migrant interdiction became a policy concern.

Argo Merchant Oil Spill, 1976

The *Argo Merchant*, a Liberian tanker, grounded off Nantucket in December, carrying 7.3 million gallons of fuel oil. Coast Guard cutters *Sherman*, *Vigilant*, *Spar*, and *Bittersweet* were on the scene and prepared to use the air deployable antipollution transfer system on the vessel. However, deteriorating weather, thirty-knot winds, and heavy seas prevented removal of its cargo before the hull began to buckle. The bow was wrenched from the hull and opened the cargo to the sea. This was the largest oil spill up until then in American waters. Northwesterly winds dispersed the oil out to sea. The *Argo Merchant* accident and fourteen more tanker accidents in or near American waters over the next ten weeks caused great concern about tanker safety.

Mariel Boatlift, 1980

Boats with Cuban migrants on board began departing Mariel, Cuba, in April 1980. The first two boats arrived in Miami on the same day, marking the largest Cuban migration to the United States. Cuban leader Fidel Castro then declared the port of Mariel "open," increasing the number of boats involved in the exodus and giving the exodus its name. By the time the boatlift came to an end in October 1980, more than 125,000 Cubans had made the journey to the United States. It was the largest search-and-rescue operation conducted by the Coast Guard since World War II and gave the service an indication of the increasing role the service would play regarding migrant interdiction in upcoming years.

Exxon Valdez Oil Spill, 1989

The *Exxon Valdez* oil spill led to the creation of the Oil Pollution Act of 1990 (OPA-90). In this act Congress addressed tanker construction, personnel licensing, and the emergency rapid-response capability. The act called for mandatory double hulls on new tankers and the gradual phasing out of noncomplying vessels. The licensing requirements for ship's officers were strengthened in the area of drug and alcohol testing. The rapid-response capability was expanded nationwide, and new emphasis was placed on oil pollution research. The act required the Coast Guard to create response groups (known as strike teams) capable of responding to spills and other disasters.

Hurricanes Katrina and Rita, 2005

Hurricane Katrina struck the Gulf of Mexico states on 29 August 2005, devastating a 90,000-square-mile area from Grand Isle, Louisiana, to Mobile, Alabama. It was one of the worst natural disasters in American history, with more than 1,200 deaths and $80 billion in property damage. The enormous storm surge and subsequent multiple levee breaches flooded more than 60 percent of New Orleans. Tens of thousands of men, women, and children were left stranded on rooftops and in city buildings with no food, water, or electricity.

The massive response that followed was the largest search-and-rescue operation in U.S. history. More than 5,000 Coast Guardsmen from around the country responded to more than 33,000 people. They also responded to thousands of oil spills totaling more than 9 million gallons, repaired navigational aids, and restored waterways in and around some of the country's most vital ports.

The Coast Guard has proven agility and significant experience in leading integrated joint-agency operations. Due largely to this, the Coast Guard's chief of staff was chosen to be the principal federal official in the early days of the disaster to coordinate the massive federal, state, and local responses to Hurricane Katrina and, subsequently, Hurricane Rita, which hit nearly the same area just one month later.

Hurricanes Katrina and Rita tested the Coast Guard's organizational character. These events once again illustrated our service's unique national value: Complementary missions and skills, core-value competencies, and deeply ingrained principles of operation enabled the Coast Guard to rise to the moment.

Haiti Earthquake, 2010

On 12 January 2010 a devastating 7.0 earthquake struck Haiti. The Coast Guard cutter *Forward* was diverted from a law enforcement patrol in the Caribbean and arrived in Port-au-Prince, Haiti, within seventeen hours of the disaster. The cutter *Forward* provided the world with firsthand images and reports of the devastation that had struck the impoverished nation. The first helicopters on scene were from Coast Guard units deployed to the Turks and Caicos Islands for counternarcotics operations. More than eight hundred Coast Guard personnel, cutters, and aircraft were involved in response and relief efforts, including operations to evacuate injured personnel and U.S. citizens, and in delivery of aid. Coast Guard cutters and personnel also completed port assessments and developed plans to reconstitute port operations, allowing for safe transit of humanitarian supplies from relief ships to coastal logistics staging points.

Deepwater Horizon Oil Spill, 2010

On 20 April 2010 the Coast Guard responded to a distress call following an explosion on the Deepwater Horizon oil rig in the Gulf of Mexico. During the initial search-and-rescue response more than one hundred people were rescued. As the scope and focus of the incident expanded and shifted to the ensuing oil spill, the Coast Guard led the monumental federal response to the first ever Spill of National Significance and one of the worst environmental disasters in U.S. history. The scope of the unprecedented response involved more than seven thousand Coast Guard personnel, forty-six cutters and boats, and twenty-two aircraft. Throughout the incident, the Coast Guard led a whole of government and industry response to secure the wellhead, contain and clean up the spill, and protect the Gulf Coast ecosystem.

Figure 2-1.
USCGC *Aquidneck* (WPB-1309), homeported in Atlantic Beach, North Carolina,
patrols up the Khawr Abd Allah River in Iraq, 28 March 2003.

CHAPTER 2

Missions

By law the Coast Guard has eleven statutory missions that are defined as either "homeland security" or "non–homeland security." These missions clearly demonstrate how the Coast Guard's assets, competencies, capabilities, authorities, and partnerships bring forth a unique ability to serve as a leading maritime responder and incident manager within government.

HOMELAND SECURITY MISSIONS

Defense Readiness

As one of the nation's five armed services, the Coast Guard provides unique authorities and capabilities to support our nation's war plans. These unique authorities and capabilities provided under the Defense Operations program encompass eight activities: maritime interdiction operations, combating maritime terrorism, port operations security and defense, military environmental response operations, coastal sea control operations, maritime operational threat response, rotary wing air intercept operations, and support for theater security cooperation initiatives. They are essential military tasks assigned to the Coast Guard as a component of joint and combined forces in peacetime, crisis, and war.

Moreover, outside of U.S. coastal waters, the Coast Guard assists foreign naval and maritime forces through training and joint operations. Many of the world's maritime nations have forces that operate principally in the littoral seas and conduct missions that resemble those of the U.S. Coast Guard and not the U.S. Navy. Since it has such a varied mix of assets and missions, the Coast Guard is a powerful role model that is in ever-increasing demand abroad. The service's close working relations with these nations not only improve mutual cooperation during specific joint operations in which the Coast Guard is involved but also support U.S. diplomatic efforts in general: the promotion of democracy, economic prosperity, and trust between nations.

Ports, Waterways, and Coastal Security (PWCS)

The PWCS mission entails the protection of the U.S. maritime domain and the U.S. Marine Transportation System (MTS) and those who live, work, or recreate near them; the prevention and disruption of terrorist attacks, sabotage, espionage, or subversive acts; and the response to and recovery from those acts that do occur. Conducting PWCS deters terrorists from using or exploiting the MTS as a means for attacks on U.S. territory, population centers, vessels, critical infrastructure, and key resources. PWCS includes the employing awareness activities; conducting counterterrorism, antiterrorism, preparedness, and response operations; and establishing and overseeing a maritime security regime. PWCS also includes the national defense role of protecting military outload operations.

Drug Interdiction

The Coast Guard is the lead federal agency for maritime drug interdiction and shares lead responsibility for air interdiction with the U.S. Customs Service. As such, it is a key player in combating the flow of illegal drugs to the United States. The Coast Guard's mission is to reduce the supply of drugs from the source by denying smugglers the use of air and maritime routes in the transit zone, a 6-million-square-mile area that includes the Caribbean, Gulf of Mexico, and Eastern Pacific. In meeting the challenge of patrolling this vast area, the Coast Guard coordinates closely with other federal agencies and countries within the region to disrupt and deter the flow of illegal drugs.

Migrant Interdiction

The Coast Guard is the lead agency for the enforcement of U.S. immigration laws at sea. It conducts patrols and coordinates with other federal agencies and foreign countries to interdict undocumented migrants at sea, denying them entry via maritime routes to the United States, its territories, and possessions. Thousands of people try to enter this country illegally every year using maritime routes, many via smuggling operations. Interdicting migrants at sea means they can be quickly returned to their countries of origin without the costly processes required if they were to successfully enter the United States. Coast Guard migrant-interdiction operations in the Caribbean, the Gulf of Mexico, and the Eastern Pacific are as much humanitarian efforts as they are law enforcement missions. In fact, the majority of migrant-interdiction cases handled by the Coast Guard actually begin as search-and-rescue missions, usually on the high seas rather than in U.S. coastal waters.

Other Law Enforcement

Preventing illegal foreign fishing vessel encroachment in the exclusive economic zone is a primary Coast Guard role vital to protecting the integrity of the nation's maritime borders and ensuring the health of U.S. fisheries. The Coast Guard also enforces international agreements to suppress damaging illegal, unreported, and unregulated fishing activity on the high seas.

NON–HOMELAND SECURITY MISSIONS

Marine Safety

The Marine Safety program ensures the safe operation and navigation of U.S.- and foreign-flagged vessels. The Coast Guard is responsible for

Figure 2-2. Personnel from Coast Guard Sector Miami and Aids to Navigation Team Fort Lauderdale, Florida, conduct a boom-deployment exercise at Virginia Key Beach, Florida.

providing safe, efficient, and environmentally sound waterways for commercial and recreational users. Domestic vessel inspections and port state control (foreign vessel) examinations are conducted in order to safeguard maritime commerce and international trade.

Search and Rescue

Search and rescue (SAR) is one of the Coast Guard's oldest missions and perhaps the Coast Guard's best-known mission area. Minimizing the loss of life, injury, and property damage by rendering aid to persons in distress and property at risk in the maritime environment has always been a Coast Guard priority. Coast Guard SAR response involves multimission stations, cutters, aircraft, and boats linked by communications networks.

The National SAR Plan divides the U.S. area of SAR responsibility into internationally recognized inland and maritime SAR regions. The Coast Guard is the maritime SAR coordinator. To meet this responsibility, the Coast Guard maintains SAR facilities on the East, West, and Gulf coasts, in Alaska, Hawaii, Guam, and Puerto Rico, and on the Great Lakes and inland U.S. waterways. When the rescue alarm sounds, the Coast Guard is ready to confront the inherently dangerous maritime environment, frequently going into harm's way to save others. The Coast Guard works closely with other federal, state, and local agencies, and with foreign nations, to provide the world's fastest and most effective response to distress calls. It also maintains a vessel-tracking system called AMVER (automated mutual assistance vessel rescue) that allows it to divert nearby commercial vessels to render assistance when necessary.

Aids to Navigation

The aids-to-navigation mission is a means for the Coast Guard to mark the waters of the United States and its territories to assist boaters in navigation and alert them to obstructions and hazards.

Living Marine Resources

The Living Marine Resources program's mission is to provide effective and professional at-sea enforcement of federal fisheries regulations and other regulations to advance national goals for the conservation and management of living marine resources and their environments through the detection and deterrence of illegal fishing activity. Beginning with the protection of the Bering Sea fur seal and sea otter herds and continuing through the vast expansion following World War II in the size and efficiency of global fishing fleets, Coast Guard responsibilities in this mission area now include enforcement of laws and treaties in the 3.36-million-square-mile U.S. exclusive economic zone, the largest in the world.

Marine Environmental Protection

The Marine Environmental Protection program falls under the Coast Guard's stewardship role. It is concerned with averting the introduction of invasive species, stopping unauthorized ocean dumping, and preventing the discharge of oil or hazardous substances into the navigable waters of the United States.

Ice Operations

The ice operations mission encompasses icebreaking activities in the Great Lakes, Saint Lawrence Seaway, and Northeast. These activities facilitate the movement of bulk cargoes carried by regional commercial fleets during the winter months. In addition to domestic ice operations, the Coast Guard operates the only U.S.-controlled icebreakers capable of operations in the polar regions.

Coast Guard Intelligence

While intelligence collection is not a statutory mission of the Coast Guard, it is vital to the conduct of the Coast Guard's missions. The Coast Guard is singular among seventeen agencies that make up the Intelligence Community (IC) for its national and law enforcement intelligence statutory authorities in the maritime domain. Thus, the Coast Guard is uniquely positioned to provide actionable, timely, and fused intelligence to its operational commanders and IC partners.

Figure 3-1.
ADM James Loy served as the twenty-first commandant of the
U.S. Coast Guard from May 1998 to May 2002.

CHAPTER 3

Leadership and Organization

As a military, multimission maritime service within the Department of Homeland Security, the Coast Guard is composed of more than 48,000 active-duty and reserve members, 8,500 civilian employees, and more than 30,000 members of the Coast Guard Auxiliary.

SENIOR COAST GUARD LEADERSHIP

Leaders are individuals who guide or direct in a course by showing the way. The Coast Guard's senior leadership strives to ensure that the Coast Guard can complete its missions and remain Semper Paratus, Always Ready. The Coast Guard's senior leadership includes the following.

Commandant of the Coast Guard

The commandant is the top service official, responsible for all world-wide Coast Guard activities, and for oversight of all personnel that compose Team Coast Guard. The commandant has the rank of four stars.

Vice Commandant of the Coast Guard

Second in command is the vice commandant, who directly oversees the Coast Guard's senior operational and mission support commanders as well as Coast Guard Headquarters staff, and who serves as service chief in absence of the commandant. The vice commandant became a four-star billet on 1 June 2016.

Master Chief Petty Officer of the Coast Guard

The master chief petty officer of the Coast Guard is the service's senior enlisted member, who advises the commandant on enlisted military workforce policies, is an advocate for military benefits and entitlements, enlisted mentor, and sounding board for select enlisted administrative actions.

ORGANIZATION

Coast Guard Headquarters

The Douglas Munro Coast Guard Headquarters Building is an office building of 1.2 million square feet that houses approximately 3,700 military personnel and civilian employees in Washington, D.C. When it opened in 2013 the Munro Building became the first permanent home in the U.S. Coast Guard's history. From headquarters, the commandant plans, supervises, and coordinates the overall activities of the Coast Guard and directs the policy, legislation, and administration of the Coast Guard. The commandant is aided by the vice commandant and assisted by deputy commandants and assistant commandants.

Headquarters Units

Coast Guard Headquarters units report directly to Coast Guard Headquarters in Washington, D.C. They may be geographically distant from Washington, D.C., but they are under the direct control of the commandant and operate independently of the district commander. These units were established to provide certain support services for the Coast Guard as a whole or to satisfy a requirement in a specific geographic area. Headquarters units include:

> U.S. Coast Guard Academy, New London, Connecticut
> Aviation Logistics Center (ALC), Elizabeth City, North Carolina
> Air Station Washington, D.C. (CG-7)
> Asset Project Office (APO), Baltimore, Maryland (CG-9)
> Command, Control, Communications, Computers, and Information Technology Service Center (C4ITSC), Alexandria, Virginia (CG-6)
> Counterintelligence Service (CG-CI), Washington, D.C. (CG-2)
> Cryptologic Group (CG-CG), Fort Meade, Maryland (CG-2)
> Cyber Command (CGCYBERCOM), Arlington, Virginia (CG-2)

Director of Operational Logistics (DOL), Norfolk, Virginia
  (DCMS)

Finance Center (FINCEN), Chesapeake, Virginia (CG-8)

Health, Safety, and Work-Life Service Center (HSWLSC),
  Norfolk, Virginia (CG-1)

Hearing Office (CGHO), Arlington, Virginia (CG-094)

Intelligence Coordination Center (ICC), Washington, D.C.
  (CG-2)

Investigative Services (CGIS), Arlington, Virginia (CG-2)

Legal Services Command (LSC), Norfolk, Virginia (CG-094)

Marine Safety Center (MSC), Washington, D.C. (CG-5)

Marine Safety Laboratories, Groton, Connecticut (CG-5)

National Data Buoy Center, Bay St. Louis, Mississippi (CG-5)

National Maritime Center (NMC), Martinsburg, West Virginia
  (CG-5)

National Pollution Funds Center (NPFC), Arlington, Virginia
  (CG-8)

National Vessel Documentation Center (NVDC), Falling Waters,
  West Virginia (CG-5)

Navigation Center (NAVCEN), Alexandria, Virginia (CG-5)

Personnel Service Center (PSC), Arlington, Virginia (CG-1)

Research & Development Center (RDC), New London,
  Connecticut (CG-9)

Surface Forces Logistics Center (SFLC), Baltimore, Maryland
  (CG-4)

Shore Infrastructure Logistics Center (SILC), Norfolk, Virginia
  (CG-4)

Uniform Distribution Center (UDC), Woodbine, New Jersey
  (CG-1)

Area Commands

The two largest operational commands in the Coast Guard are Coast
Guard Atlantic Area (LANTAREA) and Coast Guard Pacific Area
(PACAREA). LANTAREA is headquartered in Portsmouth, Virginia.
LANTAREA's area of responsibility includes all U.S. Coast Guard missions from the Rocky Mountains, the Great Lakes, Gulf of Mexico, and
the western rivers of North America and Central American isthmus eastward through Europe and Africa to the Arabian Gulf, spanning across five

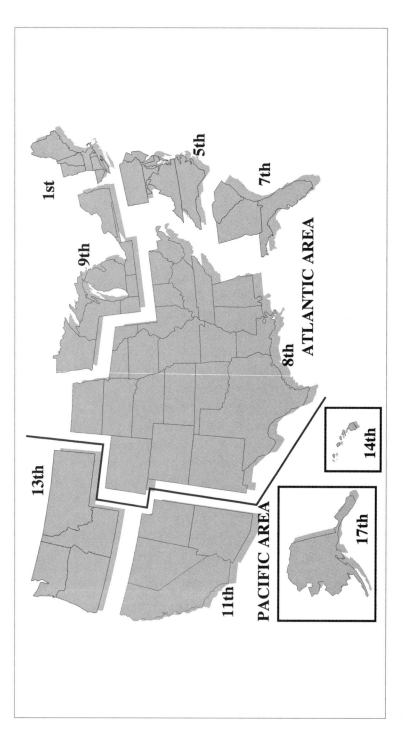

Figure 3-2. The nine Coast Guard districts.

Coast Guard Districts and forty states. PACAREA is headquartered on Coast Guard Island in Alameda, California. PACAREA's area of responsibility includes six of the seven continents, seventy-one countries, and more than 74 million square miles of ocean—from the U.S. Western states to Asia, and from the Arctic to Antarctica.

Area commanders are vice admirals who oversee the daily operations in support of the Coast Guard's eleven statutory missions in their respective area of responsibility. Nine Coast Guard districts report to the two area commands. PACAREA is made up of the 11th, 13th, 14th, and 17th Coast Guard Districts while LANTAREA is composed of the 1st, 5th, 7th, 8th, and 9th Coast Guard districts.

Coast Guard Districts

Coast Guard districts are under the command of a district commander, who is a rear admiral, directly representing the area commander in all matters pertaining to Coast Guard operations within the districts. The district commanders are responsible for the administration of their districts and for the efficient, safe, and economical performance of Coast Guard duties within their districts. The Coast Guard presently has nine numbered districts:

1st Coast Guard District (D1)

The 1st Coast Guard District, headquartered in Boston, Massachusetts, covers northern New Jersey, New York, Connecticut, Rhode Island, Massachusetts, New Hampshire, Vermont, and Maine.

5th Coast Guard District (D5)

The 5th Coast Guard District, headquartered in Portsmouth, Virginia, covers the mid-Atlantic region from central New Jersey through North Carolina. This area encompasses 156,000 square miles of ocean, bays, and rivers that are unique and filled with historical significance, several major mid-Atlantic ports, the largest naval base in the world, and our nation's capital.

7th Coast Guard District (D7)

The 7th Coast Guard District, headquartered in Miami, Florida, covers the Southeast United States and the Caribbean basin including Florida, Georgia, South Carolina, Puerto Rico, and the U.S. Virgin Islands. The 7th District encompasses an area of 1.8 million square miles and shares operational borders with thirty-four foreign nations and territories.

8th Coast Guard District (D8)

The 8th Coast Guard District, headquartered in New Orleans, covers all or part of twenty-six states throughout the Gulf Coast and heartland of America. It stretches from the Appalachian Mountains and Chattahoochee River in the east, to the Rocky Mountains in the west, and from the border between the United States and Mexico and the Gulf of Mexico to the Canadian border in North Dakota. The 8th District is home to two of the nation's busiest ports, New Orleans and Houston. More than 2 million barrels of oil and 1 million tons of cargo are imported daily.

9th Coast Guard District (D9)

The 9th Coast Guard District, headquartered in Cleveland, Ohio, covers the five Great Lakes, the Saint Lawrence Seaway, and parts of the surrounding states including 6,700 miles of shoreline and 1,500 miles of the international border with Canada.

11th Coast Guard District (D11)

The 11th Coast Guard District, headquartered in Alameda, California, covers the states of California, Arizona, Nevada, and Utah as well as the coastal and offshore waters out over thousand miles and the offshore waters of Mexico and Central America down to South America.

13th Coast Guard District (D13)

The 13th Coast Guard District, headquartered in Seattle, Washington, covers Washington, Oregon, Idaho, and Montana as well as the entire Pacific Northwest coast. D13 is home to the largest domestic ferry system, moving 24 million passengers and 11 million vehicles annually across ten routes; the third-largest domestic port; and the third-largest cruise ship industry in the United States. It shares a border with Canada to the north and California to the south.

14th Coast Guard District (D14)

The 14th Coast Guard District, headquartered in Honolulu, Hawaii, covers more than 12.2 million square miles of land and sea, from the Hawaiian Islands and across most of the Central and Western Pacific.

17th Coast Guard District (D17)

The 17th Coast Guard District, headquartered in Juneau, Alaska, covers the entire state of Alaska and portions of the North Pacific Ocean, Arctic Ocean, and Bering Sea. It encompasses an area of nearly 4 million square miles of the Pacific and Arctic Oceans.

U.S. Coast Guard Sector Commands

A sector is a shore-based operational unit of the United States Coast Guard. Each sector is responsible for the execution of all Coast Guard missions within its area of responsibility with operational support from Coast Guard cutters and air stations. Sub-units of a sector include stations and aids-to-navigation teams. There are thirty-five sectors across the nine Coast Guard districts:

Sector	Location of Sector Command
Sector Northern New England	South Portland, Maine
Sector Boston	Boston, Massachusetts
Sector Southeast New England	Woods Hole, Massachusetts
Sector Long Island Sound	New Haven, Connecticut
Sector Delaware Bay	Philadelphia, Pennsylvania
Sector Baltimore	Baltimore, Maryland
Sector Hampton Roads	Portsmouth, Virginia
Sector North Carolina	Atlantic Beach, North Carolina
Sector Charleston	Charleston, South Carolina
Sector Jacksonville	Jacksonville, Florida
Sector Miami	Miami, Florida
Sector Key West	Key West, Florida
Sector St. Petersburg	St. Petersburg, Florida
Sector San Juan	San Juan, Puerto Rico
Sector Ohio Valley	Louisville, Kentucky
Sector Upper Mississippi	St. Louis, Missouri
Sector Lower Mississippi	Memphis, Tennessee
Sector New Orleans	New Orleans, Louisiana
Sector Houston-Galveston	Houston, Texas
Sector Corpus Christi	Corpus Christi, Texas
Sector Buffalo	Buffalo, New York
Sector Detroit	Detroit, Michigan
Sector Lake Michigan	Milwaukee, Wisconsin
Sector Sault Ste. Marie	Sault Ste. Marie, Michigan

Sector	Location of Sector Command
Sector San Francisco	San Francisco, California
Sector Humboldt Bay	McKinleyville, California
Sector Los Angeles–Long Beach	San Pedro, California
Sector San Diego	San Diego, California
Sector Puget Sound	Seattle, Washington
Sector Columbia River	Astoria, Oregon
Sector North Bend	North Bend, Oregon
Sector Honolulu	Honolulu, Hawaii
Sector Guam	Santa Rita, Guam
Sector Anchorage	Anchorage, Alaska
Sector Juneau	Juneau, Alaska

U.S. Coast Guard Air Stations

Most Coast Guard aircraft operate from bases called air stations. The mission of an air station is to operate and provide maintenance support for the aircraft assigned. This may include providing crews and maintenance support for helicopters assigned to Coast Guard ships. Air stations have facilities such as hangars, maintenance shops, and medical facilities. The Coast Guard operates twenty-one Coast Guard Air Stations around the country:

USCG Air Station Cape Cod, Massachusetts
USCG Air Station Atlantic City, New Jersey
USCG Air Station Elizabeth City, North Carolina
USCG Air Station Savannah, Georgia
USCG Air Station Clearwater, Florida
USCG Air Station Borinquen, Puerto Rico
USCG Air Station New Orleans, Louisiana
USCG Air Station Houston, Texas
USCG Air Station Corpus Christi, Texas
USCG Air Station San Diego, California
USCG Air Station San Francisco, California
USCG Air Station Humboldt Bay, California
USCG Air Station Sacramento, California
USCG Air Station North Bend, Oregon

USCG Air Station Astoria, Oregon
USCG Air Station Port Angeles, Washington
USCG Air Station Detroit, Michigan
USCG Air Station Traverse City, Michigan
USCG Air Station Sitka, Alaska
USCG Air Station Kodiak, Alaska
USCG Air Station Barbers Point, Hawaii

Coast Guard Reserve

The Coast Guard Reserve, as a fully integrated partner in carrying out the Coast Guard's missions, adds significant capability and flexibility in meeting the service's military mobilization requirements (under Title 10) as well as domestic contingency and planned operational surge requirements (under Title 14). The Coast Guard Reserve also serves as a ready force to fill short-term active-component personnel gaps and help mitigate personnel tempo stress.

The role of the Coast Guard Reserve is to provide trained units and qualified personnel to assist the Coast Guard in meeting its national defense responsibilities, in responding to domestic emergencies, and in performing routine operations. These activities support the primary roles of the U.S. Coast Guard, which are national defense, maritime law enforcement, marine safety, and marine environmental protection.

Overall responsibility for the Coast Guard Reserve lies with the commandant of the Coast Guard, subject to regulations prescribed by the secretary of Homeland Security. Within the Coast Guard, the director of reserve and leadership, a rear admiral, formulates plans, programs, and policies of the Coast Guard Reserve and monitors and reviews the Coast Guard Reserve program.

The chain of military command for both operational and administrative control of Coast Guard Reserve training programs extends from the commandant to area/deputy commanders and to the commanders of the nine Coast Guard districts and then to the commanders or commanding officers of active-duty commands, port security units, naval coastal warfare groups and units, and harbor defense command units.

The historic 1994–95 integration of most Coast Guard selected reservists into active-component commands signaled a new era for the Coast Guard Reserve and made it unique among our nation's seven reserve components. Under Team Coast Guard, the majority of Coast Guard reservists are assigned directly to the active-component unit where they train. In the field, respective active-component commanders exercise operational control over assigned reservists.

With most Coast Guard reservists assigned to the same active-duty command that they would augment upon mobilization, they are better prepared both administratively and operationally to report, in most cases, within twenty-four hours of call-up. The exceptions to this are the Coast Guard port security units and harbor defense commands, which are nearly exclusively reserve staffed. Under the Title 14 recall authority, the secretary of Homeland Security may involuntarily recall reservists to serve in domestic emergencies, in which case the local district commander determines the specialties and number of personnel to recall.

United States Coast Guard Auxiliary

The United States Coast Guard Auxiliary is a volunteer service organization that has members in the fifty-four states and territories and the District of Columbia. Membership is open to men and women seventeen years or older, U.S. citizens of all states and territories, civilians or active-duty or former members of any of the uniformed services and their reserve components, including the Coast Guard. Facility (radio station, boat, or aircraft) ownership is desirable but not mandatory. As volunteers, members receive no pay. When auxiliarists wear the uniform or fly the auxiliary ensign, they are a member of Team Coast Guard, and their actions as auxiliarists reflect directly on the U.S. Coast Guard. In many areas they will be the Coast Guard's only link with the public.

The purpose of the Coast Guard Auxiliary is to assist the Coast Guard in any of its missions except where prohibited by statute, such as in direct law enforcement and military actions. Therefore, the primary missions of the U.S. Coast Guard Auxiliary are as follows:

> **Recreational Boating Safety:** This mission is accomplished by qualified members delivering public safety boating education, training, and classes; by providing voluntary vessel safety checks to boaters; by visiting with and educating local marine-related industries; and by general outreach to the boating public through boat shows and other public venues and events.
>
> **Operations and Marine Safety:** For those interested in boating, the Coast Guard Auxiliary offers a rigorous level of hands-on

training and qualification as boat crew and coxswain. Qualified auxiliarists perform regular safety patrol missions in their local area and support local boating activities on the water such as regattas, fireworks, and fleet visits. In addition, auxiliarists work side by side with their active-duty USCG counterparts in many other mission areas, including environmental protection, commercial vessel safety inspections, port security and planning, licensing and documentation, and other vital operational roles. Auxiliarists receive training virtually identical to that of their active-duty and reserve counterparts.

Mission Support: The Coast Guard Auxiliary needs people with leadership, administrative, and technical skills (such as Web site design, computer server administration, graphic design, photography, videography, communications, public relations and public affairs, instruction and instructional design, and personnel services) to support those auxiliarists involved in the recreational boating safety and operations and marine safety missions.

Although under the authority of the U.S. Department of Homeland Security via the commandant of the U.S. Coast Guard, the Coast Guard Auxiliary is internally autonomous, operating on the following four organizational levels:

National: The Coast Guard Auxiliary has national officers who are responsible, along with the commandant, for the administration and policymaking for the entire auxiliary program. These officers make up the National Executive Committee, which is composed of the chief director of auxiliary (an active-duty officer), national commodore, and the national vice commodores.

District/Region: Flotillas and divisions are organized in districts comparable to the Coast Guard districts and must be assigned the same district number. Some districts are further divided into regions. The district/region provides administrative and supervisory support to divisions, and promotes the policies of both the district commander and the National Auxiliary Committee. All districts and regions are governed by a district commodore, district vice commodore, and district rear commodore under the guidance of the Coast Guard district commander. At this level Coast Guard officers are assigned to oversee and promote the auxiliary programs.

Division: For maximum administrative effectiveness in carrying out auxiliary programs, flotillas in the same general geographic area are grouped into divisions. The division provides administrative, training, and supervisory support to flotillas and promotes district policy. Each division is headed by a division captain and division vice-captain and usually consists of five or more flotillas.

Flotilla: The flotilla is the basic organizational unit of the Coast Guard Auxiliary and is composed of at least fifteen qualified members who carry out auxiliary program activities. Every auxiliarist is a member of a local flotilla. Each flotilla is headed by a flotilla commander.

II
PLATFORMS

Figure 4-1.
USCGC *Rockaway* (WAVP-377, WAGO-377, WHEC-377, WOLE-377) sometime
before the Coast Guard's adoption of the "racing stripe."

CHAPTER 4

Cutters

A cutter is defined as any Coast Guard vessel sixty-five feet in length or more with adequate accommodations for the crew to live on board.

CUTTERS

418-foot Legend-class National Security Cutter (WMSL)

The national security cutter (NSC) is the flagship of the fleet, capable of meeting all maritime security mission needs. It is the largest and most technically advanced class of cutter in the Coast Guard, with robust capabilities for maritime homeland security, law enforcement, and national defense missions.

Length:	418 feet
Beam:	54 feet
Displacement:	4,306 tons

378-foot *Hamilton*-class High Endurance Cutter (WHEC)

The 378-foot *Hamilton*-class high endurance cutters are the largest cutters ever built for the Coast Guard, aside from the three major icebreakers and NSCs. Diesel engines and gas turbines provide power, and the cutters have controllable-pitch propellers. Equipped with a helicopter flight deck, retractable hangar, and facilities to support helicopter deployment, this class of cutters was introduced to the Coast Guard inventory in the 1960s. Beginning in the 1980s and ending in 1992, the entire class was modernized through the Fleet Renovation and Modernization program. The first of the class was the *Hamilton* (WHEC-715), commissioned in 1967.

Length:	378 feet
Beam:	43 feet
Displacement:	3,300 tons

282-foot *Alex Haley* Medium Endurance Cutter (WMEC)

The U.S. Coast Guard cutter *Alex Haley* (WMEC-39) is a former U.S. Navy vessel (ex–USS *Edenton* [ATS-1]) that was recommissioned for Coast Guard duty in 1999. The cutter was named after author, journalist, and Chief Petty Officer Alex Haley, USCG.

Length:	282 feet
Beam:	50 feet
Displacement:	3,000 tons

270-foot Famous-class Medium Endurance Cutter (WMEC)

Each cutter in the Famous class is named for a prominent cutter in Coast Guard history. Their primary missions include law enforcement, search and rescue, and defense readiness. They are equipped with a helicopter landing deck, retractable hangar, and the facilities to support helicopter deployments. The Famous-class cutters are the Coast Guard's primary tools for law enforcement, counterdrug, and search-and-rescue missions.

Length:	270 feet
Beam:	38 feet
Displacement:	1,852 tons

210-foot *Reliance*-class Medium Endurance Cutter (WMEC)

The *Reliance*-class cutters are older than those in the Famous class and completed a major maintenance availability (MMA) rehabilitation program. The purpose of the MMA is to improve machinery and equipment and to keep the cutters mission-capable during the second half of their service lives. Their primary missions include law enforcement, search and rescue, and defense readiness. They are equipped with a helicopter landing deck and the facilities to support helicopter deployments, but they have no hangar.

Figure 4-2. A national security cutter of the Legend class.

Figure 4-3. A high endurance cutter (WHEC) of the *Hamilton* class.

Figure 4-4. The crew of USCGC *Alex Haley* (WMEC-39) prepares to tie up at the Base Support Unit Kodiak Pier following a recent deployment.

Figure 4-5. A medium endurance cutter (WMEC) of the Famous class.

Length:	210 feet
Beam:	34 feet
Displacement:	1,000 tons

154-foot Sentinel-class Fast Response Cutter (FRC)

The Sentinel-class cutters are able to deploy independently to conduct port, waterway, and coastal security; fishery patrols; drug and illegal migrant law enforcement; search and rescue; and national defense operations. The Sentinel-class FRCs are replacing the 110-foot Island-class patrol boats.

Length:	153.5 feet
Beam:	25.4 feet
Displacement:	353 tons

110-foot Island-class Patrol Boat (WPB)

The 110-foot Island-class patrol boats are a Coast Guard modification of a highly successful British-designed patrol boat. With excellent range and sea-keeping capabilities, the Island class, all named after U.S. islands, replaced the older 95-foot Cape-class patrol boats. These cutters are equipped with advanced electronics and navigation equipment.

Length:	110 feet
Beam:	21 feet
Displacement:	168 tons

87-foot Marine Protector–class Coastal Patrol Boat (WPB)

The 87-foot coastal patrol boat has improved mission sea-keeping abilities (up to sea state 5), significantly upgraded habitability, and compliance with all current and projected environmental protection laws. It also employs an innovative stern launch and recovery system using an aluminum-hulled, inboard, diesel-powered, water-jet small boat. The larger pilot house is equipped with an integrated bridge system including an electronic chart display and information system, which interfaces with the Coast Guard's new surface-search radar.

Length:	87 feet
Beam:	19 feet
Displacement:	91 tons

Figure 4-6. A medium endurance cutter (WMEC) of the *Reliance* class.

Figure 4-7. A fast response cutter of the Sentinel class.

Figure 4-8. A patrol boat (WPB) of the Island class.

Figure 4-9. A patrol boat of the Marine Protector class.

295-foot Barque *Eagle* (WIX)

The *Eagle* is a three-masted sailing barque with 21,350 square feet of sail. It is homeported at the Coast Guard Academy, New London, Connecticut. It is the only active (operational) commissioned sailing vessel in the U.S. maritime services (and one of only five such training barques in world). Sister ships include *Mircea* of Romania, *Sagres II* of Portugal, *Gorch Frock* of Germany, and *Tovarich* of Russia.

The *Eagle* bears a name that goes back to the early history of the United States' oldest continuous seagoing service. The first *Eagle* was commissioned in 1792, just two years after the formation of the Revenue Marine, the forerunner of today's Coast Guard. Today's *Eagle*, the seventh in a long line of proud cutters to bear the name, was built in 1936 by the Blohm & Voss Shipyard, Hamburg, Germany, as a training vessel for German naval cadets. It was commissioned *Horst Wessel* and, following World War II, was taken as a war prize by the United States. On May 15, 1946, the barque was commissioned into U.S. Coast Guard service as the *Eagle* and sailed from Bremerhaven, Germany, to New London, Connecticut.

Eagle serves as a seagoing classroom for approximately 175 cadets and instructors from the U.S. Coast Guard Academy. It is on the decks and rigging of the *Eagle* that the young men and women of the Academy get their first taste of salt air and life at sea. From this experience they develop a respect for the elements that will be with them throughout their lifetimes. They are tested and challenged, often to the limits of their endurance. Working aloft, they meet fear and learn to overcome it. The training that cadets receive under sail has proven to be an invaluable asset during their subsequent Coast Guard careers.

On *Eagle*, cadets have a chance to practically apply the navigation, engineering, and other training they receive in classes at the Academy. As upper-class cadets, they perform the leadership functions normally handled by junior officers. As under-class cadets, they fill positions normally taken by the enlisted crew of the ship, including helm watch at the huge brass and wood wheels used to steer the vessel.

Sailing in *Eagle*, cadets handle more than 20,000 square feet of sail and 5 miles of rigging. Over 200 lines must be coordinated during a major ship maneuver, so cadets must learn the name and function of each line.

The ship readily takes to the task for which it was designed. *Eagle*'s hull is built of steel, four-tenths of an inch thick. It has two full length steel

Figure 4-10. Coast Guard barque *Eagle* (WIX-327).

decks with a platform deck below and a raised forecastle and quarterdeck. The weather decks are three-inch-thick teak over steel.

When at home, *Eagle* rests alongside a pier at the Coast Guard Academy on the Thames River. Key characteristics of *Eagle* include:

Length:	295 feet
Maximum Speed:	1–16 knots (under full sail)
Maximum Range:	5,450 miles
Complement:	12 officers/38 enlisted/150 cadets (average)
Commissioned:	1946

ICEBREAKERS

The largest cutters operated by the Coast Guard are icebreakers. These cutters, specifically designed for open-water icebreaking, have reinforced hulls, special icebreaking bows, and a system that allows a rapid shifting of ballast to increase the effectiveness of their icebreaking capabilities.

420-foot *Healy*-class Icebreaker (WAGB)

USCGC *Healy* (WAGB-20) is a multimission icebreaking research vessel capable of performing operations satisfying a broad spectrum of scientific and icebreaking requirements in all polar regions. *Healy* is capable of breaking four and a half feet of ice continuously at three knots and can operate in temperatures as low as minus fifty degrees Fahrenheit.

Length:	420 feet
Beam:	82 feet
Designated Draft:	29 feet

399-foot Polar-class Icebreaker (WAGB)

USCGC *Polar Star* (WAGB-10) deploys to the polar regions, supporting science and research as well as providing resupply to remote stations in places most other ships cannot reach. It is capable of breaking ice six feet thick at a speed of three knots, and at a slower speed it can break much thicker ice.

Length:	399 feet
Beam:	83.5 feet
Displacement:	13,194 tons

240-foot Great Lakes Icebreaker (WLBB)

USCGC *Mackinaw* (WLBB-30) assumed the primary duties of icebreaking and aids to navigation (ATON) from USCGC *Mackinaw* (WAGB-83), decommissioned the same day. *Mackinaw* is homeported in Cheboygan, Michigan.

Length:	240 feet
Beam:	58 feet
Displacement:	3,350 tons
Commissioned:	10 June 2006

140-foot Bay-class Icebreaking Tugs (WTGB)

Eight 140-foot icebreaking tugs make up the Coast Guard's Bay class. In addition to icebreaking, they are used for search and rescue, enforcement of laws and treaties, deployment of marine environmental protection equipment, and support for ATON activities. The Bay class uses a low-pressure-air hull lubrication, or bubbler, system that forces air and

Figure 4-11. USCGC *Healy* (WAGB-20) is the United States' most technologically advanced polar icebreaker.
Official National Oceanic and Atmospheric Administration photograph

water between the hull and ice. This system improves icebreaking capabilities by reducing resistance against the hull, thereby reducing horsepower requirements.

Length:	140 feet
Beam:	37.5 feet
Displacement:	662 tons
Commissioned:	1979–88

BUOY TENDERS

Buoy tenders are Coast Guard cutters responsible for maintaining ATON, such as fixed structures and navigational buoys.

225-foot *Juniper*-class Seagoing Buoy Tender (WLB)

The 225-foot *Juniper* was commissioned in 1996 as the lead ship in the Coast Guard's buoy tender replacement project, a major acquisition to replace the World War II–era 180-foot buoy tenders. The 225-foot WLB,

Figure 4-12. USCGC *Polar Star* (WAGB-10) is a USCGC heavy icebreaker.

Figure 4-13. USCGC *Mackinaw* (WLBB-30) is the only U.S. heavy icebreaking resource assigned to the Great Lakes.

along with the 175-foot WLM, represents the latest in shipbuilding, propulsion, and ship control technology.

The 225-foot WLB is equipped with a single controllable-pitch propeller and bow and stern thrusters, which give the cutter the maneuverability it needs to tend buoys offshore and in restricted waters. A sophisticated machinery plant control and monitoring system and an electronic chart display and information system enable the 225-foot cutters to reduce the watch standing complement compared to the 180-foot cutters. A dynamic global positioning system can hold the vessel within a ten-meter circle, allowing the crew to service and position floating ATON more efficiently than before, in winds to 30 knots and 8-foot seas.

Length:	225 feet
Beam:	46 feet
Draft:	13 feet
Displacement:	2,000 tons

175-foot Keeper-class Coastal Buoy Tender (WLM)

The 175-foot Keeper-class coastal buoy tenders, along with the 225-foot *Juniper*-class seagoing buoy tenders, represent the new wave in buoy tending. They are the first Coast Guard cutters equipped with Z-drive propulsion units, instead of the standard propeller and rudder configuration, which can independently rotate 360 degrees. Combined with a thruster in the bow, these propulsion units give the Keeper-class cutters unmatched maneuverability. With state-of-the-art electronics and navigation, these buoy tenders maneuver and position aids more accurately and efficiently with fewer crew.

Length:	175 feet
Beam:	36 feet
Draft:	7.9 feet
Displacement:	845 tons

Figure 5-1.
An HC-130J Super Hercules.

CHAPTER 5

Aircraft

COAST GUARD AVIATION

There are more than two hundred aircraft in the Coast Guard's inventory. The exact figure fluctuates operationally due to maintenance schedules. Major missions include search and rescue, law enforcement, environmental response, ice operations, and air interdiction. Fixed-wing aircraft operate from large and small Coast Guard air stations. Rotary-wing aircraft operate from flight-deck-equipped cutters, air stations, and air facilities.

Coast Guard aviation began officially from an act of Congress dated 29 August 1916 that authorized the Treasury Department to establish ten Coast Guard air stations along the coasts of the United States. Provisions were soon made to train pilots and develop rescue procedures for using airplanes in search and patrol duties. The Coast Guard also became active early in the evolution of helicopters. The Coast Guard developed rescue techniques and equipment to pioneer the use of helicopters as rescue craft. In addition to search-and-rescue duties, Coast Guard helicopters today are used for a variety of missions and are deployed on many Coast Guard cutters.

Coast Guard aircraft types vary and have changed through the years to meet mission requirements. Some of the aircraft and their designs were obtained from the Department of Defense while others were designed specifically to Coast Guard specifications.

Aircraft Markings

Aircraft operated by the Coast Guard are readily identifiable by their paint schemes and markings. Fixed-wing aircraft are painted with an overall glossy white, solar-reflective finish. The wingtips, tails, and noses of the airplanes are painted red, and the fuselages sport red and blue Coast Guard emblem stripes.

Designations

The Coast Guard uses the standard Department of Defense system to designate aircraft. A typical example follows:

HC-130H

H: modified mission symbol
C: basic mission symbol
130: model number
H: modification

Long-Range Surveillance Aircraft

HC-130H Hercules

The HC-130H is a long-range surveillance and transport aircraft that effectively serves the Coast Guard missions of search and rescue, enforcement of laws and treaties (including illegal-drug interdiction), marine environmental protection, and international ice patrols over the North Atlantic as well as cargo and personnel transport and military readiness. The HC-130Hs can exceed 2,600 nautical miles in low-altitude flight and have a mission endurance of up to fourteen hours.

HC-130J Super Hercules

The Coast Guard accepted the six "missionized" fully operational HC-130Js in July 2010. Missionized HC-130Js are equipped with the Elta-2022 surface-search radar, an infrared/optical camera system, and an advanced communications suite.

Medium-Range Surveillance Aircraft

HC-144A Ocean Sentry

The Ocean Sentry has a range of more than 1,500 nautical miles and a maximum airspeed of 230 knots. It is the Coast Guard's newest addition to the aviation fleet. The Coast Guard will use the HC-144 to perform homeland security and search-and-rescue missions, enforce laws and treaties (including illegal-drug interdiction), marine environmental protection, military readiness, and international ice patrol missions as well as cargo and personnel transport.

Figure 5-2. An HC-144 Ocean Sentry.

The size, range, and reconfiguration capabilities will fully enable the execution of the multiple missions performed by the Coast Guard. The Ocean Sentry is equipped with a state-of-the-market Rockwell-Collins Flight 2 glass cockpit instrument panel, autopilot, and avionics suite for a two-person aircrew. It also uses a mission equipment pallet that is interoperable with that of the HC-130J long-range surveillance aircraft and includes command, control, communications, computers, intelligence, surveillance, and reconnaissance (C4ISR) equipment for enhanced situational awareness, improved surveillance through radar and electro-optical/infrared sensors systems, mission data recording, a first-responder/law enforcement and marine communications suite, and enhanced secure-data encryption capabilities.

HC-27J Spartan

The Coast Guard is integrating fourteen C-27J Spartans into its medium-range surveillance aircraft fleet to operate along with its HC-144 Ocean Sentries performing drug and migrant interdiction, disaster response, and search-and-rescue missions. Following regeneration and missionization, the aircraft's range, endurance, speed, and payload will make it a valuable asset in addressing the Coast Guard's maritime flight hours gap. Key characteristics of these aircraft, transferred from the United States Air Force, include:

Length:	74 feet 6 inches
Wingspan:	94 feet 2 inches
Height:	31 feet 8 inches
Maximum Weight:	70,000 pounds

Figure 5-3. An HC-27J Spartan sits on the runway at Coast Guard Aviation Logistics Center in Elizabeth City, North Carolina, 31 March 2016.

Cruise Speed:	290 knots true airspeed
Range:	2,675 nautical miles
Endurance:	12 hours

Rotary-Wing Aircraft

Medium-Range Recovery MH-60T Jayhawk

The MH-60J, formerly the HH-60, was upgraded following the attacks of 9/11 to include aerial use-of-force capabilities. A further upgrade is under way that will change the Jayhawk to the T-model, or Thunderhawk. The MH-60 is a medium-range recovery helicopter with a maximum speed of 170 knots and a range of over 400 nautical miles. The MH-60 is the Coast Guard's "workhorse," able to perform rescues in the harshest of weather and sea states. When fully

Figure 5-4. An MH-60T Jayhawk.

upgraded, the HH-60 helicopters will be modernized and used as medium-range responders for offshore operations and will be able to provide shore-based aviation surveillance capability and transport. This project will upgrade and extend the HH-60J's service life, adding new avionics, new radar/forward-looking infrared sensors, upgrades to gas turbine power plant, and chemical, biological, and radiological environmental hazard detection and defense.

A new, modernized cockpit includes five multifunctional display screens, full-screen radar display, forward-looking infrared and hoist camera images, primary flight instruments, and an integrated traffic collision avoidance system. As previously mentioned, it will be modified with an airborne use-of-force package—including weapons for firing warning and disabling shots, and armor to protect the aircrew from small-arms fire.

Short-Range Recovery MH-65D/E Dolphin

The MH-65 Dolphin helicopter has a range of 300 nautical miles and a maximum speed of 170 knots. As the Coast Guard's main "cutter aircraft," the Dolphin has been invaluable to the fleet for search and rescue, law enforcement, and homeland security. These aircraft are currently being upgraded. Once fully upgraded, the Coast Guard's MH-65s will be redesignated as the multimission cutter helicopter (MCH) and will perform search-and-rescue, law enforcement, and homeland security missions. A major upgrade is the reengineering effort that provides 40 percent more power and higher performance. The MCH will feature enhanced C4ISR equipment. Also featured is improved vertical insertion and vertical delivery capability, which is the ability to deliver a three-person interagency counterterrorism or response team fifty nautical miles from a shore or from a Coast Guard flight-deck-equipped cutter.

C-37A Gulfstream V Long-Range Command-and-Control Aircraft

The Coast Guard operates a single, ultra-long-range Gulfstream V as its principal command-and-control transport for travel by the secretary of Homeland Security, the Coast Guard commandant, and other U.S. officials. The C-37A enjoys commonality of parts and supplies with more than a dozen C-37As operated by Department of Defense.

Figure 5-5. An MH-65 Dolphin.

Figure 6-1.
Students and instructors at the Coast Guard's National Motor Lifeboat School in Ilwaco, Washington, operate forty-seven-foot motor lifeboats in heavy surf.

CHAPTER 6

Boats

All vessels under sixty-five feet in length are classified as boats and usually operate near shore and on inland waterways. All Coast Guard small boats, whether assigned to a cutter or not, have service numbers of five digits. The first two digits give the length of the boat and the last three its individual number. Boats assigned to cutters are numbered in accordance with their location on board the cutter. The lower numbers indicate forward stowage, with odd numbers being on the starboard side and even numbers on the port side.

Sizes range from sixty-four feet in length down to twelve feet. Just some of the Coast Guard's boats include:

47-FOOT MOTOR LIFEBOAT (MLB)

The 47-foot MLB is primarily designed to operate in high seas, surf, and heavy weather environments up to 30-foot seas, 20-foot surf, and winds up to 50 knots. With safety in mind, 13 water-tight compartments are included. The MLB can self-right in only 30 seconds. With state-of-the-art electronically controlled engines, fuel management systems, and integrated electronics suite, including 4 coxswain control stations, the MLB has become the ideal platform for operations in extreme at seas weather conditions.

45-FOOT RESPONSE BOAT, MEDIUM (RB-M)

The 45-foot response boat, medium has revitalized the Coast Guard's shore-based boat fleet, delivering improved speed, maneuverability, and ergonomics over the 41-foot utility boat and other nonstandard boats it

Figure 6-2. A heavy-weather coxswain and boat crew from Station Monterey operate a forty-seven-foot motor lifeboat.

has replaced. The RB-M makes boat crews more effective in performing multiple Coast Guard missions, including search and rescue; ports, waterways, and coastal security; law enforcement; and drug and migrant interdiction.

35-FOOT LONG-RANGE INTERCEPTOR-II (LRI-II)

The LRI-II deploys from the national security cutter and was designed to extend the national security cutter's mission range. Its size allows it to carry more passengers and equipment, move through rougher seas with increased stability, and extend its range. The LRI-II also features shock-mitigating seats for 15 passengers and crew, some of which fold to facilitate crew movement and create extra storage. Characteristics include speed of 35 knots with a range of 240 nautical miles.

33-FOOT SPECIAL-PURPOSE CRAFT, LAW ENFORCEMENT (SPC-LE)

The 33-foot SPC-LE is a multimission craft designed and built primarily for counterdrug and migrant missions along the U.S. maritime border. Powered by three 300 horsepower outboard engines, the SPC-LE's speed and agility make it a very capable interceptor. The SPC-LE is outfitted with six shock-mitigating crew seats to mitigate crew fatigue.

26-FOOT TO 64-FOOT AIDS-TO-NAVIGATION BOATS

These boats were designed primarily to service aids to navigation within the inland waters of the United States.

25-FOOT AND 32-FOOT TRANSFERABLE PORT SECURITY BOAT (TPSB)

The 32-foot TPSB is designed as a multimission-capable boat. Primary missions include landside security and waterside security operations, providing waterborne security and point defense force protection in addition to standard Coast Guard missions. The TPSB is a shore-based asset.

The 25-foot TPSB (Boston Whaler) can travel at more than 40 knots and carry a .50 caliber machine gun and two M60 machine guns mounted on each side. The TPSB and port security unit (PSU) personnel are located in various points across the United States to be readily deployed overseas.

26-FOOT OVER THE HORIZON-IV (OTH-IV)

The OTH-IV is interoperable with multiple cutter classes equipped with stern launch. The OTH-IV is longer than its predecessor and features improved endurance, upgraded electronics, enhanced shock mitigation, and a bimini top that provides sun protection for the crew. Characteristics include speed of 40 knots with a range of 200 nautical miles.

25-FOOT DEFENDER-CLASS BOATS (RB-HS/RB-S)

Developed in a direct response to the need for additional homeland security assets in the wake of the September 11 terrorist attacks, the Defender-class boats were procured under an emergency acquisition authority. The response boat, small (RB-S) is a high-speed, easily deployable asset designed to operate year-round in shallow waters along coastal borders.

18-FOOT TO 24-FOOT SPECIAL-PURPOSE CRAFT, AIRBOATS (SPC-AIR)

There are several types of special-purpose crafts. These crafts are unique in their performance of an authorized mission requiring specialized capability that cannot be met within the standardized boat fleet.

The SPC-AIR are used by various Coast Guard units in the execution of ice rescue missions and flood response, and are capable of operating over water, ice, and short distances on land. Airboats are also used to get responders into shallow-water areas and marsh areas where deep-draft vessels cannot go. They range in length from 18 feet to 24 feet.

BOAT EQUIPMENT AND GEAR

The equipment supplied with each boat varies with the conditions under which it must operate. Each class of Coast Guard boat comes equipped with certain gear and parts that are standard issue (i.e., spare parts, tools, bilge pump, binoculars, navigational equipment).

Boat Outfit Lists itemize the articles of equipment recommended for each standard service boat. At shore stations, conditions of use may require modification of the list, which may be accomplished by the commanding officer or officer in charge of the unit. Checks are made at frequent intervals to ensure that all equipment required for the proper operation of the boat is on board. Lives may depend on this.

LIFEBOATS

While at sea, cutters always have one boat designated as the ready lifeboat. This boat is equipped for rescue work and can be launched immediately. The people that crew lifeboats as well as the boat-lowering crew must be thoroughly familiar with their duties and ready to take their stations immediately.

Lowering Lifeboats at Sea

The crew mans the boat and dons lifejackets. Personnel on deck—who man the falls, frapping lines, sea painter, fenders, and so forth—take their stations. When all is ready, the gripes are tripped and cleared away and fenders lowered. Personnel in the boat must grasp the knotted lifelines firmly until the boat is waterborne so that they will not be lost overboard if the boat drops out from under them. Frapping lines are passed around the falls and tended on deck. (Frapping lines prevent the boat from swinging while being lowered.) Members of the boat's crew stand by the releasing hooks of the falls fore and aft to release them at the command of the coxswain. One member of the boat's crew stands by the sea painter, ready to release it.

At the command "Lower away together," crew members manning the falls lower away smartly so that the boat will be dropped onto the crest of a wave. When the boat is waterborne, the coxswain orders the *after* falls released first, and then the *forward* falls. At this point the boat is sheered away from the ship's side by use of the sea painter and steering sweep. Once the boat is clear of the ship's side, the sea painter is released and the boat proceeds on its mission.

Hoisting Lifeboats at Sea

This procedure is practically a repeat of the lowering procedure but in reverse. The sea painter is passed from the ship to the boat and secured in the boat. The coxswain veers the boat under the falls, using the sea painter and rudder. The *forward* fall is hooked first, then the *after* fall. When the falls are hooked and secured, the slack is taken up and, at the command "Hoist away," the deck crew hoists the boat with the utmost speed possible.

Prior to the boat's return to the ship, the deck crew makes full preparations to hoist in without delay; frapping lines are passed, fenders manned, sea painter made ready to pass to the boat, falls led out, and all hoisting stations manned.

It is important to keep hands off the gunwales. If the boat hits the side of the ship, a person could lose part or all of a hand caught between the ship and the gunwale.

FUEL

Most Coast Guard motorboats are powered by diesel engines. However, many gasoline outboard engines are also in use. Gasoline is a highly inflammable, volatile liquid that must be handled with great caution. Diesel fuel, though not as inflammable or as volatile as gasoline, is still dangerous.

MOTORBOAT HANDLING

The following factors should be considered in handling motorboats:

- *Normal effect of the rudder.* When the steering wheel is turned to the right, or the tiller moved to the left, the boat's head is swung to the right if the boat is moving ahead.
- *Screw current.* With the screw turning ahead, the normal rudder effect is greatly increased due to increased pressure exerted by the water expelled from the screw. When the screw is turning astern, the water expelled moves in the reverse direction; hence, the result

is to slightly lessen the effectiveness of the rudder if the boat is still going ahead.

- *Sidewise pressure of screw blades.* As the blades turn, they tend to push the stern in the opposite direction from that in which they are moving. Fortunately, upper and lower blades exert this force in opposite directions. However, there is a difference in water pressure at the top and the bottom of the screw equal roughly to a half-pound for each foot of the diameter of the screw. The lower blades are moving in greater water pressure and therefore exert more sidewise thrust. For this reason, when standard service boats are going ahead, the stern tends to swing to starboard. The reverse effect is noted when going astern. The effect of the sidewise pressure of the blades is hardly noticeable when going ahead. Going astern, there is a strong tendency for the stern to swing to port. This is why single-screw boats should always make portside landings if possible.

- *Effect of wind.* The wind has a decided effect on a boat, especially at slow speed. If the bow is high out of the water, it will be difficult to turn into the wind, as the sail area of the bow will have a "wind rudder" effect opposite to that of the rudder. This situation can be helped by increasing the bow weight—a process called "trimming." Despite trim, if the wind is strong and the boat is going astern, her stern will come up into the wind in spite of anything that can be done.

- *Effect of current.* If stopped, the boat will float with the current like a cork. When running, the course made good is a result of this motion and the motion imparted to the boat by the effects of screw, rudder, and wind.

III
MILITARY BASICS

Figure 7-1.
Chief Engineer James M. MacDougall, USRCS, a New York native,
in his service dress uniform, circa 1871. At this time officers were appointed
specifically as "engineering officers" to distinguish them from those of the "line."

CHAPTER 7

Uniforms
and Awards

The uniform identifies Coast Guardsmen as professional members of a military service that is more than 220 years old. It shows, at a glance, that they are members of the U.S. Coast Guard.

Because many of the Coast Guard's missions require close contact with the public, it is important that the public have faith and confidence in, and respect for, the Coast Guard. Presenting a proper appearance is a critical but easy way of developing that attitude in the public.

The *Uniform Regulations Manual*, COMDTINST M1020.6 (series), can answer just about any question there may be about uniforms, grooming, and insignia.

COMMON TYPES OF UNIFORMS

Operational Dress Uniform

Coast Guardsmen wear the operational dress uniform when engaged in work that would soil the dress uniform. This uniform is not a liberty uniform but is authorized for wear when traveling between work and residence. It may also be worn to service-type facilities on base, such as commissaries, exchanges, snack bars, and dispensaries during working hours.

Service Dress Blue

Service dress blue is the basic dress uniform for all officers and enlisted personnel. It is appropriate any time and in any season when a dress uniform is required. It is an authorized liberty uniform.

Tropical Blue

Tropical blue is also the standard Coast Guard uniform of the day, regardless of season or location. It is an authorized liberty uniform.

Winter Dress Blue

Winter dress blue is an optional cold-weather dress uniform. Normally this uniform may be worn from 1 November to 31 March or as local prescribing authority dictates.

Maternity Uniform

The maternity uniform is worn when the regular Coast Guard uniforms no longer fit. The maternity uniform is actually an Air Force maternity uniform.

Other Uniforms

The Coast Guard personnel also wear full dress, formal dress, and dinner dress uniforms. These are prescribed for special occasions, such as changes of command, ceremonial visits, and formal evening functions. Refer to the *Uniform Regulations Manual* for a listing of the items that are worn with these uniforms.

MARKING AND TAILORING

Ownership markings and tailoring take place when the service member enters the service. A complete description of markings and tailoring requirements are found in the *Uniform Regulations Manual*.

ORGANIZATIONAL CLOTHING

Local commands can authorize and issue clothing, such as flotation coats and law enforcement coveralls, if a need exists. Local directives will dictate when to wear these items.

CIVILIAN CLOTHING

Officers and enlisted personnel are permitted to possess civilian clothing on board ship and at units ashore. Personnel may wear civilian clothing while leaving or returning to ships or stations, while on leave or liberty, or in any off-duty status ashore. Civilian clothing should be appropriate for the occasion and in good taste so as not to bring discredit upon the Coast

Guard, and it should not conflict with accepted attire in the geographic area.

RESTRICTIONS

There are several restrictions on the wearing of the uniform:

- Coast Guardsmen should not wear the uniform if their actions in uniform would bring discredit upon the Coast Guard.
- Do not wear the uniform if wearing it indicates that the Coast Guard approves of an action or a product. This includes political activities and civilian employment.
- Women may not wear skirts when serving as crew members on cutters, boats, or aircraft except for certain official ceremonies.
- As a general rule, do not wear uniform items with civilian clothing. Some uniform items that do not present a distinctive Coast Guard appearance, such as raincoats, gloves, and shoes may, however, be worn with civilian clothes.

REPLACING UNIFORMS

Use Coast Guard–approved items when it is time to replace uniforms. Do not substitute items with similar civilian material. Uniforms have met certain exacting standards for safety. For example, shirts and trousers meet fire retardancy standards, and the steel in safety shoes meets specific impact resistance standards.

After six months of active duty, enlisted members will receive a clothing maintenance allowance. This is a monetary allowance to replace and maintain uniform items.

If your unit does not have access to a uniform store, you may mail-order items from the Uniform Distribution Center in Cape May. Your unit's personnel or supply office can provide you with information and forms for ordering.

APPEARANCE

The wearing of the Coast Guard uniform must be a matter of personal pride. Members of the Coast Guard are representatives of the United States, and their dress must reflect credit upon themselves, the Coast Guard, and our nation.

Uniforms must be kept clean and pressed. Refer to the labels in the uniform items for instructions on cleaning and pressing. Shine shoes and polish brass. Wear headgear squarely on the head. Button or zip up jackets at least two-thirds of the way. You may carry items in your pockets, but they must not be exposed. Keep hands out of pockets. Refer to the *Uniform Regulations Manual* for specific information regarding jewelry, sunglasses, and accessories such as backpacks and cell phones.

Men's Grooming

Hair should present a neat, clean, military appearance, tapered from the lower hairline upward. It should not touch the ears. Its length (distance from the scalp) should not exceed one and one-half inches (bulk). The style should not interfere with the wearing of a cap or safety device. Sideburns must be straight, end in a horizontal line, and not extend below the lowest part of the external ear opening.

Mustaches are permitted if they present a neat, clean, military appearance. They must be kept off the lip, above the corners of the mouth, and no more than one-quarter inch horizontally past the corners. Beards are not authorized.

Women's Grooming

Hair should present a neat, clean, military appearance. It may touch but not extend below the lower edge of the collar. The style should not interfere with the wearing of a cap or safety device. Hair ornaments such as pins and barrettes should be functional rather than decorative and approximate the color of the hair. Cosmetics should be conservative in color and amount. Gold, silver, pearl, or diamond ball earrings (one-eighth to one-quarter inch in diameter) may be worn centered on the earlobe, one per ear.

IDENTIFICATION OF PERSONNEL

Awards, devices, insignia, and emblems are displayed on the uniform to identify the rate, rank, rating, and experience of each individual.

Combination Cap Insignia

Nonrated members and petty officers (E-1 through E-6) wear a Coast Guard emblem on their caps. The chin strap is black. Chief petty officers (E-7 through E-9) wear a device consisting of an anchor and shield. E-8s and E-9s also wear one or two stars. The chin strap is black. Chief warrant officers and commissioned officers wear a device that is predominantly an eagle. The chin strap is gold. Senior and flag officers (O-5 through O-10) wear acorns and oak leaves on the brim.

Dress Jacket Sleeve Insignia

Nonrated members (E-1 through E-3) wear a left-arm sleeve insignia consisting of red, green, or white stripes that designate their rate. Petty officers (E-4 through E-6) wear left-arm sleeve insignia consisting of an eagle and red chevrons. Chief petty officers (E-7 through E-9) wear left-arm sleeve insignia consisting of an eagle, gold chevrons, and a gold rocker. E-8s wear one star above the eagle, and E-9s wear two stars.

Chief warrant officers and commissioned officers wear gold/blue or gold stripes around the lower part of both sleeves. Chief warrant officers wear a gold insignia designating their specialty. Commissioned officers have a gold shield above the stripe(s).

Enlisted Service Stripes

For every four years of service, enlisted members wear a service stripe on the lower left sleeve of the dress jacket: nonrated members and petty

Figure 7-2. Coast Guard sleeve insignia.

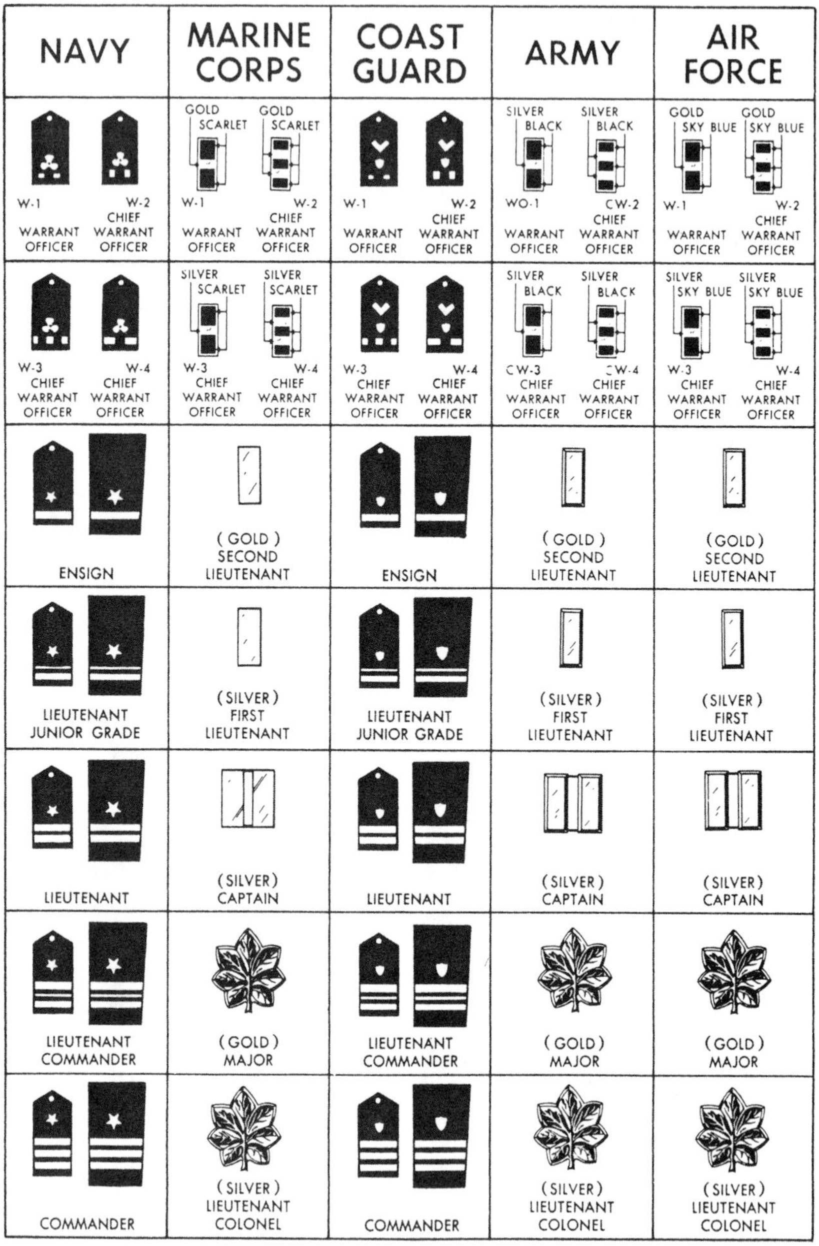

Figure 7-3. Officers' insignia.

NAVY	MARINE CORPS	COAST GUARD	ARMY	AIR FORCE
CAPTAIN	COLONEL	CAPTAIN	COLONEL	COLONEL
REAR ADMIRAL (LOWER HALF)	BRIGADIER GENERAL	REAR ADMIRAL (LOWER HALF)	BRIGADIER GENERAL	BRIGADIER GENERAL
REAR ADMIRAL	MAJOR GENERAL	REAR ADMIRAL	MAJOR GENERAL	MAJOR GENERAL
VICE ADMIRAL	LIEUTENANT GENERAL	VICE ADMIRAL	LIEUTENANT GENERAL	LIEUTENANT GENERAL
ADMIRAL	GENERAL	ADMIRAL	GENERAL	GENERAL
FLEET ADMIRAL	NONE	NONE	GENERAL OF THE ARMY	GENERAL OF THE AIR FORCE
NONE	NONE	NONE	AS PRESCRIBED BY INCUMBENT GENERAL OF THE ARMIES	NONE

	**Aviation Maintenance Technician (AMT)** A two-bladed winged propeller.	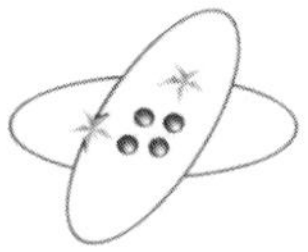	**Electronics Technician (ET)** A helium atom about which revolve two electrons.
	**Aviation Survival Technician (AST)** A winged flaming spherical shell with parachute.		**Food Service Specialist (FS)** Crossed quill and wheat spike with a key across top.
	**Avionics Electrical Technician (AET)** A winged helium atom about which revolves two electrons.		**Gunner's Mate (GM)** Crossed gun barrels with muzzles up.
	**Boatswain's Mate (BM)** Crossed anchors with crowns down.		**Health Services Technician (HS)** A caduceus.
	**Damage Controlman (DC)** Crossed fire axe and maul with handles down and axe blade to the front.	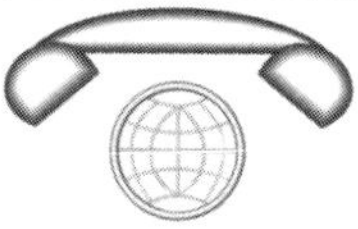	**Information System Technician (IT)** A globe with lines of embroidery representing lines of latitude and longitude, with a French type telephone above it.
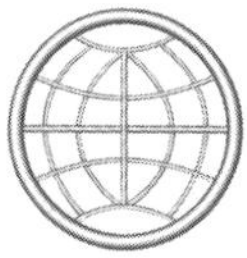	**Electrician's Mate (EM)** A globe with lines of embroidery representing lines of latitude and longitude.		**Intelligence Specialist (IS)** Crossed quill and lightning bolt, both pointing down; quill pen on top with bolt to the back.

Figure 7-4. Coast Guard rating badge specialty marks.

	Machinery Technician (MK) Face of an eight-toothed gear.		**Operations Specialist (OS)** An A-Scope superimposed on an arrow. Arrow is pointing diagonally upward and to the front
	Marine Science Technician (MST) A trident rising through waves		**Public Affairs Specialist (PA)** A camera surcharged with a quill pen
	Maritime Enforcement (ME) Law enforcement shield with a Coast Guard shield embedded.		**Storekeeper (SK)** Crossed keys with stems down and webs outward
	Musician (MU) A lyre.		**Yeoman (YN)** Crossed quill pens with nibs pointing down.

Reserve-Specific Ratings

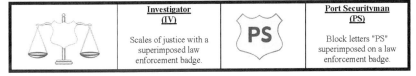

	Investigator (IV) Scales of justice with a superimposed law enforcement badge.		**Port Securityman (PS)** Block letters "PS" superimposed on a law enforcement badge.

NAVY	MARINES	ARMY	AIR FORCE	
MASTER CHIEF P.O.	SGT MAJOR / MASTER GUNNERY SGT	STAFF SGT MAJOR / COMMAND SGT MAJOR / SPEC 9	CHIEF MASTER SGT / CHIEF MASTER SGT OF THE AF	E-9
SENIOR CHIEF P.O.	1ST SGT / MASTER SGT	1ST SGT / MASTER SGT / SPEC 8	SENIOR MASTER SGT	E-8
CHIEF P.O.	GUNNERY SGT	SGT 1ST CLASS / SPEC 7	MASTER SGT	E-7
P.O. 1ST CLASS	STAFF SGT	STAFF SGT / SPEC 6	TECHNICAL SGT	E-6
P.O. 2ND CLASS	SGT	SGT / SPEC 5	STAFF SGT	E-5
P.O. 3RD CLASS	CORPORAL	CORPORAL / SPEC 4	SENIOR AIRMAN	E-4
SEAMAN	LANCE CORPORAL	PRIVATE 1ST CLASS	AIRMAN 1ST CLASS	E-3
SEAMAN APPRENTICE	PRIVATE 1ST CLASS	PRIVATE	AIRMAN	E-2
SEAMAN RECRUIT	PRIVATE	PRIVATE	BASIC AIRMAN	E-1

Figure 7-5. Enlisted ratings and pay grades—all five military services. (The Navy and the Coast Guard use the same ratings and pay grades.)

officers (E-1 through E-6), red stripes; chief petty officers (E-7 through E-9), gold.

Shoulder Boards

Chief warrant officers and commissioned officers wear shoulder boards on certain uniforms. The stripes are similar to the sleeve insignia.

Collar Devices

Nonrated members (E-1 through E-3) do not wear collar devices. Petty officers (E-4 through E-6) wear collar devices consisting of a Coast Guard shield and chevrons. Chief petty officers (E-7 through E-9) wear collar devices consisting of an anchor and shield along with the appropriate number of stars. Chief warrant officers and commissioned officers wear collar devices appropriate to their rank. Chief warrant officers wear a device indicating their specialty.

Enlisted Specialty Marks

Enlisted specialty marks are located on the dress jacket sleeve insignia between the eagle and chevrons. These marks identify the member's rating (job specialty).

Awards

Awards are displayed on the uniform as ribbons or medals. Awards have an order of rank to them. That order can be found in the *Medals and Awards Manual*, COMDTINST M1650.25 (series). This manual also lists awards earned in other branches of the armed forces that are authorized on the Coast Guard uniform.

Ribbons are centered one-quarter inch above the left pocket on a dress uniform. Ribbons are displayed in order from top to bottom, from center out. In other words, they would be displayed as if someone else was reading lines of print on your chest. Medals are displayed in a similar fashion.

Breast Insignia

Breast insignias are worn to indicate a special qualification or designation such as coxswain, cutterman, or air crewman. When ribbons or medals are worn, the insignia is worn centered one-quarter inch above the ribbons or medals. On uniforms without ribbons or medals, the insignia is worn centered one-quarter inch above the left pocket.

Figure 8-1.
Lowering the national ensign during evening colors on board the Coast Guard
cutter *Harriet Lane*. Colors is a traditional ceremony that takes
place on board all U.S. military vessels and bases every day at 0800 and sunset.

CHAPTER 8

Customs, Traditions, and Ceremonies

The customs, traditions, courtesies, and ceremonies observed in the Coast Guard have developed over many years. Some are common to all military services, and some are naval in nature because the Coast Guard is essentially a seagoing service. Coast Guard personnel should learn and observe all the customs, traditions, and ceremonies of the service and take pride in carrying them out properly. Like wearing the uniform, they are a mark of the select group of citizens who have taken a special oath to support and defend their country.

In one sense, service etiquette is simply the Coast Guard way of observing the rules of good manners that prevail in some form everywhere. To reduce these rules to their essence, well-mannered people always show respect and politeness toward each other, and it is always the duty of the junior to take the initiative. But the senior is equally obligated to respond.

Coast Guard personnel have a special reason for observing the rules of courtesy: their uniform makes them representatives of the United States and their service, wherever they are.

Customs and ceremonies of the armed forces show courtesies of several kinds. Some of the most important are

- Respect toward the emblems of our nation and toward its officials;
- Courteous behavior on board a ship of the Coast Guard or Navy;
- Mutual respect and courtesy between enlisted personnel and officers; and
- Respect and courtesy toward the flags, ships, officials, and other personnel of friendly nations.

Honors and ceremonies are prescribed in *United States Coast Guard Regulations.*

NATIONAL ENSIGN AND ANTHEM

The national ensign—the Stars and Stripes—is the flag of the United States. The national anthem is the song "The Star-Spangled Banner." These symbols represent the nation and are always treated with great respect. Similarly, the anthems and flags of countries whose governments are formally recognized by the United States (and not at war with the United States) are treated with respect. The same salutes should be rendered for other national flags and anthems as described here for our own. Our national flag is called "ensign" when displayed on a vessel or facility, "colors" when carried by foot, and "standard" when displayed on a vehicle or aircraft.

Colors

Every Coast Guard shore command and every ship not under way performs the ceremony of colors twice a day—at 0800 and at sunset. This ceremony consists of paying honor to the national ensign as it is hoisted at 0800 and lowered at sunset.

At five minutes before 0800 and five minutes before sunset, "First call" is sounded on the bugle (if the ship or station has a bugler), or a recording of first call is played. If not, the boatswain's mate pipes or whistles and passes the word, "First call to colors." Preparative Pennant is hoisted close-up.

At 0800, "Attention" is sounded on the bugle. Commands without buglers or records will pass the word, "Attention to colors," over their loudspeakers, or the quartermaster will blow a blast with a whistle. PREP is executed.

The national ensign is hoisted smartly to the top of the flagstaff. While it is being hoisted, the band, if there is one, plays the national anthem. If there is no band but there is a bugler, the bugler plays "To the colors." Everyone within sight or hearing renders honors as follows:

- If in ranks, you will be called to attention or to present arms.

- If you are in uniform but not in ranks, you stop whatever you are doing, face the colors, and salute until "Carry on" is sounded.

- If you are in a vehicle and traffic safety permits, you stop and sit at attention but do not salute. If conditions permit, the senior person in the vehicle gets out and salutes. The driver remains seated.

Figure 8-2. Folding the national ensign.

- If you are a passenger in a boat, you remain at attention, seated or standing. The boat officer or coxswain salutes for the boat.
- If you are in civilian or athletic gear at colors, stop and face the colors at attention. If you have a hat, hold it in your right hand, over your heart. If you have no hat, salute by holding your right hand over your heart. Members of the armed forces and veterans not in uniform may render the military salute.

At sunset a similar ceremony takes place. Five minutes before sunset, "First call" is sounded. At sunset, "Attention" is sounded and the band, if there is one, plays the national anthem. The flag is lowered slowly, so that it reaches the bottom and is balled up within the field of blue on the last note of the music. If there is no band but there is a bugler, then the bugler plays "Retreat." Your behavior during evening colors is the same as that for morning colors. You face the ensign at attention and hold your salute until "Carry on" is sounded.

Commissioned vessels not under way also hoist and lower the jack at morning and evening colors. This is a square flag with white stars on a blue background, the same as the small square in the national ensign. It is hoisted on the jackstaff, a small flagpole at the bow of a ship. Ships that are under way do not hold morning or evening colors. They hoist the national ensign at the gaff as they get under way, but the jack is not flown.

Folding the Ensign

The ensign is folded in half the long way so that the crease parallels the red and white stripes. It is folded in half again so that the new crease parallels the red and white stripes and the blue field is to the outside. The fly end (away from the blue field) is folded up to the top so that the single edge lies perpendicular to the stripes. By repeatedly folding the thick triangle thus formed about the inboard edge of the triangle, the ensign is folded into the shape of a cocked hat.

Shifting Colors

On unmooring, the instant the last mooring line leaves the pier or the anchor is aweigh, the quartermaster will blow a long whistle blast over the ship's intercom system and pass the word "Shift colors." The jack and ensign, if flying, will be hauled down smartly. At the same instance the "steaming" ensign will be hoisted on the gaff and the ship's call sign and other signal flags will be hoisted or broken. The jack is not flown. On mooring, the instant the anchor is let go or the first mooring line is made fast on the pier, the quartermaster passes the same word as for unmooring, the ship's call sign and the "steaming" ensign are hauled down smartly, and the jack and ensign are hoisted.

Half-Masting the Ensign

The ensign is half-masted as a tribute to the dead. Whenever the national ensign is displayed at half-mast, it is first hoisted to the peak and then lowered to the half-mast position. Before lowering, the ensign is again raised to the peak. On Memorial Day, the ensign is half-staffed from 0800 until completion of the twenty-one-gun salute at 1200, or until 1220 if no salute is fired. During burial at sea, the ensign is at half-mast from the beginning of the funeral service until the body is committed to the deep.

Dipping the Ensign

Merchant ships salute naval ships by dipping their ensign. They seldom fly an ensign at sea, so this will normally occur in or near a port. A merchant ship of any nation formally recognized by the United States salutes a ship of the U.S. Navy or Coast Guard by lowering its national colors to half-staff. The Navy ship, at its closest point of approach, will lower the ensign to half-mast for a few seconds, then close it up, after which the merchant ship raises its own colors. If the salute is made when the ensign is not displayed, the Navy ship will hoist her colors, dip for the salute, close them up again, and then haul them down after a suitable interval. Coast Guard and Navy vessels dip the ensign only to answer a salute; they never salute first.

COAST GUARD ENSIGN

The initial job of the first revenue cutters was to guarantee that the maritime public was not evading taxes. Import taxes were the lifeblood of the new nation. Smuggling had become a patriotic duty during the revolution. If the new nation under the Constitution were to survive, this activity needed to be stopped. Working within a limited budget, cutters needed some symbol of authority. Neither officers nor crew members had uniforms. How could a revenue cutter come alongside a merchant ship during an age of pirates and privateers and order it to heave to? The solution was to create an ensign unique to the revenue cutter to fly in place of the national flag while in American waters.

Nine years after the establishment of the Revenue Cutter Service, the Congress passed the Customs Administration Act on 2 March 1799. This

Figure 8-3. Coast Guard ensign.

act provided that "the cutters and boats employed in the service of the revenue shall be distinguished from other vessels by an ensign and pendant, with such marks thereon as shall be prescribed and directed by the President of the United States."

On 1 August 1799, Secretary of the Treasury Oliver Wolcott issued an order announcing that the distinguishing ensign and pennant would consist of, "16 perpendicular stripes, alternate red and white, the union of the ensign to be the arms of the United States in a dark blue on a white field."

The ensign was poignant with historical detail inasmuch as in the canton of the flag there are thirteen stars, thirteen leaves to the olive branch, thirteen arrows, and thirteen bars to the shield. All corresponded to the number of states constituting the union at the time the nation was established. The sixteen vertical stripes in the body are symbolic of the number of states in the union when this ensign was officially adopted. This ensign soon became very familiar in American waters and served as the sign of authority for the Revenue Cutter Service until the early twentieth century.

The ensign was intended to be flown only on revenue cutters and boats connected with the Customs Service. But over the years it was found flying atop customs houses as well. However, President William Howard Taft issued an executive order on 7 June 1910 adding an emblem to the ensign flown by the revenue cutters, to distinguish it from the ensign flown from the customs houses. The order read:

> By virtue of the authority vested in me under the provisions of Section 2764 of the revised Statutes, I hereby prescribe that the distinguishing flag now used by vessels of the Revenue Cutter Service be marked by the distinctive emblem of that service, in blue and white, placed on a line with the lower edge of the union, and over the center of the seventh vertical red stripe from the mast of said flag, the emblem to cover a horizontal space of three stripes. This change to be made as soon as practicable.

At about this time, cutters began flying the U.S. flag as their naval ensign, and the revenue ensign became the service's distinctive flag. When the service adopted the name Coast Guard in 1915, the Revenue Cutter Service's ensign became the distinctive flag on all Coast Guard cutters as it had been for the revenue cutters.

The colors used in the Coast Guard ensign today, as in the Revenue Cutter Service, are all symbolic. The color red stands for our youth and sacrifice of blood for liberty's sake. The color blue not only stands for justice but also for our covenant against oppression. The white symbolizes our desire for light and purity.

As it was intended in 1799, the ensign is displayed as a mark of authority for boardings, examinations, and seizures of vessels for the purpose of enforcing the laws of the United States. The ensign is never carried as a parade or ceremony standard.

There is no set procedure on how to fold the Coast Guard ensign, and one unit may vary from another. If a unit does not have a specific folding procedure, the ensign should be folded neatly and stored with care.

THE STANDARD

The origins of the Coast Guard standard are very obscure. It may have evolved from an early jack; at least one contemporary painting supports this theory. In an 1840 painting, the revenue cutter *Alexander Hamilton* flies a flag very similar to today's Coast Guard standard as a jack. This flag, like the union jack, which is the upper corner of the U.S. flag, appears to be the canton or upper corner of the revenue cutter ensign.

An illustration in 1917 shows the Coast Guard standard as a white flag with a blue eagle and thirteen stars in a semicircle surrounding it. At a later date the words "United States Coast Guard—Semper Paratus" were added. After 1950 the semicircle of stars was changed to the circle containing thirteen stars.

The Coast Guard standard is used during parades and ceremonies and is adorned by the Coast Guard battle streamers. The Coast Guard is the only branch of the U.S. military to have two official flags, the Coast Guard standard and the Coast Guard ensign.

The Battle Streamers

The Coast Guard cherishes its many peacetime activities. But it is also proud of its services in the wars of the United States. The "system of cutters" was only seven years old when several of its fleet fought in the Quasi-War with France. In this war and the War of 1812, these small, lightly armed cutters proved their worth against experienced European warships.

Embroidering the names of battles on flags may be traced to the early days of the republic. By the end of the nineteenth century, embroidery was

Figure 8-4. A Coast Guard color guard presents the national ensign and the Coast Guard ensign during the national anthem at a retirement ceremony.

discontinued in favor of inscribed silver bands around the color staffs. This too was changed in World War I in favor of small ribbons bearing battle names; the Coast Guard adopted battle streamers in 1968 following the practice established by the U.S. Marine Corps.

Streamers are attached to the Coast Guard standard, replacing cords and tassels. They are carried in all ceremonies representing heroic actions in all naval encounters from 1790 to the present. Only major headquarters commands may display a complete set of battle streamers. Individual units may only display those they have earned. The Coast Guard has authorized a total of forty-three battle streamers:

Unit
Coast Guard Presidential Unit Citation
Navy Presidential Unit Commendation
Department of Transportation—Secretary's Outstanding Unit Award
Coast Guard Unit Commendation
Navy Unit Commendation
Coast Guard Meritorious Unit Commendation
Navy Meritorious Unit Commendation
Army Meritorious Unit Commendation

Campaign
Maritime Protection of the New Republic
French Naval War (Quasi-War with France)
War of 1812
African Slave Trade Patrol
Operations against West Indian Pirates
Indian Wars
Mexican War
Civil War
Spanish-American War
World War I Victory
China Service
Yangtze Service
American Defense Service
American Campaign
European-African-Middle Eastern Campaign
Asiatic-Pacific Campaign
World War II Victory

Navy Occupation Service
Korean Service
National Defense Service
Armed Forces Expeditionary
Vietnam Service
Southwest Asia Service
Kosovo Campaign
Global War on Terrorism Expeditionary
Global War on Terrorism Service
Afghanistan Campaign
Iraq Campaign

Foreign

Croix de Guerre, French, World War II
Philippine Defense
Philippine Liberation
Philippine Independence
Philippine Presidential Unit Citation
Republic of Vietnam (RVN) Armed Forces Meritorious Unit Citation,
        Gallantry Cross with Palm
RVN Meritorious Unit Citation, Civil Actions Medal First-Class Color,
        with Palm

The Jack

During its early years, the Revenue Cutter Service flew the canton (the upper corner of the flag nearest the staff) of the revenue cutter ensign as its jack. This practice persisted at least into the 1830s. Prior to the Civil War, the Revenue Cutter Service adopted as its new jack the canton of the United States flag (the union jack), and this continues to this day. Now the jack is flown from the jackstaff only while at anchor. During the early years of the service it was frequently flown on special occasions either at the jackstaff or atop the mainmast while under way as well as when at anchor.

The Commission Pennant

The Coast Guard commission pennant was created in 1799 at the same time as the ensign. The original commission pennant bore the same-style American eagle as the ensign, sixteen vertical red and white stripes and a white-over-red vertical tail.

Prior to the Civil War the Revenue Cutter Service adopted a commission pennant that had thirteen blue stars on a white field, thirteen vertical red and white stripes, and a red swallowtail. Sometime after the war, the service adopted the same commission pennant as the U.S. Navy. This pennant has thirteen white stars on a blue field, thirteen vertical red and white stripes, and a red swallowtail. The pennant is flown from the top of the mainmast.

Seals and Emblems

The creation of an official Coast Guard seal confirmed the existence of a symbol that had evolved over the decades. The Revenue Cutter Service, the Life-Saving Service, and the Lighthouse Service all had their own unique distinguishing devices. The Bureau of Marine Inspection and Navigation used drawings of ships and marine equipment on licenses and stationery. But it was not until 1927, after Treasury Secretary Andrew W. Mellon approved a design, that the Coast Guard had its first official seal or emblem. At that time, the seal and emblem were the same. This seal/emblem was designed by civilian Coast Guard draftsman Oscar H. Kee.

Over the years the seal and emblem became two different devices. The emblem, a simplification of the seal, began to appear throughout the service. Consequently, in 1957 Assistant Secretary of the Treasury David W. Kendall signed an order prescribing the distinctive emblem of the Coast Guard. This order specified that the emblem be used on the Coast Guard ensign but did not indicate any additional use. The emblem apparently continued in wide use.

The seal is used for official documents and records of the Coast Guard. It may also be used for jewelry, stationery, and so forth, at the discretion of the commandant. The official seal is also used on invitations, programs, certificates, diplomas, and greetings.

The Coast Guard emblem, a simplified version of the seal, was created as a visual identifier for the Coast Guard. It not only appears on the Coast Guard ensign but also inside the distinctive slash on the sides of cutters, craft, aircraft, and at units. It is also used on medals and plaques where space is limited.

The Slash

Our familiar and distinctive red slash or "racing stripe" did not appear on cutters, boats, and aircraft until the 1960s. In 1964 an industrial design firm recommended that the Coast Guard adopt a symbol or mark that could be easily distinguished from other government agencies and easily applied to ships, boats, aircraft, stations, vehicles, signs, and printed forms.

Their design was a wide red bar to the right of a narrow blue bar, both canted at 64 degrees. Centered on the red bar was a new emblem. Studies were done with experimental markings for their impact on the public as well as for their long-run compatibility with the Coast Guard's mission and traditions. The reaction was overwhelmingly favorable. Three years later, on 6 April 1967, the now-famous slash appeared throughout the Coast Guard. In the final design only the emblem changed. The traditional Coast Guard emblem was selected for centering on the red stripe over the new design.

Semper Paratus

No one seems to know exactly how "Semper Paratus" was chosen as the Coast Guard's motto, but there is no doubt as to who put the famous motto to words and music. Capt. Francis Saltus Van Boskerck wrote the words in 1922 while he was commanding officer of USCGC *Yamacraw* (CG-21). He wrote the music to "Semper Paratus" five years later on a beat-up old piano in Unalaska, Alaska.

For probably as long as Captain Van Boskerck could remember, "Semper Paratus" had been a Revenue Cutter Service and Coast Guard watchword. The words themselves—Latin for "always ready" or "ever ready"—date back to ancient times. Captain Van Boskerck hoped to give "Semper Paratus" as much recognition as "Semper Fidelis" of the Marines and "Anchors Aweigh" of the Navy.

Personal Flags and Other Flags and Pennants

The ship carrying an officer of flag rank who commands a fleet or a unit of a fleet flies his personal flag from the main truck at all times unless he is absent for more than seventy-two hours. This is a blue flag carrying five white stars for a fleet admiral, four for admiral, three for vice admiral, two for rear admiral (upper half), and one for rear admiral (lower half).

An officer below flag rank, when in command of a force, flotilla, squadron, carrier division, or aircraft wing, flies a broad command pennant, white with blue stripes top and bottom. An official in command of any other division, such as a mine division or destroyer division, flies a burgee command pennant, white with red stripes top and bottom.

Absence Indicators

When the commanding officer, or any flag officer, is absent from the command for less than seventy-two hours in port, an "absentee pennant" is flown as follows:

- First Substitute (starboard yardarm): Flown when the admiral or unit commander whose personal flag or pennant is flying is absent.
- Second Substitute (port yardarm): Flown when the chief of staff is absent.
- Third Substitute (port yardarm): Flown when the commanding officer is on an official absence of more than seventy-two hours. The executive officer, serving as acting commanding officer, is accorded the use of this pennant as if the executive officer were the commanding officer.
- Fourth Substitute (starboard yardarm): Flown when the civil or military official whose flag is flying (such as the secretary of Defense) is absent.

Church Pennant and Jewish Worship Pennant

The church pennant and the Jewish worship pennant are the only flags authorized to be flown over the national ensign at the same point of hoist, when divine services are conducted afloat by a chaplain. When divine services are conducted ashore by a chaplain, these two pennants may be flown from the same halyard as a personal flag would be flown. If a personal flag is already flying, it is moved to the halyard one down in precedence.

Red Cross Flag

Hospital ships in wartime fly the Red Cross flag instead of the commission pennant. Boats engaged in sanitary service and landing parties from hospital boats fly the Red Cross flag from a staff in the bow. Under the 1949 Geneva Conventions and previous international agreements, this flag is the recognized symbol of protected medical and religious persons and activities. Ships and installations flying these flags or otherwise marked with these symbols are exempt from attack. Their misuse in order to injure, kill, or capture the enemy is an act of perfidy under the laws of warfare.

Other Flags and Pennants

Both in port and at sea, ships fly many single flags or pennants with special meanings. The senior officer present afloat may prescribe certain flag hoists for local use, such as a request for a garbage or trash lighter, or a water barge. At anchor, ships awarded the Presidential Unit Citation, Coast Guard Unit Commendation, Navy Unit Commendation, or Coast Guard Meritorious Unit Commendation should fly the pennant at the foretruck from sunrise to sunset.

Other Honors to the Flag and Anthem

There are many occasions other than colors when you will render honors to the ensign or national anthem. The usual rules are face the flag if it is displayed; face the music if the flag is not displayed. Hold your salute until the anthem ends, or, if there is no anthem, until the flag has been hoisted or lowered or has passed your position.

Anthem Played Outdoors, Flag Not Displayed

Men and women not in formation stand at attention and face the music. If in uniform render a hand salute; if armed, present arms. In civilian attire, face the music and place your right hand over your heart. Members of the armed forces and veterans not in uniform may render the military salute. Hold the salute until the last note of music is played. Formations are brought to a halt and the officer in charge faces the music and renders the salute; the ranks remain at attention facing the direction they were in at the halt.

Anthem Played Outdoors and Flag in View

The salutes are the same, except that the officer in charge of a formation faces the flag instead of the music and salutes; the ranks remain at attention facing in the direction they were in at the halt.

Flag Ceremony Inside a Building

If the anthem is played during a ceremony inside a building in which the flag is brought forward and presented to the audience or is then retired, the procedures are as follows:

- All personnel face the flag.
- Personnel in civilian attire stand, face the flag, and render the right-hand-over-heart salute from the first note of the anthem to the last. Members of the armed forces and veterans not in uniform may render the military salute.
- Uncovered military personnel stand at attention with hands at sides.
- Personnel with rifles present arms.

- Personnel covered or with side arms in uniform render the hand salute.
- Military formations stand at attention; their officers in charge render the salutes.
- If the audience is all or almost all military, the officer in charge will call "Attention." The officer may order all to salute, or the officer may salute for the audience.
- Salutes are held until the flag is placed and its bearer steps aside.

Anthem Played Inside a Building, Flag Not Displayed

All persons stand and face the music. Men and women in uniform, if covered, render the hand salute; if not covered, stand at attention. In civilian attire, face the music and place your hand over your heart. Members of the armed forces and veterans not in uniform may render the military salute.

Personnel in Boats

During the playing of the national anthem, only the boat officer (or coxswain, if there is no boat officer) stands and salutes. Crew and passengers remain standing or seated at attention. If in civilian clothing, remain at attention and do not salute. This is an exception to the general rule.

Passing in Parades

If you are in a formation, obey orders. Your officer in charge will render the salute for the formation. If you are not in formation and the flag is hoisted, lowered, or passes in a parade, obey the following procedures:

- If seated, rise, come to attention, face the flag, and salute.
- If standing, come to attention, face the flag, and salute.
- If walking, halt at attention, face the flag, and salute.
- If riding in a vehicle, remain seated at attention.

The rules for saluting the anthem apply only if it is played or broadcast as a part of a public ceremony. If you hear it in private or as you walk past a radio, you need not stop or salute.

Pledge of Allegiance

Coast Guard personnel in uniform but uncovered or in civilian clothes render the pledge of allegiance to the flag by facing the flag and standing at attention with the right hand over the heart. Personnel in uniform and covered render the military salute and repeat the pledge.

Hail to the Chief

The song "Hail to the Chief" is played when the president of the United States is being honored. When "Hail to the Chief" is so played, render the same honor given during the playing of the national anthem.

Boarding and Leaving Ship

The officer of the deck (OOD) or a representative—who may be an officer or enlisted—will always be on duty to greet persons leaving or boarding the ship.

The procedure for boarding your own ship is as follows:

- At the gangway, if the national ensign is flying, turn aft and salute the ensign.
- Then turn to the OOD or a representative, salute and say, "I report my return aboard sir/ma'am."
- The OOD will return your salute and say, "Very well," or "Very good."

The procedure for leaving your own ship is as follows:

- Step to the OOD, salute, and request, "Permission to leave the ship, sir/ma'am?" or "I request permission to go on the pier to check the after mooring lines, sir/ma'am."
- After the OOD has said "Very well" or "Permission granted" and has returned your salute, drop your salute and step to the gangway.
- If the ensign is flying, salute in its direction and leave.

When you go on board a ship other than your own, you must obtain permission from that ship's OOD. Stand at the gangway and salute the ensign, if it is flying, turn to the OOD or a representative, salute, and say, "I request permission to come aboard, sir/ma'am."

On leaving the ship you visited, turn to the OOD or a representative, and say "With your permission, sir/ma'am, I shall leave the ship." After the OOD has said "Very well" or "Permission granted" and has returned your salute, step to the gangway, and, if the ensign is flying, salute in its direction and leave.

If you are in a work party, only the person in charge makes the requests to the OOD to board and leave the ship. However, each member salutes the ensign (if it is flying) and the OOD as they file by, both coming and going. If you are making many trips bringing stores on board, then you salute only the first and last times over the gangway.

Shipboard Customs

Quarterdeck Customs

The quarterdeck is not a specified deck like the flight deck or the gun deck. It is an area designated by the commanding officer to serve for official and ceremonial functions. Therefore, the quarterdeck is treated as the "sacred" part of a ship, and you must obey the following rules:

- Do not be loud or sloppy in its vicinity.
- Never appear on the quarterdeck unless wearing the uniform of the day or as a member of a working party.
- Never smoke on the quarterdeck.
- Never cross or walk on the quarterdeck except when necessary.
- Do not lounge on or in the vicinity of the quarterdeck.

Sick Bay Customs

In the old days, medical and sanitary conditions were so bad that sick bay usually meant a place for dying rather than for getting well, particularly after a battle. Accordingly, it became customary to remove one's cap when entering sick bay, out of respect for the dying and the dead. Today medicine is so highly developed that we less commonly think of death in connection with sick bay. However, the custom of removing one's cap remains. Of course, one should also maintain quiet, and smoking is prohibited.

Mess Deck and Living Areas

The mess deck is where enlisted personnel eat; the wardroom is where officers eat. If you enter any of these areas while a meal is in progress, uncover. Even if you are on watch and wearing the duty belt, uncover while a meal is in progress.

Officers' country is the part of a ship where the officers' staterooms and the wardroom are. CPO berthing is where the chief petty officers have their living spaces and mess. Enter these areas only on official business. Never use their passageways as thoroughfares or shortcuts. If you enter the wardroom or any compartment or office of an officer or CPO, remove your cover. If you are on watch and wearing the duty belt or sidearm, remain covered unless divine services or a meal is in progress in one of the spaces. Always knock before entering any officer's or chief petty officer's stateroom.

Divine Services

When divine services are in progress, the ship flies the church pennant above the national ensign and the word is passed, "Divine services are in progress in [location]. Maintain quiet about the decks during divine services."

If you enter the area where divine services are in progress, uncover. Even if you are wearing the duty belt and sidearm, uncover. There is one exception: for a Jewish ceremony, remain covered.

The Salute

The military custom you will learn first and use most is the hand salute. This is a courtesy that has been observed for centuries by military personnel of every nationality. It exists by more than force of custom: the occasions and methods of saluting are specified by detailed orders in *United States Coast Guard Regulations*.

Whom to Salute

All uniformed members of the armed forces recognize and greet each other. However, military organization custom requires that this recognition take the form of the official hand salute or rifle salute to any of the following:

- Commissioned and warrant officers of the Coast Guard, Navy, Marine Corps, Air Force, and Army
- Officers of the National Oceanic and Atmospheric Administration and the Public Health Service
- Officers of foreign armed services whose governments are recognized by the United States (in practice, all foreign officers, unless we are actually at war with the country in question)

If in doubt, salute. It is better to be overcourteous than to not salute when you should have. Always hold your salute until it is returned or acknowledged.

Saluting Rules

The hand salute is given with the right hand. If you have an injury that makes this impossible, or if you are using a boatswain's pipe, salute with the left hand. Personnel in the Coast Guard, Navy, and Marine Corps render the left-hand salute when necessary; personnel in the Air Force and Army do not.

Accompany your salute with a cheerful, respectful greeting. "Good morning, sir/ma'am," "Good afternoon, Commander," "Good evening, Miss White," as appropriate. When meeting two or more officers, direct your greeting toward the most senior officer. "Sir/ma'am" should not be made plural.

Always come to attention. If double timing, slow to a walk when saluting a passing officer. You need not stop walking, but hold yourself erect. Do not bow your head or stare off in the distance; look directly at the officer as you salute. If both hands are occupied and you are unable to salute, face the officer as though you were saluting and greet him or her as described above. Do not salute with a pipe, cigar, or cigarette in your mouth.

If you are accompanying a commissioned officer, do not salute another officer until the officer with you salutes. Then salute at the same time your officer does. If you are accompanying a commissioned officer who is saluted by an enlisted person, return the salute at the same time your officer does.

Salute an officer even if the officer's hands are engaged (for example, holding packages) and the salute cannot be returned. The officer will acknowledge your salute by saying "Good morning," "Good afternoon," or "Good evening." Salute officers who are in civilian or athletic dress. Prisoners do not salute.

Distances for Saluting

Since the salute is basically a courtesy, it should be used in a manner similar to civilian greetings. An officer should be recognized and saluted at about the same distance and at about the same time as one would recognize and greet a civilian acquaintance; six paces away is a good general rule. Remember—an officer must pass near enough to be identified as rating a salute, and the person giving the salute should start while far enough away for the officer to have time to see and return it.

Salutes in Civilian Dress

When outdoors in uniform, Coast Guard personnel use the military salute when recognizing officers in civilian or athletic dress. If you have occasion to salute the flag while in civilian clothes, remove your hat with your right hand and hold it over your chest, with your hand over your heart. Members of the armed forces and veterans not in uniform may render the military salute.

If you are not wearing a hat, place your right hand over your heart. Women in civilian dress salute in this manner, except they do not remove their hat but just place their right hand over their heart.

Etiquette for Boats, Vehicles, and Passageways

The basic rule in service etiquette as in civilian etiquette is to make way for a senior quickly, quietly, naturally, and without fuss. Thus, the rule for entering boats and vehicles is seniors in last and out first. The idea is that the captain should not have to wait in a boat for a seaman to amble down the accommodation ladder. When reaching a destination, seniors are allowed to get out first because normally their business is more important and more pressing than that of subordinates.

Remember, juniors enter a boat or vehicle first; that is, they do not make last-minute dashes to reach the boat but as soon as the OOD says the boat is ready. Similarly, they are ahead of time and enter buses, cars, and other vehicles well in advance of the arrival of a senior. When the boat docks or the vehicle arrives at its destination, the juniors wait until their seniors have disembarked before disembarking themselves.

Generally, seniors take the seats farthest aft. If officers are present, do not sit in the stern seats unless invited to do so. Also, enlisted personnel maintain silence while officers are in the boat. (For reasons of safety, never become noisy and boisterous in a boat, regardless of the hour, condition of sea, or who is present.)

If boats with officers on board pass within view, the senior officer and the coxswain of each boat render salutes. Other crew members and passengers remain seated at attention. The coxswain stands and salutes all officers entering or leaving the boat.

Personnel seated in boats in which there is no officer, petty officer, or acting petty officer in charge should rise and salute all officers passing near. When an officer, petty officer, or acting petty officer is in charge of a boat, that person alone renders the salute.

Enlisted personnel seated well forward in a large boat do not rise and salute when officers enter or leave the stern seats. However, passengers in the aft section of a boat always rise and salute when a commissioned officer enters or leaves.

A boat takes rank according to the rank of the highest grade of officer embarked. A barge with an ensign is junior to a motor whaleboat with a lieutenant. When junior boats pass seniors, the junior boats salute first.

The coxswain and senior officer in each boat render the hand salute while other members of the crew not concealed by canopies stand at attention facing the senior boat. Passengers sit at attention. Boats passing U.S. or foreign men-of-war during "colors" must lay to, and their crews stand, face the colors, and salute.

For entering and leaving through doors, the rule is to let the senior go first. If possible, hold the door for the senior. If, however, the junior is a woman and the senior motions her to go first, she should do so.

On meeting a senior in a passageway, step aside and let the senior pass. If the senior is an officer and other enlisted personnel or junior officers are present, call "Gangway" so the others can make way.

Do not overtake and pass an officer without permission. When necessary to walk past, overtake on the officer's left side, salute when you are abreast, and ask "By your leave, sir/ma'am?" After the officer returns the salute and has said "Very well" or "Permission granted," drop your salute and continue past.

When walking with a senior, always walk on the left. That is, put the senior on *your* right. This rule also applies when seated in an automobile.

Addressing Officers

Senior officers—admirals, captains, and commanders (who wear gold oak leaves or "scrambled eggs" on their cap visors)—are always addressed and referred to by their titles of rank, "Admiral," "Captain," or "Commander." If several officers of the same rank are seated or working together, it is proper to use both rank and name, as "Admiral Thomas" or "Commander McMahon," to avoid confusion.

Junior officers—lieutenant commanders, lieutenants, lieutenants (junior grade), and ensigns—are addressed and referred to by "Mister" or "Miss" and their last name, as "Mister Betz," or "Miss Sinclair."

When speaking to a junior officer whose name is not known, use "sir" or "ma'am." Warrant officers, cadets, and midshipmen are treated in the same manner as junior officers. By tradition, the commanding officer of any ship or station, regardless of rank, is addressed and referred to as "Captain." Other captains or commanders in the same command should be addressed by rank and name.

Officers in the Medical and Dental Corps are addressed and referred to by rank or as "Doctor" if they are commanders and above; if of junior officer rank, they are addressed as "Doctor" instead of "Mister" or "Miss." A chaplain may be called "Chaplain," regardless of rank.

Officers of the Army, Air Force, and Marine Corps of and above the rank of captain are addressed and referred to by rank, as "General," "Colonel Jessup," "Major," and "Captain Hernandez"; others are addressed and introduced as "Mister" or "Miss" unless circumstances make it advisable to use the rank to inform other persons present of their status.

When replying to an order, the only correct reply is "Aye, aye, sir/ma'am." This means you heard the order, you understand it, and you will carry it out to the best of your ability.

"Ma'am" or "sir" is a military expression always used with "Yes" or "No" in addressing officers. Also use "sir" or "ma'am" in addressing enlisted personnel when they are performing a military duty, such as junior officer of the deck or in reporting a muster to a chief petty officer. At such times, you also exchange salutes.

Addressing Enlisted Personnel

A chief petty officer is addressed as "Chief Petty Officer Strauch" or more informally as "Chief Strauch" or as "Chief" if you do not know the chief's name. Master chief petty officers and senior chief petty officers are customarily addressed and referred to as "Master Chief O'Neill," or "Senior Chief Snyder." Chief petty officers are introduced by rate, rating, and last name: "Chief Avionics Electrical Technician Strauch." Introduce petty officers by stating the rate and rating first, followed by the last name: "Yeoman First Class Pascucci" or "Petty Officer Mullins."

Introduce "nonrated" personnel by rate and name: "Seaman Brummer" or "Fireman Truitt." When introducing first-, second-, and third-class petty officers to civilians, use the term "petty officer": "This is Petty Officer Flores." Civilians generally address enlisted people as "Mr.," "Miss," or "Mrs."

Seniors may call juniors by first name, but juniors never address seniors by first name. If a petty officer tells you to use his or her first name, do so only in privacy or on liberty. While on duty in the presence of others, use only the last name with the appropriate rank.

In civilian life, it is customary to introduce men to women and youth to age. The same general rules are followed in military life, except that in most cases, rank establishes the order of introduction. Introduce the junior to the senior, whether male or female, except that everyone, no matter the rank or sex, is introduced to a chaplain. If one person is civilian, follow the civilian rules. In making introductions, name the honored or higher ranking person first, then the name of the person being introduced or presented: "Admiral Thomas, this is Lieutenant Birst," or "Mrs. Evelyn, this is Master Chief Whitehead."

There may be times when unofficial letters are exchanged between officers and enlisted personnel—a coxswain might write to a former skipper to congratulate him on promotion. The salutation would be "Commander Senecal" and the complimentary closing should always be "Very respectfully." An officer would not address an enlisted person in writing as "Dear Pace," but as "Dear Chief Petty Officer Pace."

All officers should be addressed by their title of rank, except that officers in the Medical, Dental, Nurse, and Medical Service Corps with doctoral degrees may be addressed as "Doctor." Modifiers may be dropped; you may address a lieutenant commander as "Commander," a lieutenant (junior grade) as "Lieutenant." When a captain or lieutenant is not in uniform, he would be introduced as "of the Coast Guard" or "of the Navy," in order to avoid confusion with Army, Air Force, or Marine Corps ranks.

Honors

Gun Salutes

In olden days, it took as much as twenty minutes to load and fire a gun, so a ship that fired her guns first did so as a friendly gesture, making herself powerless for the duration of the salute. Gun salutes prescribed by regulations are fired only by such ships and stations as are designated. A national salute of twenty-one guns is fired on Independence Day and on Memorial Day and to honor the president of the United States and heads of foreign nations. Salutes for naval officers are as follows: fleet admiral, seventeen guns; admiral, seventeen guns; vice admiral, fifteen guns; rear admiral (upper half), thirteen guns; rear admiral (lower half), eleven guns. Salutes are fired at intervals of five seconds and always in odd numbers.

Manning the Rail

This custom evolved from that of "manning the yards," which is hundreds of years old. In sailing ships men stood evenly spaced on all the yards and gave three cheers to honor a distinguished person. Now the crew is stationed along the rails and superstructure of a ship when honors are rendered to the president, the head of a foreign nation, or a member of a reigning royal family. Personnel so stationed do not salute.

Dressing and Full-Dressing Ship

Commissioned ships are full-dressed on Presidents' Day and Independence Day and dressed on other national holidays. When dressing ship, the national ensign is flown from the flagstaff and usually from each masthead. The Coast Guard ensign is flown from the outboard starboard yardarm. When a ship is full-dressed, in addition to the ensigns, a "rainbow" of signal flags is displayed from bow to stern over the mastheads or as nearly so as the construction of the ship permits. Ships not under way are dressed from 0800 to sunset; ships under way do not dress until they come to anchor during that period.

Side Boys

Side boys are a part of the quarterdeck ceremonies when an important person or officer boards or leaves a ship. The custom of having side boys originated centuries ago in Britain's Royal Navy when officers were hoisted on board in a sort of basket, and since senior officers usually weighed more than juniors, more boys were needed on the line. When the side is piped by the boatswain's mate of the watch, from two to eight side boys (men or women), depending on the rank of the officer, will form a passageway at the gangway. They salute on the first note of the pipe and finish together on the last note. Side boys must be particularly smart in appearance and grooming, with polished shoes and immaculate uniforms.

Arrival and Departure Honors

On board any Coast Guard vessel capable of doing so, the arrival or departure of visiting officers, commander and above, is normally announced. Commanding officers below commander (O-5) are announced if their identity is known to the OOD. High-ranking civilian officials are announced in this manner. Accompanying the announcement of the arrival or departure is the toning of the bell. The number of "bongs" should be determined from the list below. For comparison, the number is the same as the number of side boys that the person would rate when receiving official honors. The bongs are sounded in pairs, in the same manner as done to denote the passing of time.

Officer	Number of "Bongs"
Vice Admiral (O-9 and above)	8
Rear Admiral (O-8 or O-7)	6
Captain (O-6)	4

Officer	Number of "Bongs"
Commander (O-5)	4
Other officers (OOD's discretion)	2

Senior officers are normally announced on board using the name of their command or title, as in the following examples:

CO of USCG barque *Eagle*	EAGLE
CO of Coast Guard District One	FIRST DISTRICT
Superintendent of USCGA	COAST GUARD ACADEMY
Commandant, USCG	COAST GUARD
Sector Commander of Sector New York	SECTOR NEW YORK
Senators	SENATOR, STATE OF ——
Representatives	REPRESENTATIVE, (#) DISTRICT, STATE OF ——

The following procedure should be adhered to: As the person approaches the bow or begins up the accommodation ladder, the correct number of bongs is sounded in pairs and the announcement is passed. For example, if the captain of USCG barque *Eagle* is returning, the announcement is "Bong, Bong . . . Bong, Bong . . . EAGLE (half pause) ARRIVING." Only in the case of the captain or an officer whose pennant is flown on board the ship is a final "bong" sounded as the person steps on board or ashore. No other voice announcement is made.

Passing Honors

When ships pass each other, all hands who are topside or who are visible from outboard and are free to do so should face the ship being passed, stand at attention, and salute on signal, whether in ranks or not.

Coast Guard and Navy ships use the following whistle signals when exchanging passing honors:

Ship is passing to starboard—1 blast
Ship is passing to port—2 blasts
Salute—1 blast
End salute—2 blasts
Carry on—3 blasts

Ceremonies

Change of Command

When a new commanding officer or officer in charge takes over any ship or station, all hands fall in at quarters in dress uniform. The departing officer reads the detachment orders, and the relieving officer reads the orders to take over the command. Both officers then inspect the crew and the ship.

Christenings and Commissionings

The tradition of christening and commissioning a ship started well before the first ten revenue cutters entered the service. The practice dates back to ancient times when the Greeks, Romans, Egyptians, and Vikings called upon the gods to protect their ships and crew from the perilous sea. Religion played an important role in these ceremonies. In fact, christenings originated as a way to appease the gods of the elements.

Christening gives a ship its identity. Over the years, different cultures and people changed and shaped the way ceremonies were performed, and some of these traditions were carried over into modern times.

A ship is traditionally christened or given its name at the time it is launched into the water. When a ship is christened, it is tradition to break a bottle across the ship's bow. This practice began in Britain in the late seventeenth century. Previously, an official would sip wine from a "standing cup," a large loving cup made of precious metal, then pour out the remaining wine onto the deck or over the ship's bow. The cup was then tossed overboard. This practice soon became too costly, and a net was used to catch the cup so that it could be reused at other launchings. Wine was the traditional liquid used to christen a ship, although other liquids were used such as whiskey, brandy, and water. At the close of the nineteenth century, champagne became the popular liquid. However, during Prohibition ships were christened with water.

Ships' sponsors were generally royalty or senior naval officers. In the nineteenth century, women became ship sponsors for the first time. Women sponsored ships more and more frequently, although it was not the rule.

The actual physical process of launching a new ship from a building site to the water involves three principal methods. The oldest, most familiar, and most widely used is the "end-on" launch in which the vessel slides, usually stern first, down an inclined shipway. The "side launch," whereby the ship enters the water broadside, came into use in the nineteenth century on inland waters, rivers, and lakes. It was given major impetus by the World War II shipbuilding program. Another method involves ships built in basins or graving docks. When ready, ships constructed in this manner are floated by admitting water into the dock.

The commissioning ceremony completes the cycle from christening and launching to full status as a cutter in the U.S. Coast Guard. Commissioning ceremonies are typically hosted by a local Navy League Council.

Dining-In and Mess Night

The tradition of Dining-In or Mess Night can be traced to the days when the Roman legions held great banquets to celebrate their victories or paraded the prizes of their conquests. The Vikings had a tradition of celebrating great battles and feats by formal ceremony. It is from these customs of celebrating special events that we have formal dinners today.

The Dining-In is a formal dinner function for the officers of a military organization or unit. Originally such functions provided an excellent setting to recognize both individual and unit achievements as well as to bid farewell to departing officers and welcome new ones. Today they are primarily occasions to gather socially at a formal function. But the protocol and traditional amenities remain intact.

The Chief Petty Officer Academy continues the tradition, calling it "Chiefs' Mess Night." Chiefs' Mess Night is a formal military function like the Dining-In, but the mess members are chiefs and guests. It provides an occasion for chiefs to meet socially and give recognition to a dignitary, individual, or unit. It also may simply be a pleasant way for individuals to become better acquainted.

In early-eighteenth-century Europe, various regiments of the established monarchies would gather for an evening of good food, drinking, and fellowship to honor individuals and organizations. More recently this custom can be traced to the "messes" of the Royal Navy and regimental "messes" of the British Army. Our early leaders gained background and training from either British regulars or colonial militia in the French and Indian Wars. It is most likely that they became indoctrinated in the formal aspects of military life as practiced by the men of that period.

Prior to World War II, Mess Night had reached its greatest prominence. While the occasion became less of a celebration for individual achievements, the protocol became more formalized. The uniforms prescribed were evening dress with medals.

Dining-In and Mess-In are strictly for military people, while Mess-Out and Dining-Out are for military as well as invited guests.

Decommissioning Ceremony

When a ship is taken out of service, either to go into "mothballs" or to be transferred to a foreign navy, the ceremony is smaller and simpler. The commanding officer reads the decommissioning orders to officers and crew at quarters and the ensign, jack, and commission pennant are hauled down. By tradition, the captain keeps the commission pennant.

Crossing the Line

When a ship crosses the equator, all "pollywogs"—those who have never crossed—are initiated into the "Ancient Order of the Deep" and become shellbacks. All hands are issued a certificate and a wallet-sized card with the name of the ship, date, and longitude. On large ships, shellbacks may spend days planning an elaborate "Neptune party" in which King Neptune, Davy Jones, and a group of "royal" police, judges, surgeons, barbers, bears, and other characters initiate the pollywogs. A Neptune party is an all-hands affair during which the shellbacks take over the ship to a certain extent and initiate all pollywogs, whether they are seamen or admirals.

Golden Dragons

When a ship crosses the 180th meridian—the International Date Line—going west, it enters "The Realm of the Golden Dragon," and all hands are issued a card to mark the occasion.

Blue-Nose Polar Bear

When a ship crosses the Arctic Circle, all hands are issued a wallet-sized card, similar to the shellback card, stating the circumstances, and naming them "Blue-Nose Polar Bears."

New Year's Log

It is traditional that the log for the first midwatch of the new year—1 January—must be written in verse. In recent years some such log entries have been published and are widely circulated on the Internet.

Ship's Bell

On some ships the oldest member of the crew strikes the old year out with eight bells, and the youngest member of the crew strikes the new year in with eight bells.

Homeward-Bound Pennant

No ship carries this flag; it is made up only when needed. It is flown by a ship that has been overseas for nine months or more when she gets under way to return to the United States. The homeward-bound pennant flies until sunset of the day of arrival in a U.S. port. The pennant next to the hoist is blue and has one white star for the first nine months overseas and an additional star for each six months. The rest is divided lengthwise; the upper section is red, the lower is white. The pennant is one foot long for each person on board who has been outside the United States for at least nine months but is never longer than the length of the ship.

After being hauled down the pennant is cut up. The blue part goes to the commanding officer. The remainder is divided equally among all hands.

Burial at Sea

During combat or at other times when operations prevent holding a body for services ashore, the dead are buried at sea. When schedules permit, retired persons may be buried at sea. The ceremony may be very simple in combat or more elaborate when time permits. The body is covered by a U.S. flag, which is removed as the body is committed to the deep and is presented to the next of kin.

Figure 9-1.
A smart appearance at ceremonies requires each person to be well grounded
in basic military fundamentals.

CHAPTER 9

Military Drill Fundamentals

There are a multitude of reasons for military drills: to teach the fundamentals of military bearing; to provide experience in giving and following commands; to prepare for military operations on land; to facilitate movements of companies from one place to another; and to promote teamwork The primary reference document for this chapter is the *U.S. Marine Corps Drill and Ceremonies Manual.*

DRILL COMMANDS

Preparatory commands are indicated in this chapter beginning with a capital letter followed by lowercase letters, and those of execution by CAPITAL LETTERS. A military drill command has two parts:

1. The preparatory command, such as Hand, which indicates the movement that is to be executed.
2. The command of execution, such as SALUTE, HALT, or ARMS, which causes the desired movement, or halt, or element of the manual to be executed.

When appropriate, the preparatory command includes the name or title of the group concerned, such as "Bravo Company, Hand, SALUTE." In certain commands, the preparatory command and the command of execution are combined. Examples would be AT EASE, REST, FALL IN, and so forth.

To call back or revoke a command or to begin again a movement that started wrong, the command AS YOU WERE is given, at which the movement stops and the former position is taken.

The Positions

Position of Attention

Command: AT-TEN-TION or FALL IN. Heels close together, feet turned out to form an angle of 45 degrees, knees straight, hips level, body erect, with the weight resting equally on the heels and balls of the feet. Shoulders squared, chest arched, arms hanging down without stiffness so that thumbs are along the seams of the trousers, palms and fingers relaxed. Head erect, chin drawn in, and eyes straight to the front. In coming to attention, the heels are brought together smartly and audibly.

The Rests

Commands: FALL OUT, REST, AT EASE; and 1. Parade, 2. REST, 3. FALL OUT. Personnel break rank but remain nearby. Personnel return to places and come to attention at the command FALL IN.

REST: Right foot is kept in place. Personnel may talk and move.

AT EASE: Right foot is kept in place. Personnel keep silent but may move about.

1. Parade, 2. REST ("Parade, REST"): Move the left foot smartly twelve inches to the left from the right foot; at the same time, clasp the hands behind the back, palms to the rear, the right hand clasping the left thumb, arms hanging naturally. Keep silent and motionless.

To resume attention from any rest other than FALL OUT, the commands are, for example, 1. Detail, 2. ATTENTION ("Detail, ATTENTION").

Eyes Right or Left

Commands: 1. Eyes, 2. RIGHT or LEFT ("Eyes, RIGHT/LEFT"), 3. Ready, 4. FRONT ("Ready, FRONT"). At the command RIGHT, each person turns his head and eyes smartly 45 degrees to the right. The personnel on the extreme right file keep the head and eyes to the front. At the command FRONT, the head and eyes are turned smartly to the front. The opposite is carried out for "Eyes, LEFT."

Hand Salute

Commands: 1. Hand, 2. SALUTE ("Hand, SALUTE"), 3. Ready, 4. TWO ("Ready, TWO"). The command TWO is used only when saluting by command. At the command SALUTE, salute smartly, looking toward the person saluted.

At the command TWO, drop the arm to its normal position by the side in one movement and turn the head and eyes to the front. In passing in review, execute the hand salute in the same way. During a review the salute is held until six paces beyond the person saluted. Under other circumstances the salute is dropped when returned or upon passing.

Facings

Right or Left Face

Commands: 1. Right (or Left), 2. FACE ("Right/Left, FACE"). At the command FACE, slightly raise the left heel and right toe; face to the right, turning on the right heel, putting pressure on the ball of the left foot. Hold the left leg straight. Then place the left foot smartly beside the right.

About Face

Commands: 1. About, 2. FACE ("About, FACE"). At the command, place the toe of the right foot about a half-foot length to the rear and slightly to the left of the left heel without moving the left foot. Put the weight of the body mainly on the heel of the left foot, right leg straight. Then turn to the rear, moving to the right on the left heel and on the ball of the right foot. Place the right heel beside the left to complete the movement.

Steps and March Commands

All movements executed from the halt, except right step, begin with the left foot. Forward, Half Step, Halt, and Mark Time may be executed one from the other in Quick or Double Time.

The following table prescribes the length in inches and the cadence in steps per minute of steps in marching:

Step	Time	Length	Cadence
Full	Quick	30	120
Full	Double	36	180
Full	Slow	30	—★
Half	Quick	15	120
Half	Double	18	180
Side	Quick	12	120
Back	Quick	15	120

★ This is a special step executed only as a funeral escort is approaching the place of interment. The cadence, in accordance with that set by the band, varies with different airs that may be played.

All commands of execution are given on the foot, right or left, in the direction the movement is going. For example, if the march is to be to the right, as: 1. By the Right Flank, 2. MARCH ("By the Right Flank, MARCH"), the command MARCH is given on the right foot.

Quick Time

All steps and movements are executed in Quick Time, which is what most people understand as normal marching pace, unless the unit is marching Double Time, or unless Double Time is added to the command. Example: "Double Time, MARCH."

Marching

At Halt, to march forward to Quick Time, the commands are 1. Forward, 2. MARCH ("Forward, MARCH"). At the command Forward, shift the weight of the body to the right leg. At the command MARCH, step off smartly with the left foot and continue the march with thirty-inch steps taken straight forward without stiffness or exaggeration of movements. Swing the arms easily in their natural arcs six inches straight to the front and three inches to the rear of the body.

Double Time

To march in Double Time, the commands are 1. Double Time, 2. MARCH ("Double Time, MARCH").

1. If at halt, at the command Double Time, shift the weight of the body to the right leg. At the command MARCH, raise the forearms, fingers closed, knuckles out, to a horizontal position along the waistline and take up an easy run with the step and cadence of Double Time, allowing the arms to take a natural swinging motion across the front of the body. Keep the forearms horizontal.

2. If marching in Quick Time, at the command 1. Double Time, 2. MARCH, given as either foot strikes the ground, take one more step in Quick Time and then step off in Double Time.

3. To resume the Quick Time from Double Time, the commands are 1. Quick Time, 2. MARCH. At the command MARCH, given as either foot strikes the ground, advance and plant the other foot in Double Time, resume the Quick Time, dropping the hands by the sides.

Halt

Commands: 1. Company, 2. HALT ("Company, HALT").

1. When marching in Quick Time, at the command HALT, given as either foot strikes the ground, execute the halt in two counts; advance and plant the other foot, then bring up the rear foot.

2. When marching in Double Time, first bring the Squad (Company) to Quick Time, then halt as above.

3. When executing Right Step or Left Step, at the command HALT, given as the heels are together, plant the foot next in cadence and come to the halt when the heels are next brought together.

Mark Time

Commands: 1. Mark Time, 2. MARCH ("Mark Time, MARCH").

1. Being in march, at the command MARCH, given as either foot strikes the ground, advance and plant the other foot, then bring up the rear foot, placing it so that both heels are in line, and continue the cadence by alternately raising and planting each foot. When raised, the ball of the foot is two inches above the ground.

2. Being at a halt, at the command MARCH, raise and plant first the left foot, then the right as described above.

3. Mark Time may be executed in either Quick-Time cadence or Double-Time cadence. While marking time, any errors in alignment should be corrected.

4. The HALT is executed from Mark Time as from Quick Time or Double Time. Forward MARCH, HALT, and Mark Time may be executed one from the other in Quick Time or Double Time.

Half Step

Commands: 1. Half Step, 2. MARCH ("Half Step, MARCH").

1. Being in march, at the command MARCH, take steps of fifteen inches in Quick Time instead of the normal thirty inches. The half step is executed in Quick Time only.

2. To resume the full step from half step, the commands are: 1. Forward, 2. MARCH ("Forward, MARCH").

Right Step

Commands: 1. Right Step, 2. MARCH ("Right Step, MARCH"). At the command MARCH, carry the right foot twelve inches to the right, then place the left foot beside the right, left knee straight. Continue in the cadence of Quick Time. The right step is executed in Quick Time from a halt for short distances only.

Left Step

Commands: 1. Left Step, 2. MARCH ("Left Step, MARCH"). At the command MARCH, carry the left foot twelve inches to the left, then place the right foot beside the left, right knee straight. Continue in the cadence of Quick Time. The left step is executed in Quick Time from a halt for short distances only.

Back Step

Commands: 1. Backward, 2. MARCH ("Backward, MARCH"). At the command MARCH, take steps of fifteen inches straight to the rear (begin with left foot and swing your arms). The back step is executed in Quick Time for short distances only.

To Face to the Right (or Left) in Marching

Commands: 1. By the Right (or Left) Flank, 2. MARCH ("By the Right/ Left Flank, MARCH").

1. To face to the right (left) in marching and advance from a halt, at the command MARCH, turn to the right (left) on the ball of the right (left) foot; at the same time, step off with the left (right) foot in the new direction with a half or full step in Quick Time or Double Time, as the case may be.

2. To face to the right (left) in marching and advance, being in march, at the command MARCH, given as the right (left) foot strikes the ground, advance and plant the left (right) foot; then face to the right (left) in marching, and step off with the right (left) foot in the new direction with a half or full step in Quick or Double Time, as the case may be.

To Face to the Rear in Marching

Commands: 1. To the Rear, 2. MARCH ("To the Rear, MARCH").

1. Being in march at Quick Time, at the command MARCH, given as the right foot strikes the ground, advance and plant the left foot,

then turn to the right all the way about on the balls of both feet, and immediately step off with the left foot (keep arms pinned to your side while you are turning, continue to swing arms after you step off).

2. Being in march at Double Time, at the command MARCH, given as the right foot strikes the ground, advance two steps in the original direction; turn to the right all the way about while taking four steps in place, keeping the cadence, and step off.

To Change Step
Commands: 1. Change Step, 2. MARCH ("Change Step, MARCH").

1. Being in march in Quick Time, at the command MARCH, given as the right foot strikes the ground, advance and plant the left foot; then plant the toe of the right foot near the heel of the left, and step off with the left foot.

2. The same movement may be executed on the right foot by giving the command MARCH as the left foot strikes the ground and planting the right foot; then plant the toe of the left foot near the heel of the right and step off with the right foot.

To March at Ease
Commands: 1. At Ease, 2. MARCH ("At Ease, MARCH"). At the command MARCH, adopt an easy natural stride, without any requirement to keep step or a regular cadence. Keep silent and maintain interval and distance.

To March at Route Step
Commands: 1. Route Step, 2. MARCH ("Route Step, MARCH"). At the command MARCH, adopt an easy natural stride, without any requirement to keep step or a regular cadence, or to maintain silence (must keep interval and distance).

MANUAL OF ARMS

The manual of arms described here is performed with the M16 rifle. For instruction purposes, it may be taught by the numbers. When marching at Quick Time, the only movements that may be executed are Right/Left Shoulder Arms, Trail Arms, and Port Arms. The cadence of all motions is Quick Time. Recruits learning the manual of arms should concentrate their attention on the details of the motion; the cadence will be acquired as they become accustomed to handling their rifles. The instructor may require them to count aloud in cadence with the motions.

General

FALL IN is executed with the rifle at Order Arms. When companies are formed, rifles are immediately inspected. (See commands for Inspection Arms, following.) Prior to executing any movements with the rifle, the magazine is removed and the sling is positioned on the left side of the rifle, drawn tight, the keeper lying flat and on top of the pistol grip just below the selector lever. Before starting any movement from the halted position, except for movements requiring the position of Trail Arms, rifles will be at Right Shoulder Arms, Port Arms, or Sling Arms. While at a position other than Sling Arms, the rifle is brought to Port Arms for marching at Double Time.

Position of Order Arms

This is the basic rifle position (see figure 9-2). The rifle butt rests on deck, with the toe of the rifle in line with the toe of the right shoe. The left side of the stock is along the outer edge of the right shoe. The magazine well is to the front and the barrel is in a vertical position. The rifle is held with the fingers extended and joined, the junction of the front sight assembly and the barrel rests in the "V" formed by thumb and fingers of the right hand. The right thumb is on the trouser seam and the entire right arm is behind the rifle (this may cause a slight bend in the arm). The body is at the position of Attention, as it is executed without arms.

Trail Arms from Order Arms

Commands: 1. Trail, 2. ARMS ("Trail, ARMS"). This command is given at the position of Order Arms only. At the command of execution, raise the rifle vertically three inches off the deck. Do not change the grasp of the right hand, keep the right thumb on the trouser seam and the right arm behind the rifle; keep the left hand in position.

Order Arms from Trail Arms

Commands: 1. Order, 2. ARMS ("Order, ARMS"). At the command, lower the butt of the rifle to the deck and assume the position of Order Arms. The rifle is kept at Trail Arms during any movement of Back Step, Extending, or Closing ranks; the position of Order Arms is assumed automatically on halting.

Figure 9-2. Order arms.

Figure 9-3. From order arms to port arms.

Port Arms from Order Arms

Commands: 1. Port, 2. ARMS ("Port, ARMS"). At the command of execution, slide the right hand to the barrel with fingers joined and wrapped around it, raise and carry the rifle diagonally across in front of and slightly to the left of your face. The right wrist and forearm are straight, elbow held down without strain. The right arm does not need to be parallel to the deck. Rifle muzzle is up, bisecting the angle formed by neck and shoulder, with magazine well to the left, butt in front of right hip. At the same time, smartly grasp the handguard, fingers joined and wrapped around it, little finger above the slip ring, thumb inboard, centered on the chest. Left wrist and forearm are straight, elbow against the body.

TWO. Release grasp of right hand on the barrel and regrasp the small of the stock. Fingers are joined, wrapped around the small of the stock, parallel to the deck, elbow against the body, upper arm in line with the back. The rifle should be four inches from the belt. (See figure 9-3.)

Order Arms from Port Arms

Commands: 1. Order, 2. ARMS ("Order, ARMS"). At the command of execution, release the grasp of the right hand from the stock and smartly regrasp the barrel. Fingers are joined, wrapped around the barrel, palm to the rear, little finger just above the bayonet stud. The right wrist and forearm are straight, elbow held down without strain. The right arm does not need to be parallel to the deck.

TWO. Release the grasp of the left hand from the handguard. With the right hand, lower the rifle to the deck, magazine in front, muzzle in a

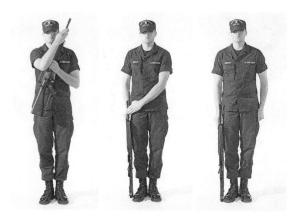

Figure 9-4. From port arms to order arms.

vertical position. At the same time, guide the weapon into the right side with left hand until thumb is on the trouser seam, left fingers extended and joined, thumb along the forefinger. The first joint of the forefinger should touch the metal below the flash suppressor. The left wrist and forearm are straight, elbow against the body.

THREE. Smartly return the left hand to the left side in the position of Attention, and gently lower the rifle to the deck with the right hand, with the toe of the rifle in line with the toe of the right shoe. The magazine is to the front and in a vertical position. The position is the same as described for Order Arms, above (see figure 9-4).

Present Arms from Order Arms

This is used as a salute to persons and colors during parades and ceremonies, and by sentries. Commands: 1. Present, 2. ARMS ("Present, ARMS"). On the first count, raise and carry the rifle to a vertical position centered on the body, magazine well to the front.

The fingers of the right hand are joined and wrapped around the barrel, thumb on the left side. At the same time, smartly grasp the rifle at the handguard with the left hand, fingers joined and wrapped around the handguard, thumb just above the slip ring, which should be four inches from the body. The left wrist and forearm are straight, parallel to the deck, elbow against the side and upper arm in line with the back.

TWO. Release the grasp of the right hand from the barrel and regrasp the small of the stock, fingers extended and joined, and the charging

Figure 9-5. Present arms.

handle resting in the "V" of thumb and forefinger. Wrist and forearm are straight, forming a straight line from fingertips to elbow, which is held against the body (see figure 9–5).

Order Arms from Present Arms

Commands: 1. Order, 2. ARMS ("Order, ARMS"). On the command of execution, release the grasp of the right hand from the small of the stock and smartly regrasp the barrel, fingers joined and wrapped around the barrel, palm facing left. Right wrist and forearm are straight, elbow held down without strain. The right arm does not need to be parallel to the deck.

TWO. Release the grasp of the left hand from the handguard and lower the weapon to the right side until the rifle butt is about three inches from the deck, muzzle in front and in a vertical position, magazine well to the front. At the same time, guide the weapon into the right side with the left hand until the thumb is on the trouser seam. Fingers are extended and joined, thumb along the forefinger, palm to the rear, first joint of the forefinger touching the metal just below the flash suppressor. The left wrist and forearm are straight, elbow held against the body. The entire right arm is behind the rifle.

THREE. Smartly, in the most direct manner, return the left hand to the left side in the position of Attention. At the same time, lower the rifle to the deck with the right hand so the toe of the rifle is in line with the toe of the right shoe; the magazine should be well to the front, the muzzle in

Figure 9-6. From present arms to order arms.

a vertical position. With the butt of the rifle on deck, the right hand takes the position described earlier for Order Arms (see figure 9-6).

Right Shoulder Arms from Order Arms

Commands: 1. Right Shoulder, 2. ARMS ("Right, Shoulder, ARMS"). At the command of execution, slide the right hand to the muzzle, fingers joined and wrapped around it, and in continuous motion raise and carry the rifle diagonally across the front of the body until the right hand is in front of and slightly left of the face. The right wrist and forearm are straight and slightly to the left of the face. The muzzle is up, bisecting the angle formed by the neck and left shoulder, magazine well to the left, butt of the rifle in front of the right hip. At the same time, smartly grasp the handguard with the left hand, fingers joined and wrapped around the handguard. The little finger is just above the slip ring, thumb on the inboard side and centered above the chest. The left wrist and forearm are straight and the elbow is against the body.

TWO. Release the grasp on the muzzle, regrasp the butt of the rifle, thumb and forefinger joined on the comb of the stock, remaining fingers joined and wrapped around the butt. The right arm is almost extended; the elbow is against the right side of the body.

THREE. Relax the left-hand grip on the handguard, and with the right hand carry the rifle to the right side. At the same time, rotate the rifle so the magazine is well to the rear, and place it on the right shoulder with the pistol grip in the right armpit. Guide the rifle into the right shoulder with the left hand by sliding it down the rifle until the first joint of the left index finger touches metal at the base of the charging handle. The left

Figure 9-7. From order arms to right shoulder arms.

fingers should be extended and joined with the thumb along the forefinger, palm to the rear. The left wrist and forearm are straight, the elbow held down without strain. The right hand's grasp remains unchanged. The right wrist and forearm are straight, parallel to the deck, elbow against the body, and upper arm in line with the back.

FOUR. Return the left hand, in the most direct manner, to the left side at the position of Attention (see figure 9-7).

Port Arms from Right Shoulder Arms

Commands: 1. Port, 2. ARMS ("Port, ARMS"). From the position of Halt, at the command of execution, press the rifle butt down quickly with the right hand so the rifle comes off the right shoulder. At the same time, rotate the rifle with the right hand a quarter of a turn so the magazine is well to the left, and let the rifle fall diagonally across the front of the body. Bring the left hand up, grasp the handguard, left fingers joined and wrapped around it; the little finger is just above the slip ring, thumb on the inboard side and centered on the chest. The left arm and wrist are straight, elbow held against the body. Rifle muzzle is up, bisecting the angle formed by the neck and left shoulder; rifle butt is in front of the right hip. The right hand's grasp is unchanged; the right arm is nearly extended, but held against the body.

TWO. Release the grasp of the right hand, regrasp the small of the stock with fingers joined, thumb on the inboard side of the rifle. Right wrist and forearm are straight, parallel to the deck, elbow in and upper arm in line with the back. The rifle should be about four inches from the belt.

When the command "Port, ARMS" is given while troops are marching at "Right Shoulder, ARMS," it will be given as the left foot strikes the deck. On the next step, the first count of the movement is carried out as if done at a halt. With each following step, another count is executed until the movement is complete. The movement is executed in cadence. Troops continue marching in this position until another command is given.

Order Arms from Right Shoulder Arms

Commands: 1. Order, 2. ARMS ("Order, ARMS"). On the command of execution, press the rifle butt down quickly with the right hand so the rifle comes off the right shoulder. At the same time, rotate the rifle a quarter turn with the right hand so the magazine well is to the left, and let the rifle fall diagonally across the front of the body. Bring the left hand up and grasp the handguard; the fingers of the left hand are joined and wrapped around the guard, the little finger just above the slip ring. The left wrist and forearm are held straight and against the body, as is the elbow. The rifle muzzle is up, bisecting the angle formed by the neck and left shoulder. The grasp of the right hand remains unchanged. The right arm is extended, with elbow against the body.

TWO. Release the grasp of the right hand, regrasp the rifle with the palm to the rear, fingers joined and wrapped around the muzzle, thumb inboard. The wrist and forearm are straight, held down without strain.

THREE. Release the grasp of the left hand from the handguard and lower the rifle with the right hand on the right-hand side of the body until the butt is three inches from the deck, magazine to the front and the rifle in a vertical position. At the same time, guide the rifle to the right side with the left hand until the right thumb is on the trouser seam. Left fingers are joined, thumb along the forefinger. The first joint of that finger touches the metal just below the flash suppressor. The left palm faces the rear; the left wrist and forearm are straight, the elbow against the body. The entire right arm is behind the rifle.

FOUR. Return the left hand to the left side in a direct manner and assume the position of Attention. Then lower the rifle to the deck, gently, so the toe of the rifle is in line with the toe of the right shoe, magazine well to the front and rifle in a vertical position. When the butt is on deck, slide the right hand down the barrel so that the front sight assembly and barrel will rest in the "V" formed by the thumb and fingers of the right hand. The fingers of the right hand are joined and extended diagonally across the

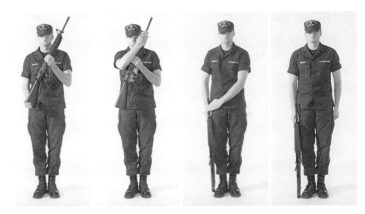

Figure 9-8. From right shoulder arms to order arms.

rifle, right thumb on the trouser seam; the entire right arm is behind the rifle. (See figure 9-8.)

Right Shoulder Arms from Port Arms

Commands: 1. Right Shoulder, 2. ARMS ("Right Shoulder, ARMS"). At the commands of execution release the right-hand grasp on the stock and smartly regrasp the butt of the rifle, thumb and forefinger joined and wrapped around the comb of the stock and the remaining three fingers joined and wrapped around the butt. The heel of the butt will be visible between the forefinger and middle finger. The right arm is nearly extended; the right elbow is against the body.

TWO. Relax the grasp of the left hand on the handguard, and with the right hand carry the rifle to the right side. At the same time, rotate the rifle a quarter turn so the magazine is well to the rear, and place the pistol grip in the right armpit; guide the rifle into the right shoulder with the left hand by sliding it down the rifle until the first joint of the index finger touches the base of the charging handle. The left fingers should be joined with the thumb along the forefinger, palm to the rear. The left wrist and forearm are straight, the elbow held down without strain. The right-hand grasp remains unchanged; the right wrist and forearm are straight and parallel to the deck, elbow against the body, upper arm in line with the back.

THREE. Return the left hand to the side in the most direct way and come to the position of Attention.

Note: This command can be executed while marching at Quick Time, in the same manner as Port Arms from Right Shoulder Arms.

Order Arms from Left Shoulder Arms

Commands: 1. Order, 2. ARMS ("Order, ARMS"). On the command of execution, bring the right hand across the body in the most direct manner and grasp the small of the stock. The fingers are joined and wrapped around the small of the stock, thumb inboard; right wrist and forearm straight, elbow held down without strain.

TWO. Release the grasp of the left hand on the butt of the rifle, and with the right hand bring the rifle from the shoulder to a position diagonally across the front of the body, rotating it a quarter turn so the magazine is well toward the left. At the same time, smartly regrasp the handguard with the left hand; the fingers should be joined and wrapped around the handguard, the little finger just above the slip ring, and the thumb inboard and centered on the chest. The left wrist and forearm are straight, elbow against the body; right wrist and forearm straight, elbow in and in line with the back. The rifle muzzle is up, bisecting the angle formed by the neck and left shoulder, the butt in front of the right hip.

THREE. Release the right-hand grasp on the stock and smartly regrasp the barrel, fingers joined, wrapped around the barrel, palm to the rear and little finger just above the bayonet stud. The right wrist and forearm are straight, elbow held down without strain.

FOUR. Release the grasp of the left hand from the handguard, and with your right hand lower the rifle to the right side until the butt is three inches from the deck. The magazine should face the front, rifle in a vertical position. At the same time, guide the rifle into the right side with the left hand until the right thumb is on the trouser seam. The left fingers are extended and joined, thumb along the forefinger, first joint of the forefinger touching the metal of the flash suppressor. The left palm faces the rear; left wrist and forearm are straight, elbow against the body. The right arm is behind the rifle.

FIVE. Return the left hand to the left side and assume the position of Attention. At the same time, lower the rifle to the deck so the toe of the rifle is in line with the toe of the right shoe, with the magazine to the front and the rifle in a vertical position. As the rifle touches the deck, slide the right hand to the front sight assembly; fingers joined and extended diagonally across the front of the barrel. The right arm is behind the rifle.

Left Shoulder Arms from Right Shoulder Arms

This movement can be made while at Right Shoulder, ARMS, Marching at Quick Time, or at a Halt. Commands: 1. Left Shoulder, 2. ARMS ("Left Shoulder, ARMS"). On the command of execution, press the rifle butt down quickly with the right hand so the rifle comes off the right shoulder; rotate the rifle a quarter turn with the right hand so that the magazine well is to the left, and let the rifle fall diagonally across the front of the body. At the same time bring the left hand up and smartly grasp the handguard with left fingers joined and wrapped around it, little finger just above the slip ring, thumb inboard and centered on the chest. The left wrist and forearm are straight, elbow against the body. The right-hand grasp is unchanged; the right arm is nearly extended, the right elbow against the body. The rifle muzzle is up and bisecting the angle formed by the neck and left shoulder, the butt behind the right hip, and the rifle four inches in front of the body.

TWO. Release the grasp of the right hand from the butt and smartly regrasp the small of the stock, fingers joined and wrapped around it, thumb inboard. The right wrist and forearm are straight, elbow against the body, upper arm in line with the back.

THREE. Release the grasp of the left hand from the handguard, and with the right hand carry the rifle to the left side, rotating it a quarter turn so the magazine is well to the rear; place the pistol grip in the left armpit. The right wrist and forearm are straight, elbow held down without strain. At the same time, regrasp the butt with the left hand; thumb and forefinger joined on the comb of the stock, remaining three fingers joined and wrapped around the butt. The heel of the butt will be visible between the forefinger and middle finger.

FOUR. Release the grasp of the right hand from the small of the stock and return it across to the right side in a direct manner for the position of Attention.

Note: This command can be executed while marching at Quick Time, in the same manner as Port Arms from Right Shoulder Arms.

MOVEMENTS OF REST

Rest movements are used to give troops a change from the position of Attention and still maintain an orderly formation. Parade Rest and At Ease consist of a single count, that of execution. Fall Out has no count at all.

Figure 9-9. Parade rest, front and side views.

Parade Rest

When the preparatory command Parade is given, shift the weight from the left leg to the right without any body movement. When the command of execution, REST, is given, move the left foot smartly twelve inches to the left of the right foot measured from inside the heels. Heels should be on the same line, legs straight but not stiff, body weight equal on both legs.

The left hand is placed in the small of the back, just below the belt, fingers extended and joined, thumb along the forefinger, palm to the rear and elbow in line with the body. The butt of the rifle is kept on deck, in line with the feet, and against the right foot. When the left foot is first moved, relax the grasp of the right hand and move it to a point just below the flash suppressor. The fingers are joined and curled with the forefinger touching the thumb. At the same time, straighten the right arm directly to the front so the rifle muzzle points upward.

While at Parade Rest, maintain silence. The only command given to troops in this position is Attention. When that command is given, bring the left foot smartly back against the right foot, lower the left arm back to the side, move the rifle to the position of Order Arms and regrasp the rifle for that position (see figure 9-9).

At Ease

In this position the right foot remains stationary, and the rifle butt remains on deck. The rifle is held as in Parade Rest, but the grip and body stance may be more relaxed. Equipment may be adjusted, but silence must be maintained. The only command given while in this position is Attention, the movement for which is described above.

Rest

This is exactly the same as At Ease, except that troops may talk in conversational tones. The only command given to troops in this position is Attention. On the preparatory command, come to Parade Rest. On the command of executions, come to Attention and Order Arms.

Fall Out

When this order is given, troops may leave the present position and go to a designated area or remain in the general vicinity. The only command that can be given after FALL OUT is FALL IN, at which troops assume the position of Order Arms.

RIFLE SALUTES

Troops in formation, or individuals who find themselves in a saluting situation, may execute the rifle salute. Salutes are made from Order Arms, Trail Arms, Right Shoulder Arms, or Left Shoulder Arms.

At Order Arms

Commands: 1. Rifle, 2. SALUTE ("Rifle, SALUTE"). Carry the left hand across the body in the most direct manner until the first joint of the forefinger touches the metal just below the flash suppressor, left fingers extended and joined, thumb along the forefinger, palm down. The wrist and forearm are straight, elbow against the body.

Ready, TWO. Smartly return the left hand to the left side and assume the position of Order Arms (see figure 9-10).

At Trail Arms

On the command of execution, carry the left hand across the body in the most direct manner until the first joint of the forefinger touches the metal at the base of the flash suppressor. Left fingers extended and joined, palm down, wrist and forearm straight, elbow against the body.

Ready, TWO. Smartly bring the left hand to the side and assume the position of Order Arms.

Figure 9-10. Hand salute at order arms.

Figure 9-11. Hand salute at right shoulder arms.

At Right Shoulder Arms

Smartly carry the left hand across the body until the first joint of the left forefinger touches the metal at the charging handle. The left fingers are extended and joined, palm down, arm parallel to the deck.

Ready, TWO. Smartly return the left hand to the side and assume the position of Order Arms (see figure 9-11).

At Left Shoulder Arms

On the command of execution, smartly carry the right hand across the body until the first joint of the forefinger is touching the base of the charging handle. Fingers of the right hand are extended and joined, with thumb along the forefinger, palm down, arm parallel to the deck.

TWO. Smartly return the right hand to the side and assume the position of Order Arms.

Inspection Arms (Without Magazine)

Commands: 1. Inspection, 2. ARMS. On the command of execution, slide the right hand to the muzzle, fingers wrapped around the barrel. Raise and carry the rifle diagonally across the front of the body until the right hand is in front of and slightly to the left of the face. The right wrist and forearm are straight and held down without strain. The rifle muzzle is up and bisecting the angle formed by the neck and left shoulder, magazine well to the left, butt in front of the right hip. Smartly grasp the handguard

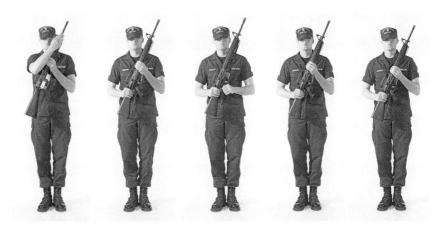

Figure 9-12. Inspection arms.

with the left hand, little finger just above the slip ring, thumb centered on the chest. Left wrist and forearm are straight, elbow against the body.

TWO. Release the grasp of the right hand on the barrel; regrasp the small of the stock with the thumb wrapped around the inboard portion. Right wrist and forearm are straight and parallel to the deck; elbow against the body, upper arm in line with the back. The rifle should be four inches from the body.

THREE. Release the grasp of the left hand from the handguard and regrasp the rifle at the pistol grip. The thumb of the left hand is over the lower portion of the bolt catch.

FOUR. Release the grasp of the right hand from the stock and use the thumb and forefinger to unlock the charging handle; at the same time pull it sharply to the rear, apply pressure on the bolt catch, and lock the bolt to the rear.

FIVE. With the thumb and forefinger of the right hand, push the charging handle until it is locked in the forward position. Regrasp the small of the stock with the right hand.

SIX. Elevate the rifle to the left with both hands until the rear sight is in line with the left shoulder. At the same time, rotate it a quarter turn clockwise so the chamber is visible, and turn the head and eyes toward the ejector port to inspect the chamber.

SEVEN. After seeing the chamber clear, or clearing it, turn head and eyes back to the front and lower the rifle to the position of Port Arms. As it is lowered, release the grasp on the pistol grip and regrasp the handguard as in Port Arms (see figure 9-12).

Port Arms from Inspection Arms (Without Magazine)

Commands: 1. Port, 2. ARMS. At the preliminary command of Port, release the grasp of the left hand and regrasp the rifle at the magazine well and trigger guard, fingers extended and joined, thumb inboard, thumb and forefinger forming a "V" at the magazine well and trigger guard. At the same time press the bolt catch and allow the bolt to slide forward. With the fingertips of the left hand, push forward and close the ejection cover, then slide the left hand toward the pistol grip and place the thumb on the trigger.

At the command ARMS, push down on the trigger with the left thumb and uncock the hammer. Then release the grasp of the left hand and regrasp the rifle at the handguard, fingers extended and joined, little finger just above the slip ring, thumb inboard. Left wrist and forearm are straight, elbow against the body.

CLOSE-ORDER DRILL

A group of personnel in uniform and bearing arms must be under the direct control of a senior officer in order for them to move in an orderly fashion. Close-order drill will accomplish this. Such drill also serves to teach discipline by instilling habits of automatic and precise response to orders, developing better morale through team spirit, and making it possible for a Coast Guard unit to make a favorable appearance in a parade or other public ceremony.

The basic unit in a military formation is the squad, a group of eight to twelve people organized as a team. The squad is usually kept intact. Its normal formation is a single rank or file. This permits variation in the number of personnel in the squad. The original formation is always in line.

The squad marches in line only for minor changes of position. The squad leader, when in ranks, is posted as the right person of the squad in line or as the leading person in column.

To Form the Squad

The command is: FALL IN. At the command, the squad forms in line. Each person, except the one on the extreme left, extends the left arm to the side at shoulder height, palm down, fingers extended and joined. Each person, except the one on the extreme right, turns the head and eyes to the right and moves enough in line so that the right shoulder touches the tips of the fingers of the person on the right. As soon as proper intervals have been obtained, each person drops the arm and turns the head smartly to the front.

To Form at Close Interval

Commands: 1. At Close Interval, 2. FALL IN. At the command, the personnel fall in as above, except that close intervals are obtained by placing the left hand on the hip, fingers extended downward and joined, thumb along the forefingers, heel of the hand resting along the hip bone, elbow extended to the side.

If the squad is under arms, rifles are inspected. The squad executes the positions, movements, and manual of arms with all personnel executing the movements together.

To Dismiss the Squad

Commands: 1. Inspection, 2. ARMS, 3. Port, 4. ARMS, 5. DISMISSED. If the squad is not under arms, the single command DISMISSED is used (the squad, of course, being at attention or having been brought to attention).

To Count Off

Commands: 1. Count, 2. OFF. At the command OFF, each person except the one on the right flank, turns their head and eyes to the right. The right-flank person calls out "Zero One." Each person in succession calls out "Zero Two," "Zero Three," and so forth, turning the head and eyes to the front as the number is given.

To Align the Squad

In line, commands: 1. Dress Right (or Left), or 1. At Close Interval, Dress Right (Left), 2. DRESS, 3. Ready, 4. FRONT. At the command DRESS, each person except the one on the left, extends the left arm (or if at close intervals, places the left hand upon the hip), and all dress to the right. The squad leader is on the right flank, one pace from and on the line of the squad. From this position the squad leader checks alignment, ordering individuals forward or backward as necessary. Having checked alignment, the squad leader faces to the right in marching, and moves three spaces forward, halts, faces to the left and commands: 1. Ready, 2. FRONT. At the command FRONT, arms are dropped quietly and smartly to the side, and heads turned to the front. In column, command: COVER. At the command COVER, personnel cover from front to rear with 40-inch intervals between them.

To Obtain Close Interval

Commands: 1. Close, 2. MARCH. At the command MARCH, all personnel, except the right-flank person, face to the right in marching and form at close interval.

To Extend to Normal Interval

Commands: 1. Extend, 2. MARCH. At the command MARCH, all personnel, except the right-flank person, face to the left in marching and form at normal interval.

To March to the Flank

Commands: 1. Right (or Left), 2. FACE, 3. Forward, 4. MARCH. These movements are executed with all personnel stepping off together.

To March to the Oblique

Commands: 1. Right (or Left) Oblique, 2. MARCH. (The word "oblique" is pronounced to rhyme with "strike.") At the command MARCH, given as the right foot strikes the ground, each person advances, plants the left foot, faces half right, and steps off in a direction of 45 degrees to the right. Each remains in relative place, shoulders parallel to those of the guide, steps regulated so that the ranks stay parallel to the original front.

The command HALT is given on the left foot when halting from the right oblique march, and on the right foot when halting from left oblique march.

At the command HALT (from left oblique march), given as the left foot strikes the ground, each person advances, planting the right foot, turns to the front on the ball of the right foot, and brings the left foot to the side of the right. The opposite is done from right oblique march so that in either case the personnel end facing front.

To stop temporarily the execution of the movement for the correction of errors, the commands are 1. In Place, 2. HALT. All halt in place without facing to the front and stand fast. To resume the movement the commands are 1. Resume, 2. MARCH.

If at half step or mark time while obliquing, the full step is resumed by the command: 1. Resume, 2. MARCH.

To resume original direction, the commands are 1. Forward, 2. MARCH. At the command MARCH, each person faces half left in marching and moves straight to the front. If at half step or marking time while in oblique march, the full step is resumed by the command 1. Oblique, 2. MARCH.

To March toward the Flank

Commands: 1. By the Right (Left) Flank, 2. MARCH. At the command MARCH, each person simultaneously executes the movements as described above.

In Column, to Change Direction

Commands: 1. Column Right (or Left), (Half Right), (Half Left), 2. MARCH. At the command MARCH, the leading person executes the movement immediately; the others follow in succession as they reach the pivot point.

In Line, to Take Double-Arm Interval and Assemble

Commands: 1. Take Interval to the Left (or Right), 2. MARCH. At the command MARCH, the right-flank person stands fast, extends the left arm at shoulder height, palm down, fingers extended and joined, until the person on the left obtains the proper interval, then drops the arm. Other personnel face to the left in marching and step out until they have an interval of two arms' length from the person on the right, then halt and face to the front. Each person, except the one on the left who raises the right arm only, extends both arms laterally at shoulder height. Each person, except the one on the right flank, then turns the head and eyes to the right and moves enough for alignment so that the fingertips of the right hand touch those of the left hand of the person on the right. As soon as each person is aligned at two arms' length, they drop the right arm to the side and turn the head and eyes to the front. Each drops the left arm when the person on the left has obtained the proper interval.

To assemble, the commands: 1. Assemble to the Right (or Left), 2. MARCH. At the command MARCH, the right-flank person stands fast; all other personnel face to the right in marching and form at normal interval.

Column of Twos

When marching small groups, not at drill, the group may be marched in a column of twos by forming it in two ranks and giving the commands: 1. Right (or Left), 2. FACE.

To Form Column of Twos from Single File and Reform

The squad being in column, at halt, to form a column of twos the commands are 1. Form Column of Twos to the Right (Left), 2. MARCH. At the command MARCH, the leading person stands fast; the second person in the squad moves by the oblique to the left of and abreast of the squad leader with normal interval, and halts. The third person moves forward until behind the squad leader at normal interval, and halts. The fourth person moves by the oblique to the left of and abreast of the third person at normal interval, and halts, and so on.

To Form Single File from Column of Twos

Being in column of twos, in marching, to reform single file the squad is first halted. Commands: 1. Column of Files from the Right, 2. MARCH. At the command MARCH, the leading person of the right column moves forward. The leading person of the left column steps off to the right oblique, and then executes left oblique so as to follow the right file at normal interval. The remaining twos follow successively in the same way.

IV
PERSONAL
DEVELOPMENT

Figure 10-1.
Coast Guard Training Center Cape May is the home of the Coast Guard's
enlisted corps—it is our nation's only Coast Guard enlisted accession
point and recruit training center. More than four thousand of America's finest
young men and women arrive every year for the first chapter of
their Coast Guard careers—boot camp.

CHAPTER 10

Training and Education

Enlistment in the Coast Guard offers you a choice of many career opportunities, many of which are very marketable in the private sector. The option you finally select will depend, in part, on the training you receive, your own aptitudes, and the needs of the service. This chapter describes some of the training opportunities available to you and points out how you can start on a Coast Guard career. Remember that the amount of responsibility and self-discipline you display in your work will have a great effect on your duty assignments and promotion potential.

RECRUIT TRAINING

Your service in the Coast Guard will begin in Cape May, New Jersey. Having been accepted into the Coast Guard and sworn in, you have been ordered to report to Training Center Cape May. Here, in eight weeks, you must learn to make the difficult change from school and home life to a strict military routine. You will probably be wishing you were anywhere else but at Coast Guard boot camp at first, but so will almost everyone else. Your company commanders and instructors are among the most professional members of the U.S. Coast Guard and are there to help you get organized and learn to do things the Coast Guard way.

One of the biggest problems the average recruit has upon entering military service is learning how to live with a large group of people and become a team member. Recruit training is designed to assist you in solving those problems, but it cannot do it alone. You must help yourself by

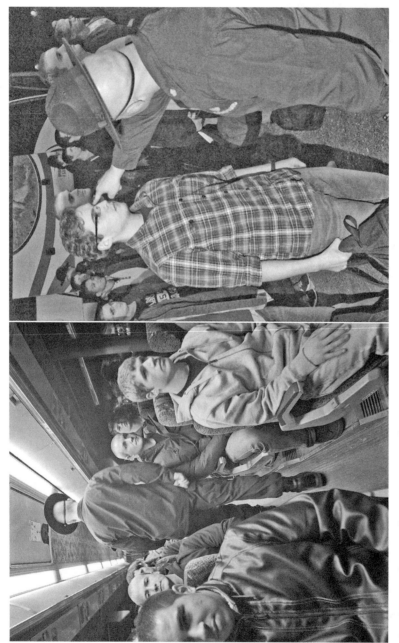

Figure 10-2. Welcome to the U.S. Coast Guard.

being tolerant of other people's peculiarities since you probably have some of your own.

While you are in recruit training, you will learn many things. Some of them are taught in formal classes; others you will gradually absorb in your daily work. Recruit training is an entry-level instructional program designed to provide you with fundamental skills and knowledge so that you can be productive in a military environment. You will learn the various skills required in the Coast Guard: military courtesies, drills and ceremonies, military justice and code of conduct, leadership and supervision, seamanship, uniforms, ranks and rates, career development, first aid and survival, fitness, wellness, and quality of life, Coast Guard history, traditions and values, safety, damage control, small arms, Coast Guard assets and missions, administration and personal finance, watch-standing, security, and communications. You will also learn the main essential of good citizenship—getting along with others.

FIRST DUTY ASSIGNMENTS

Upon completion of recruit training, each graduate is permitted to take up to five days regular leave. This "recruit" leave may provide you time to visit your family and friends back home for a few days. They may base their entire opinion of the Coast Guard on what you do and how you act and display the training you have received, so make sure you leave them with a good impression.

First duty assignments may be either to sea duty or shore duty. Sea duty could mean a cutter, icebreaker, buoy tender, patrol boat, or any of many other vessels. Shore duty could be at a small-boat station or sector. Wherever you go, you will find new friends and exciting experiences.

Training

Training consists of basic ("A") schools and specialized ("C") schools. "A" schools provide the minimum essential rating training designed to provide the basic technical knowledge and skills required for job entry-level performance for the respective rating. The "A" school graduate is an apprentice who will initially require additional on-the-job training under close supervision to carry out assigned duties. "C" school is short-term training designed to provide advanced/specialized knowledge and skills to perform a task, or group of tasks, required by a specific billet.

Figure 10-3. Company commanders are role models and mentors.

Assignment to "A" school may be made upon completion of recruit training. Most personnel gain assignment to "A" schools after serving in a ship or at a station for a specific amount of time.

The Coast Guard conducts the following "A" schools:

Aviation Technical Training Center (ATTC), Elizabeth City, North Carolina

Avionics electrical technician (AET)	(26-week course)
Aviation maintenance technician (AMT)	(24-week course)
Aviation survival technician (AST)	(24-week course)

Training Center, Petaluma, California

Electronics technician (ET)	(29-week course)
Culinary specialist (CS)	(12-week course)
Health services technician (HS)	(23-week course)
Information system technician (IT)	(29-week course)
Operations specialist (OS)	(17-week course)
Storekeeper (SK)	(8-week course)
Yeoman (YN)	(8-week course)

Training Center, Yorktown, Virginia

Boatswain's mate (BM)	(14-week course)
Damage controlman (DC)	(15-week course)
Electrician's mate (EM)	(19-week course)
Gunner's mate (GM)	(13-week course)
Intelligence specialist (IS)	(14-week course)
Machinery technician (MK)	(13-week course)
Marine science technician (MST)	(10-week course)

Coast Guard Maritime Law Enforcement Academy at the Federal Law Enforcement Training Center, Charleston, South Carolina

Maritime enforcement specialist (ME)	(10-week course)

Defense Information School, Ft. Meade, Maryland

Public affairs specialist (PA)	(12-week course)

Naval Diving and Salvage Training Center, Naval Support Activity Panama City, Florida

Diver (DV)	(17.5-week course)

Selection for Training

In order to attend an advanced school, you must have attained minimum scores in your classification tests, must have a good conduct record, and must be recommended by your commanding officer. Your commanding officer's recommendation will be based on observation of your work and attitude by the CO, the executive officer, and your division and petty officers.

Graduates of "A" schools will be advanced to petty officer third class once they have completed the enlisted professional military education (EPME) requirements for E-4, have demonstrated leadership abilities, passed the Advancement Qualification Exam (AQE) for E-4, and have the required six months' service in pay grade E-3. Those not advanced on graduation will be given a school-assigned designator and classified as a "rated nonrate." This allows them to work in their new rating. They will be promoted when the above requirements have been satisfied and they are recommended by their commanding officer.

Some of the knowledge you need can be gained only by experience. The rest comes by studying various publications, including this manual. Remember that knowledge alone will not ensure promotion. You must have a record of satisfactory performance in your present assignment, your

Figure 10-4. Class "A" schools prepare you for duty in a specific rating.

conduct record must be satisfactory, you must be reliable, and you must be recommended by your commanding officer. The officers and petty officers of your unit must be able to depend upon you.

Ratings Qualifications

After reaching pay grade E-3, you should decide what rating you prefer to enter. Most ratings are covered by "A" schools, but for some you can make your own preparations by "striking a rating."

In addition to the actual knowledge requirements for advancement in rate, you must serve a certain minimum time in the lower rate, be recommended for advancement, and have completed the enlisted performance qualifications (EPQ) for the specific rating. The *Enlisted Performance Qualifications Manual*, COMDTINST M1414.8 (series), gives details on EPQs.

Performance Qualification Guide

The Coast Guard Institute (CGI) furnishes performance qualification guides covering the requirements for advancement in practically all ratings. These guides enable you to study for a rating at your own unit. Completion of one of them is required for advancement. "A" school graduates do not have to complete the rate-specific performance qualification guide for

Figure 10-5. Graduation day at Coast Guard Training Center Cape May.

advancement to pay grade E-4, but they do complete the EPME require-ments and pass the AQE course for advancement to pay grade E-4. Note: Some ratings require "A" school attendance. Check with your educational services office.

On-the-Job Training

On-the-job training goes hand in hand with correspondence courses. What you learn in your courses in theory, you can practice in your daily work. In many of the ratings, experience is an absolute necessity. For instance, a boatswain's mate must have experience as a leading seaman. Here, while serving an apprenticeship, you actually learn while doing, under an expe-rienced petty officer. Whether you train on the job or in a formal school, both theoretical knowledge and practical experience are necessary.

Leadership

You need more than knowledge and experience to hold a rating in the Coast Guard. The device on the petty officer badges means more than that they are specialists—the eagle indicates petty officer, a leader. A petty officer is both a specialist and a military leader. Military duties and respon-sibilities as a petty officer take precedence over skill in rating. Leadership is a quality that few people are born with. The acquisition of leadership requires study and practice. Some of the study can come from a book or formal course. Much of it must come from observation and emulation of good leaders. In your daily life, you will come in contact with many leaders. Watch them, note the qualities that make them good leaders, and copy them.

A good leader must have a good conduct record. Conduct itself is an integral part of leadership. You cannot expect a higher standard of conduct from your subordinates than you yourself display.

Information and Education Program

It is possible to study almost any subject in your off-duty time through the Coast Guard's educational program. The principal sources of off-duty education are the U.S. Coast Guard Institute (CGI) and various educational and commercial institutions.

The Coast Guard Institute

The CGI and courses produced there have been reviewed by the National Home Study Council (NHSC) and by the American Council on Education (ACE). The NHSC has approved accreditation of the CGI, and the ACE has recommended educational credit for many Coast Guard correspondence courses.

This accreditation makes it easier for students to obtain college or vocational credit for their correspondence course work. In general, the Coast Guard Institute provides correspondence courses and service-wide examinations for enlisted advancements. Most of the courses are directly related to the knowledge and skills that Coast Guard personnel in the various ratings and pay grades must possess. As a general rule, Coast Guard personnel must complete a CGI correspondence course before competing for advancement.

The unit's educational services officer (ESO) has a copy of the CGI's *Correspondence Course Manual*, which contains the rules and procedures of the program and a list of the courses available. The ESO will take care of the details for your enrollment, keep track of your progress, and give instruction and guidance if you have any problems while you are completing the course. You are eligible for enrollment in any course offered by the CGI. Enrollment is usually limited to three courses at a time.

In addition to enlisted rating courses, the CGI offers a number of special courses, among which are oceanography, meteorology, boating safety, and search and rescue. All courses are furnished without cost, and you may keep most of the materials for reference. Satisfactory course completion is recorded in your service record.

Coast Guard Tuition Assistance Program

The Coast Guard tuition assistance program covers courses taken during off-duty hours at accredited colleges, universities, junior colleges, high schools, and commercial schools but not correspondence courses. (See Veterans Administration Tuition Assistance concerning these.) To qualify for assistance, the desired course must be one that would broaden your technical and academic background and increase your usefulness to the service. Generally, any course that will increase your chances for promotion—management for a yeoman, accounting for a storekeeper—will meet this requirement. See your ESO for more details.

Education Benefits

Your education is good for you and the Coast Guard. Your increased knowledge will be reflected in improved skills and performance. Therefore, the Coast Guard is ready to help you attain your educational goal, whether it is a high school equivalency or a college degree. Depending on your unit and your job, you could be eligible for off-duty tuition assistance. You will also be able to take correspondence courses through the CGI and DANTES (see below). Your educational services officer can help you order CLEP (see below) and DANTES tests that you can use for college credit.

There are several educational benefits programs that apply to service members, depending upon when you entered the service. For an explanation of the Montgomery G.I. Bill, the Post-9/11 G.I. Bill, and the Veterans Administration Educational Program, check with your unit's educational services office.

Defense Activity for Non-Traditional Education Support (DANTES)

DANTES offers a number of programs through your local ESO or from various military installation education centers. The DANTES program can help you get credit for what you have learned outside the classroom. You develop and expand your skills and knowledge through training courses and programs, self-study activities, volunteer or community work, on-the-job experiences, and through a variety of other learning experiences. These are considered nontraditional education experiences because they do not normally take place in a traditional classroom setting.

Nontraditional education programs are practical for many Coast Guard members who cannot participate in formal education programs due to their work schedules, education levels, or duty-station assignments, or for personal reasons. The DANTES program offers high school equivalency credential (such as the GED), college admission examinations (ACT and SAT), military evaluation programs (which are explained in detail under the ACE program), college credit by examination, independent study, experiential learning assessment, and college external degree programs.

College Level Examination Program (CLEP)

The College Level Examination Program (CLEP) can give you a good start on your road to college. By taking CLEP exams, you can receive college credit that will save you time and money during your college career. CLEP credit can even help you satisfy college requirements for Officer Candidate School. The CLEP consists of a series of examinations that test college-level knowledge you may have gained through military classes or technical assignments, or through your own reading, travel, and intellectual curiosity. About two-thirds of the colleges and universities in the United States give credit for CLEP exams, but each institution has its own policy for acceptable test scores. Your ESO or the CGI can provide you with more details.

American Council on Education (ACE)

Today's Coast Guard members are trained to operate the most technologically advanced equipment anywhere in the world. They are continuously and expertly trained, and many have acquired knowledge equivalent to that of college students. ACE recognizes that some military learning experiences can be equated to college course work and encourages college and university personnel to award academic credit for military learning experiences.

Servicemembers Opportunity Colleges (SOC)

The Servicemembers Opportunity Colleges (SOC), a consortium of national higher-education associations of more than 550 institution members, function in cooperation with the Department of Defense, the military services, and the Coast Guard to help meet the voluntary higher-education needs of military members and their dependents.

Hundreds of Coast Guard members and their family members enroll annually in college-level programs. Many of these programs make it possible for a Coast Guard member to earn a college degree with little or no residency requirement. Over 95 percent of the SOC institutions will accept the ACE-recommended college credits, CLEP, and DANTES tests to apply toward your degree program. For more information on the SOC and ACE programs, see your ESO.

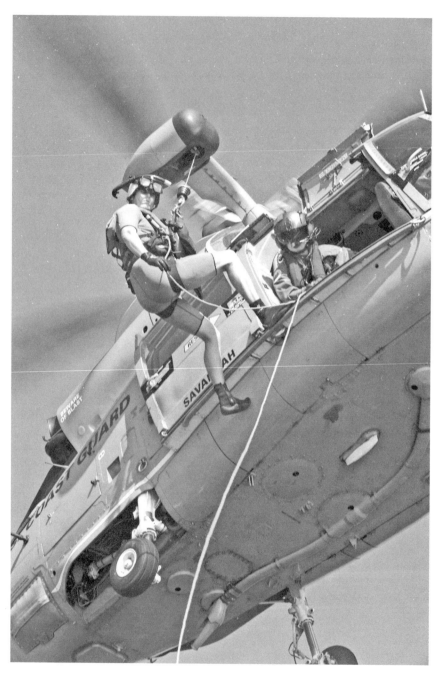

Figure 11-1.
Helicopter rescue training.

CHAPTER 11

Advancement and Promotion

The Coast Guard Enlisted Advancement System is a fully qualified, competitive system. Personnel are advanced to pay grades E-4 through E-9 to fill service-wide vacancies in total Coast Guard allowances. Opportunities exist for enlisted personnel to join the commissioned officer ranks. The *Coast Guard Personnel Manual* and the *Reserve Policy Manual* contain complete details.

COMMON CHANNELS OF ADVANCEMENT

There are three common avenues for advancement in the enlisted rates. They are nonrated, "A" school, and competitive advancements.

Nonrated advancements. Recruits are normally advanced to seaman apprentice or fireman apprentice (E-2) upon graduation from basic training. Six months later, they may be advanced to seaman or fireman (E-3) by their commanding officers if they have maintained good conduct, satisfactory performance, and completed the enlisted professional military education (EPME) requirements for E-3.

"A" school advancements. Graduates of "A" schools will be advanced to petty officer third class once they have completed the EPME requirements for E-4, have demonstrated leadership abilities, passed the Advancement Qualification Exam (AQE) for E-4, and have the required six months' service in pay grade E-3.

Competitive advancements. Advancement to E-5 and above is achieved through competition in the Servicewide Advancement System.

The Servicewide Advancement System

Candidates become eligible for advancement after they

- Earn good performance evaluations;
- Demonstrate the ability to perform at the next higher rate;
- Complete the rating-specific EPQs;
- Complete the rating-specific course; the written portion is called the performance qualification guide;
- Pass the end-of-course test for the material covered in the performance qualification guide;
- Complete the EPME requirements for the next pay grade;
- Pass the test for the EPME requirement, which is the AQE when competing for E-4, E-6, and E-8;
- Receive their commanding officer's recommendation;
- Compete in the service-wide examination;
- Fulfill sea time and other special requirements for certain ratings, serve a specified minimum number of years in the Coast Guard, and serve a specified number of months in their current rate in the Coast Guard; and
- Fulfill additional eligibility requirements for personnel in E-7/E-8/ E-9 examinations (i.e., Chief Petty Officer Academy, Officer in Charge course).

Recommended candidates must compete with all other candidates in the target rate by taking the service-wide examination (SWE). Candidates will be rank ordered with other candidates. The rank order for advancement to each rate is based upon the final multiple.

The final multiple is computed from the SWE score, average performance evaluations for a given period of time, time in service (TIS), time in grade (TIG), points for certain medals and awards, and sea time.

The number of candidates to be advanced is based upon a forecast of service-wide vacancies made by the Coast Guard Personnel Service Center's Enlisted Personnel Management (EPM) Division. The forecast for each rate will vary with each examination cycle.

TIS and TIG must be completed by specified cutoff dates for each advancement cycle. Completion of performance-based qualifications requires candidates to satisfactorily demonstrate certain tasks. The performance standards must be met before the candidates may be recommended to participate in the service-wide examination.

All eligibility requirements except TIS and TIG in present rating must be met before an individual may be recommended by the commanding officer. TIS and TIG in present rating must be completed on or prior to terminal eligibility dates established.

The commanding officer's recommendation is the last and by far the most important prerequisite. Even if all other requirements are met, a candidate cannot compete without the commanding officer's recommendation.

After all prerequisites have been met, the Servicing Personnel Office completes the Advancement Recommendation and Examination Request. Upon receipt of this form at the CGI, a service-wide examination will be mailed to the candidate's examining board.

The Service-Wide Examination

Each member who wants to take the service-wide exam (SWE) must personally verify his or her Personnel Data Extract (PDE) prior to taking the exam. The next and final step is participation in the service-wide examination.

The *Enlisted Performance Qualifications Manual*, COMDTINST M1414.8 (series), is the basis for the development of the service-wide examination. This manual prescribes the minimum occupational and military standards (performance-based) for advancement in rate.

The main goal of the SWE is to identify those candidates best qualified for advancement on the basis of rate-related or rate-required standards. The SWE is a norm-referenced examination. The minimum acceptable score is set by headquarters for each examination series and may be adjusted for selected rates.

The difficulty of the SWE is set so that, in the judgment of the subject-matter expert, the minimally competent candidate will score 70 percent or higher on the SWE. Usually, 80 percent of the candidates should be minimally competent; however, the subject-matter expert will make that determination after consultation with the program, training, and rating force managers. Candidates are then ranked based on their final multiple.

TIG and TIS are calculated through 1 January and 1 July following the SWE examination. Candidates for E–8 must have ten years of cumulative service in one or more components of the U.S. armed forces as of 1 January following the SWE. Candidates for E-9 must have twelve years of cumulative service in one or more components of the armed forces as of 1 January following the SWE.

Service-Wide Schedule

The SWE administration schedule for active-duty enlisted personnel is as follows:

E-5 through E-6 November and May
E-7 through E-9 May

The Advancement Results

The results are announced eight to ten weeks after the service-wide examination is administered. All candidates who participate in the SWE will receive a profile letter. The profile letter indicates the overall performance of the candidate as well as performance by sections. Section scores are provided to identify those areas that the candidate may use to study for the next SWE, if applicable.

An advancement eligibility list is published shortly after the profile letters are mailed. This list includes the names of all candidates who are eligible for advancement. Names that fall *above* the cutoff are guaranteed advancement. Names *below* the cutoff may be advanced but should plan to compete again in the next SWE. This is not an automatic process; the candidate must remain qualified and be recommended by the commanding officer for each SWE cycle.

Advancement in the Coast Guard Reserve

The regular procedures for advancement of active-duty enlisted personnel in the Coast Guard apply also to the Coast Guard reservists with the following exceptions: an annual SWE for reservists is given in October, and TIG and TIS are calculated through 1 January following the SWE.

Reservists on extended active duty may not participate in the regular SWE. They can only compete in the reserve SWE. If they place above the cut, headquarters will determine if the person is to be advanced.

Personnel who passed an active-duty Coast Guard SWE for advancement to E-4, E-5, or E-6 within one year of their separation from active duty but were not advanced for any reason other than disciplinary may be advanced without taking a reserve SWE *if* their commanding officer recommends advancement within one year of the candidate's separation from active duty, the candidate's advancement multiple was at least as high as the lowest multiple used in advancing a reservist on the same examination, and no SWE for that rate has been taken by the candidate since passing that SWE.

Promotion to Chief Petty Officer

The rich nautical tradition of the chief petty officer (CPO) in the Coast Guard dates back to 18 May 1920, when Congress approved the Coast Guard's adoption of the Navy's rate structure. The original need for this higher authority came from the merger of the Life-Saving Service and the Revenue Cutter Service. The position of chief was established to oversee differences between the "surfman" of the Life-Saving Service and the "petty officer" of the Revenue Cutter Service. Since then, the CPO has been the leader of enlisted ranks and the conduit from the mess decks to the wardroom. The chief, who instills pride and professionalism, demands respect up and down the chain of command. The chief is responsible for bringing the concerns of the crew to the command, and for policy enforcement.

The chief's corps provides extensive training, knowledge, and experience in senior-level enlisted supervision and in administration. They bring to bear a wide range of technical expertise to manage equipment and tasking associated with the functioning of the Coast Guard. A chief is the link between the enlisted and officer corps. A CPO plans and administers on-the-job and other training programs for subordinates and junior officers serving within the same specialty.

Accordingly, many CPOs will serve as senior enlisted advisers in matters concerning Coast Guard personnel. They exemplify Coast Guard core values of honor, respect, and devotion to duty. When strenuous mental and physical conditions arise and experienced leadership is a must, all Coast Guard men and women know these three words: "Ask the Chief!"

Promotion to Chief Warrant Officer and Commissioned Grades

Every petty officer rating has a normal path of advancement to chief warrant officer rank. Selections for chief warrant officer are made from among first-class and chief petty officers on the eligibility list who apply for this advancement. An applicant must have the required service, must make satisfactory scores on warrant qualification and specialty examinations, and must be recommended by his or her commanding officers. A selection board of commissioned officers chooses those considered best qualified to fill the vacancies in the various chief warrant officer categories. Selected candidates are promoted to W–2 and can later compete for further advancement.

Officer Candidate School

Selected enlisted personnel may be ordered to the Officer Candidate School (OCS) at the U.S. Coast Guard Leadership Development Center at the Coast Guard Academy, New London, Connecticut. This training leads to appointments for temporary service as ensigns, provided that certain requirements are met. Many such temporary officers eventually receive permanent commissions after meeting service requirements and after having been selected for promotion to lieutenant. Also, at certain times enlisted personnel with the necessary educational qualifications are given the opportunity to become commissioned officers in the Coast Guard Reserve.

Coast Guard Academy Scholar Program

The Coast Guard Academy (CGA) recognizes that not every prospective cadet emerges from high school prepared for the rigorous academics, military, and athletic criteria established by this institution. The Academy's Scholar Program was designed to develop the necessary foundation for success as a cadet, an officer and, most importantly, a leader of character.

The Academy offers one-year prep school scholarships to properly motivated students interested in demonstrating their personal potential

Figure 11-2. Members of the U.S. Coast Guard Academy class of 2016 graduate and receive their commissions as officers.

and willingness to succeed academically, militarily, and physically. Most participants who successfully complete this year of focused preparation receive a full appointment to the Academy as a member of the next entering class, where they will be recognized as leaders from the beginning of Swab Summer. CGA Scholars appointees are selected from within the general applicant pool. There is not a separate admissions process for this program.

Coast Guard Academy

Enlisted personnel may compete for direct appointments to the Coast Guard Academy. Cadets are selected annually by competitive examination. To qualify for the examination, a candidate must be between the ages of eighteen and twenty-two, be of good character, be a United States citizen, and be a high school graduate with school credits in the required subjects. These examinations also are open to enlisted people in any of the armed forces branches.

Pre-Commissioning Program for Enlisted Personnel (PPEP)

This upward-mobility program enables selected enlisted personnel to attend college on a full-time basis for up to two years, receive a bachelor's degree, and attend OCS. Upon graduation from OCS, graduates receive commission as an ensign in the U.S. Coast Guard. Selections are highly competitive, and only members who have demonstrated the ability to excel will be considered for selection. Students receive full pay and allowance while attending school.

To qualify for the PPEP, applicants must have completed four years of active duty with a minimum of two years in the Coast Guard. They must have sufficient college credits to complete a bachelor's degree within twenty-four months of entering the program and maintain a cumulative minimum grade point average of 2.5. Applicants are also required to have a SAT score of 1,000 or an ACT score of 23. An ASVAB (the Armed Services Vocational Aptitude Battery) GT score of 109 is also acceptable (GT = AR + VE).

Figure 12-1.
The Coast Guard is looking for people who are "born ready," motivated to save lives, protect the environment, and defend America's coastlines and waterways.

CHAPTER 12

Career Information

As a member of the Coast Guard, you are entitled to many benefits. Upon completing your enlistment, either through discharge or retirement, many of these benefits will still be available to you. This chapter briefly describes some of the most important features of the benefit package.

PAY AND ALLOWANCES

Your pay consists of basic pay, basic allowances (one for housing and one for subsistence or meals), incentive pay or hazardous duty pay, and miscellaneous pay and allowances.

You will be paid twice a month, generally the first and the fifteenth. The actual amount you receive each pay day is equal to basic pay, plus certain allowances, minus taxes, and minus any allotments you may have made. Your basic pay is determined by your rank and your total years of creditable service. Basic pay and basic allowances are computed on a monthly basis.

Direct Deposit

Direct deposit is the Coast Guard's primary pay delivery method. The direct deposit method of payment allows you to have your net pay deposited directly into your savings or checking account at a designated financial institution of your choice. You can access a copy of your leave and earnings statement (LES) via Direct Access to verify your pay information.

Direct deposit provides reliability and convenience and avoids lost or stolen checks. Afloat cashiers on board vessels deployed outside the continental United States and away from home port more than four days over a payday are authorized to cash personal checks for members on direct deposit. The primary purpose of this check-cashing authority is to provide members on direct deposit with a source of funds to cover incidental expenses, not to provide a cash payroll.

Basic Allowances for Subsistence and Housing

As a member of the Coast Guard, you are entitled to food and shelter in addition to basic pay. Regardless of your pay grade, you are entitled to certain allowances in cash to provide housing for your spouse or legal dependents. If you have no legal dependents, you will usually be provided subsistence and housing "in kind." This means that you will eat in a government mess and will sleep on board ship or in government barracks or quarters. Your food costs you nothing, so you receive no cash allowance for it. There may be times when it is impracticable to furnish you with either subsistence or housing or both in kind. In that case you will receive a cash allowance in lieu of items not furnished.

Discounted Meal Rate

When a regular Coast Guard dining facility is available but you choose to eat elsewhere, your commanding officer may authorize payment of a daily allowance for this. You may still eat at a dining facility, but you must pay for the meal. The conditions under which the discounted meal rate may be paid will be explained by your Servicing Personnel Office (SPO).

Incentive Pay for Hazardous Duty

For pay purposes the following duties have been designated as hazardous:

- Duty involving frequent and regular participation, as a crew member, in aerial flights.
- Duty involving frequent and regular participation, not as a crew member, in aerial flights.
- Duty involving frequent and regular participation as a flight deck tie-down crew member.
- Duty involving frequent and regular participation in diving activities.

Assignment to such duties will entitle you to monthly incentive pay that varies depending on rate and time in service. The directives on hazardous duties and incentive pay change from time to time. Check with your SPO.

Clothing Allowance

Six months after graduation from recruit training, you will begin receiving a monthly allowance for clothing maintenance. Special clothing, such as arctic clothing, is furnished to you at no cost when you are required to use it.

Special-Duty Assignment Pay

Special-duty assignment pay was established as an incentive to help fill shortages in certain ratings or special-duty assignments. Each year Coast Guard Headquarters determines which ratings, skills (identified by qualification codes), and special-duty assignments will receive this additional pay.

Family Separation Allowance

A family separation allowance is paid to help offset additional expenses you might have when stationed away from your family. The rate of pay varies, and eligibility is involved. Check with your SPO.

Allotments

You may assign or allot a part of your pay regularly to a spouse, parent, bank, or insurance company. You can also use this method to buy savings bonds. The money you allot is paid each month and is deducted from your pay.

Thrift Savings Plan

The Thrift Savings Plan (TSP) is a retirement savings and investment plan for federal employees and members of the uniformed services. Congress established the TSP in the Federal Employees' Retirement System Act of 1986. The purpose of the TSP is to provide retirement income. The TSP offers federal employees and members of the uniformed services the same type of savings and tax benefits that many private corporations offer their employees under 401(k) plans. More information concerning the TSP for members of the uniformed services can be found on the TSP Web site: http://www.tsp.gov. The contributions that you make to your TSP account are voluntary and are separate from your contributions to other retirement plans.

Travel, Transportation, and Per Diem

When traveling under orders, you are entitled to do so at government expense. You may be authorized to travel by automobile or by government or commercial transportation. In addition, a per diem (daily) allowance may be paid to cover the cost of food, lodging, and other expenses.

With the exception of personnel ordered to their first duty station after recruit training, transportation of dependents and household goods at government expense is authorized on a permanent change of station. Orders to a new duty station involve many expenses. The Coast Guard recognizes this and reimburses you through allowances for dislocation, trailer, mileage, dependents' transportation, and shipment of household goods. The rules and regulations covering transportation and travel expenses are complicated and subject to frequent changes. You are only entitled, however, to allowances authorized or directed in the orders.

WORK-LIFE PHILOSOPHY

The mission of work-life is to create a network of individuals and organizations willing and able to provide support for our members and their families. Through the enhancement of customer service and service delivery, the right information will be in the hands of the right people at the right time. The Coast Guard recognizes that work-life issues are critical to individual and organizational success. In response, a beneficiary guide was developed and distributed to every active-duty Coast Guard member. It introduces the programs and benefits available to you.

BENEFITS

Veterans Administration Educational Programs

The Montgomery G.I. Bill became effective 1 July 1985 and is available to any member who first became a member of the armed forces or first began active duty on or after that date. COMDTINST M1760.6 requires that an allotment authorization be submitted for members who elect to participate in the program. Only the recruit SPO at Training Center Cape May is authorized to start this allotment, except in the case of a direct-commission officer or a member going through OCS. Detailed information on this program will be provided during recruit training.

The Post-9/11 G.I. Bill provides financial support for education and housing to individuals with at least ninety days of aggregate service on or after 11 September 2001 or individuals discharged with a service-connected disability after thirty days. You must have received an honorable

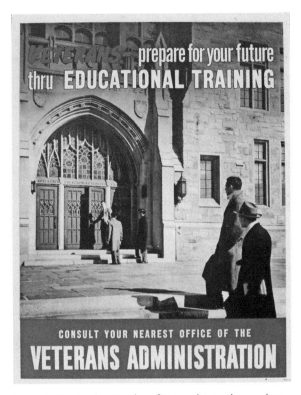

Figure 12-2. VA education benefits may be used toward traditional degrees, non-college degrees, on-the-job training, apprenticeships, and more.

discharge to be eligible for the Post-9/11 G.I. Bill. The amount of support that an individual may qualify for depends on where they live and what type of degree they are pursuing. For a summary of Post-9/11 G.I. Bill benefits, see the Department of Veterans Affairs Web site: www.benefits.va.gov/gibill/post911_gibill.asp.

Approved training under the Post-9/11 G.I. Bill includes graduate and undergraduate degrees and vocational/technical training. All training programs must be offered by an institution of higher learning and approved for G.I. Bill benefits. Additionally, tutorial assistance, licensing, and certification test reimbursement are approved under the Post-9/11 G.I. Bill. The Post-9/11 G.I. Bill will pay your tuition based upon the highest in-state tuition charged by an educational institution in the state where the educational institution is located. For more expensive tuition, a program exists that may help to reimburse the difference. This program is called the Yellow Ribbon Program. For more information on the Yellow Ribbon Program visit www.benefits.va.gov/GIBILL/yellow_ribbon.asp.

The Post-9/11 G.I. Bill also offers some service members the opportunity to transfer their G.I. Bill to dependents; for more information, see www.benefits.va.gov/gibill/post911_transfer.asp.

Social Security

All service personnel participate in the Social Security program. A portion of your pay is deducted each month for this purpose. The amount of deduction is matched by the Coast Guard. These funds will earn Social Security benefits for you and your dependents.

Insurance Programs

Dependency and Indemnity Compensation is a form of insurance payable to the survivors of personnel who die in the line of duty while on active duty. As a member of the Coast Guard you automatically receive this coverage, free.

Servicemen's Group Life Insurance is a government-sponsored term-insurance program. Upon joining the Coast Guard, you are automatically insured. Part of your pay is deducted each month to pay for this insurance. You may reduce your coverage or cancel it; however, this is seldom done because the cost is so low. This insurance is good only while you are on active duty, or for certain categories of reservists. The Survivors' Benefit Plan applies to retired personnel. Under this program you agree to a reduced retirement pay upon retirement, and in return your spouse continues to receive a portion of your retirement pay after you die.

Mortgage insurance for servicemen. Sometime during your career you may decide to purchase a residence. As a member of the Coast Guard, you are eligible to apply for a Federal Housing Administration in-service loan, or a Veterans Administration (VA) in-service loan, which allows you to purchase a home with a greatly reduced down payment. VA helps service members, veterans, and eligible surviving spouses become homeowners. VA home loans are provided by private lenders, such as banks and mortgage companies. The VA guarantees a portion of the loan, enabling the lender to provide you with more favorable terms. Check chapter 16 of the *Coast Guard Personnel Manual* about additional requirements concerning eligibility.

Medical care. Free medical and dental care is provided all members of the Coast Guard on active duty. This care includes surgery and emergencies as well as routine immunizations, physical exams, and so forth. Dependents are eligible for medical care in the facilities of any of the armed services; they pay a small daily charge when they are hospitalized. Another medical care program is TRICARE. Under this program dependents may receive outpatient treatment and, in some instances, inpatient care from civilian facilities. The government pays most of each TRICARE bill, but the individual may also pay part of it. A dental plan is also available.

Off-Duty Employment

Personnel on active duty are in a twenty-four-hour duty status, and their military duties take precedence at all times. However, you can hold a part-time job or work for yourself in off-duty hours. If you do such work, you must remember that even though you are on leave or liberty, you are subject to recall and duty at any time. The *Coast Guard Personnel Manual* covers the conditions under which civilian work or employment is prohibited.

Servicemembers Civil Relief Act

On 19 December 2003 President George W. Bush signed into law the Servicemembers Civil Relief Act (SCRA). This law is a complete revision of the Soldiers' and Sailors' Civil Relief Act (SSCRA), which provided a number of significant protections to service members.

While protecting the United States during the war on terrorism, some servicemen and servicewomen may face difficulty in meeting certain financial obligations at home, such as rent or mortgage payments, if they are activated for military duty. The SCRA can provide many forms of relief to military members. Below are some of the most common forms of relief:

- Mortgage relief
- Termination of leases
- Protection from eviction
- Cap on interest rates at 6 percent
- Stay of proceedings
- Reopening default judgments

The SCRA actually provides many more protections than those listed above, and the Supreme Court has ruled that the SCRA must be read with "an eye friendly to those who dropped their affairs to answer their country's call." Coast Guard legal assistance attorneys are available to provide guidance on the SCRA.

Coast Guard Mutual Assistance

Coast Guard Mutual Assistance (CGMA) is an independent, nonprofit, charitable organization providing short-term financial assistance to the entire Coast Guard family—active duty and retired military personnel, civilian employees, reservists, auxiliarists, commissioned officers of the Public Health Service serving with the Coast Guard, and their families. Most assistance is provided to members finding themselves in unexpected financial need due to circumstances beyond their control. All CGMA assistance is based on financial need. CGMA is not part of the U.S. Coast Guard and does not receive government or appropriated funds. CGMA is solely funded through tax-deductible contributions and returns on investments.

A wide range of financial assistance is available to our members from CGMA through six programs:

Emergency loans
Educational assistance
Medical assistance
General assistance
Housing assistance
Debt management assistance

Contact your local CGMA representative or CGMA headquarters for specific program information and eligibility or to make contributions.

The American Red Cross

From the first day of enlistment, members of the U.S. Coast Guard and their families are eligible for Red Cross assistance. This organization provides many special services geared toward military personnel and their families. Among these are verification of requests for emergency leave or leave extensions based on emergencies, assistance with the unit's blood program, and individual counsel, guidance, and financial assistance in the form of a loan or grant. Any financial assistance given will depend on the type of emergency and how serious it is.

The American Red Cross Emergency Communications Center is available 7 days a week, 24 hours a day, 365 days a year, with two options for requesting assistance, online at http://www.redcross.org/about-us/our-work/military-families or by phone. To speak to a Red Cross Emergency Communications Specialist call 1-877-272-7337. When contacting the Red Cross, be prepared to provide the following information (if you do not have all of the information, please use the phone option):

Service member information

Full legal name
Rank/rating
Branch of service (Army, Navy, Air Force, Marines, Coast Guard)
Social Security number or date of birth
Military unit address
Information about the deployed unit and home base unit (for deployed service members only)

Information about the emergency

Name and contact for the immediate family member experiencing the emergency (could be spouse, parent, child, grandchild, or grandparent)
Nature of the emergency
Where the emergency can be verified (hospital, doctor's office, funeral home)

Eligibility requirements

Service members eligible to receive emergency communications regarding an immediate family member include:
 - Service members on active duty in the Army, Marines, Navy, Air Force or Coast Guard
 - An activated member of the Guard and Reserve of all branches of the U.S. armed forces
 - A civilian employed by or under contract to the Department of Defense and stationed outside the Continental United States
 - A cadet or midshipman at a service academy; ROTC cadet on orders for training
 - A Merchant Marine on board a U.S. naval ship

The American Red Cross does not authorize emergency leave for members of the U.S. military. The Red Cross role is to verify the emergency, enabling the service member's commander to make an educated decision regarding emergency leave.

The American Red Cross facilitates emergency financial assistance on behalf of military aid societies. These aid societies determine the financial assistance package that will be offered—a grant or a loan. The Red Cross is the mechanism to expedite access to these financial resources 24/7.

You should let your family members know about the American Red Cross services. Have them contact a Red Cross chapter if they are unable to reach you directly concerning a family emergency.

Nonappropriated Fund Activities

Many of the larger units have a base or ship's exchange, laundry, tailor shop, barber shop, snack bar, enlisted club, and so forth. These functions are called nonappropriated fund activities because they are supported by their own profits rather than by government funds. Some of the profits are used by the commanding officer for recreational equipment for the crew, for crew's parties, and so forth. Some of the profits are distributed to smaller units not having nonappropriated fund activities in order that they may also benefit. You are entitled to use the exchanges or stores at any Army, Navy, Marine, Air Force, or Coast Guard installation.

Legal Assistance

Coast Guard legal assistance attorneys assigned to the Legal Service Command provide advice and counsel regarding personal legal issues to thousands of service members, dependents, and retirees each year at no cost. These issues may involve family law, estate planning, consumer law, landlord–tenant relations, immigration, or many other topics. They will interview, advise, and assist members of the Coast Guard and their dependents who have legal problems and, in certain cases, refer them to competent attorneys. All matters are treated confidentially and will not be disclosed without your permission.

DUTY TOURS

In general, you may expect to get the duty or training you desire unless the needs of the Coast Guard make it impossible. Sometimes you may have to wait, but eventually your requests will be taken care of. It is important to keep up to date on changes in rules and regulations in order to take advantage of all opportunities for either duty or schools.

Sea Duty

Sea duty for pay, promotion, and rotation is defined by the commandant. The length of duty tours beyond the continental limits of the United States varies according to the location, degree of isolation, and the needs of the service.

Shore Duty

After a certain amount of continuous sea duty, overseas duty, or a combination of both, you will become eligible for transfer to shore duty. The interests of the service come first, but every effort is made to give everyone a fair share of shore duty.

Time Not Creditable for Service

You will not be paid or credited for time served if you were not performing duty. In counting time served, days in which you were absent from duty due to misconduct will be deducted from the total. The following count as time not served:

- Absence over leave, absence without leave, and desertion.
- Absence from duty due to injury, disease, or sickness resulting from intemperate use of drugs or alcohol or from other misconduct.
- Absence due to arrest and serving of sentence imposed either by civil or military authorities. If the arrest results in no conviction, the time spent under arrest is not deducted.

Duty Assignments

All enlisted personnel list their personal data and duty preferences on Direct Access via the e-résumé tool. Assignments to duty are based on the e-résumé, so be sure to have one on file and up to date.

Nonrated and petty officer assignments are handled by the Central Assignment Control program at Personnel Service Center. They are based on the billet requirements of each unit, the needs of the service, the enlisted qualification code of certain billets, and billet vacancies.

Mutual Exchange of Station and Unilateral Transfers

On your first duty assignments, it is not always possible to match duty stations with your personal preference; personnel are sent where they are needed most. Later, changes in personnel at various stations make it possible for people to be sent to stations they prefer, but because there is no advantage to the Coast Guard in such a transfer, it must be made at no expense to the government. The person wanting the transfer must find someone at the other station with the same rating and experience who will "mutual" with them—swap duty stations with both paying their own expenses. If a "mutual" exchange cannot be made, a unilateral change of station may be requested. One condition for a unilateral transfer is that the person's unit is in excess of its allowance for the member's rating, or that it has the recommendation of the commanding officer that no immediate replacement will be required if the request is granted. Other restrictions apply; check the *Coast Guard Personnel Manual*.

Humanitarian Assignments

Special humanitarian assignments are made to help a member handle a hardship of such a nature that cannot be taken care of by emergency leave. Such transfers are usually for four months, but in special cases a permanent change of station will be considered at government expense. Details are contained in the *Coast Guard Personnel Manual*.

DISCHARGES

There are six types of discharges for enlisted personnel:

Type of Discharge	Character of Separation	Length
Honorable	Honorable	Administrative action
General	Under honorable conditions	Administrative action
Undesirable	Conditions other than honorable	Administrative action
Bad Conduct	Conditions other than honorable	Court-martial
Dishonorable	Dishonorable	General court-martial
Uncharacterized	Without specific character	Administrative action

Honorable Discharge

To obtain an honorable discharge, you must fulfill certain conditions described in detail in the *Coast Guard Personnel Manual*. Your record must show high standing in performing the duties of your rate and in conduct.

General Discharge

A general discharge is given for the same reasons as the honorable discharge. If your conduct and performance of duty have been satisfactory but not high enough to warrant an honorable discharge, you will receive a general discharge.

Discharge under Other than Honorable Conditions

This type of discharge is given for reasons of misconduct, security, or for the good of the service.

Bad Conduct Discharge

A bad conduct discharge is given, only by approved sentences of special or general courts-martial, for offenses that, although serious, are not sufficiently grave to warrant dishonorable separation.

Dishonorable Discharge

A dishonorable discharge is given only by approved sentences of general courts-martial. It is one of the consequences of serious offenses.

Uncharacterized Discharge

An uncharacterized discharge, by its own connotation, is separation without characterization of service.

Importance and Benefits of an Honorable Discharge

The kind of discharge you receive may affect you for the rest of your life. Eligibility for veterans' preference in federal employment, for payments for a service-connected disability, for a pension, and for many other benefits and privileges—state and federal—will depend upon the type of discharge you receive. Civilian employers may refuse to hire a person with the "wrong" kind of discharge. Do all you can to earn a discharge under honorable conditions.

FORMAL REASONS FOR DISCHARGE

Expiration of Enlistment

Normally, enlisted personnel are promptly discharged on the date of expiration of their enlistment. However, they may continue on duty for some time after that date if they extend their enlistment, are undergoing medical treatment, are awaiting trial, or are required to make up lost time. They may be held past the discharge date if the ship is at sea, or in time of war or national emergency.

Disability

A disability discharge is given to personnel who are unable to carry out their duties because of mental or physical disability.

Convenience of the Government

This term covers such reasons as general demobilization or the acceptance of a permanent commission, plus many more reasons.

Dependency or Hardship

Discharges given for reason of dependency or hardship, sometimes called "hardship discharges," are authorized when it is shown that undue and general hardship exists at the service member's home, that the hardship is not a temporary one, and that the conditions have arisen or worsened since the person joined the Coast Guard. A dependency discharge is not authorized merely for financial or business reasons or for personal convenience. However, a discharge will not be held up because of the Coast Guard's need to keep the billet filled or because of indebtedness to the government or to an individual.

Minority

Discharge for minority is of interest only to enlisted personnel who are under eighteen years of age. Instructions regarding such discharges are given in the *Coast Guard Personnel Manual*.

Unsuitability

The commandant may authorize or direct the discharge of personnel for unsuitability. Among the good and sufficient reasons for such a discharge are inaptitude for service life, personality disorders, apathy, defective attitudes, insanitary habits, alcohol abuse, and financial irresponsibility.

Security

A discharge for reasons of security is directed only when it is determined that the enlisted person cannot continue in service without risk to the national interest.

Misconduct

The commandant may direct the discharge of a member for misconduct in any of the following cases: conviction by civil authorities (foreign or domestic), fraudulent enlistment, absenteeism, involvement with drugs, frequent involvement of a discreditable nature with civil or military authorities, sexual perversion, an established pattern of shirking, an established pattern showing dishonorable failure to pay just debts, dishonorable failure to contribute adequate support to dependents, or failure to comply with valid orders, decrees, or judgments of a civil court concerning support of dependents.

Sentence of a Court-Martial

In addition to other punishments, a court-martial can award a discharge from the Coast Guard. Not only does this rid the service of unwanted persons but it puts—by way of punishment—a mark of serious consequences forever in the person's record.

Uncharacterized

An uncharacterized discharge will be used for most recruit separations, other than that for disability, or for prior service personnel entering recruit training. Members must have demonstrated poor proficiency, conduct, or aptitude, or unsuitability for further service during the period from enlistment through recruit training, a period of less than 180 days of active service on date of discharge.

PERSONNEL DATA RECORD

Your Personnel Data Record (PDR) is a four-section folder designed to hold only those specific documents necessary to provide for your personal and pay needs. A PDR is maintained on you at each of the following locations: CG Headquarters, Pay and Personnel Center Topeka, your local Servicing Personnel Office, and your unit. In addition, you are required to maintain your own copy of your PDR.

At the end of your service career the National Personnel Records Center in St. Louis, Missouri, will consolidate all of the above, including your medical PDR, into one PDR.

Your PDR is important during your service career and later in civilian life, so keep it "clean." Even after you are discharged or retired, information from your PDR may be needed for collecting veteran's benefits, for employment, or for school credits.

COMMON ACCESS CARD

The Common Access Card, a "smart" card about the size of a credit card, is the standard identification for active-duty Coast Guard personnel,

Selected Reserve, Coast Guard civilian employees, and eligible contractor personnel. It is also the principal card used to enable physical access to buildings and controlled spaces, and it provides access to Department of Defense/U.S. Coast Guard computer networks and systems. It is not a pass, but you should carry it at all times.

"Non-Petty Officer" is typed on the cards of people in pay grades E-1, E-2, and E-3. Cards for those in higher pay grades are marked appropriately. If you lose your ID card, you will have to sign a statement explaining how it was lost. If you were neglectful, disciplinary action will follow. The card is government property; if you alter, damage, or counterfeit it, or if you use it in an unauthorized manner, you will be subject to disciplinary action. The same warning applies to lending your card or borrowing another person's card. When you are discharged, your ID card must be turned in.

SPECIAL REQUESTS

When people want a transfer to another ship or station, assignment to school, change of rating, permission to bring some piece of personal property on board, or some other consideration, they submit a special request slip, or "request chit." Such slips are submitted in a "chain of command" fashion so that the officer to take final action will know the decision of those immediately above you. If your request is reasonable and valid and your performance of duty has been good, your request will probably be granted.

Anyone may make a request or an appeal to higher authority (senior to your commanding officer). Appeal from censure or requests for dependency or hardship transfer are possible reasons for "going up the chain." The general rule is that these types of requests are usually in letter form and must go through the "chain of command" via your commanding officer, the sector commander (if in the chain of command), district, and finally to the commandant. Do not bypass anyone in the chain of command; this is a violation of a basic military principle. You are permitted to make requests; just make them through channels.

LIBERTY

Authorized absence from duty for periods up to ninety-six hours is classed as liberty. Most liberties consist of leaving the ship or station around 1630 and returning before 0730 or 0800 the next morning. The normal forty-eight-hour liberty that is granted for weekends may be extended to seventy-two hours if the period includes a national holiday. A ninety-six-hour liberty must include two consecutive non-work days and shall not, under any circumstances, extend beyond ninety-six hours.

Various ships and stations will have different types of liberty schedules. For example: "port and starboard liberty" (one-in-two, or liberty every other evening); "three-section liberty" (two liberty nights for each duty night); and even "six-section liberty" (five liberty nights and one duty night).

All liberty is controlled by your commanding officer and is granted as workloads and operating schedules permit. Remember, liberty is a privilege, not a right.

LEAVE

Leave is the authorized absence of an individual from a place of duty, chargeable against such individual in accordance with applicable law. Leave is earned at the rate of two and one-half days per month of duty, but not for periods of training duty without pay (reservists), any period of less than thirty consecutive days of training duty (reservists), while confined as sentence of court-martial, or while in an unauthorized absent status for twenty-four hours or more.

Earned Leave

This is the amount of leave you have credited "on the books" at any given date. If you have taken more leave than you were entitled to, then you have minus leave on the books. Any minus leave over that which you would normally earn during the rest of your enlistment is treated as excess leave. Excess leave is leave over your earned leave and any advance leave granted you. Advance leave is leave granted out of the amount of leave you will earn during the rest of your enlistment. Here is how advance and excess leave works: If you have ten days' earned leave on the books but are granted, because of an emergency, twenty days' leave, those twenty days are half earned leave and half advance leave. If those twenty days were then extended by your commanding officer for ten days, those last ten days are excess leave.

On a yearly basis, the thirty days you earn are figured from 1 October of one year to 30 September of the next year. As your leave builds up, it is carried over from one fiscal year to the next, to a maximum of seventy-five days. If you have seventy-eight days on the books on 30 September, you lose three days and start with just seventy-five days.

If you are discharged with leave still on the books, you will be paid a lump sum equal to your daily pay for each day on the books. If you are discharged with minus leave on the books—in other words, owing the Coast Guard for leave time—your pay will be reduced by an amount equal to a day's pay for each day's leave you owe.

Your commanding officer has the authority to grant you (on a yearly basis) all your earned leave plus up to thirty days' advance leave. However, you may not have more than seventy-five days of leave at one time—*except for reenlistment leave.*

Sick Leave

Sick leave is the term used to describe a period of authorized absence granted to persons while under medical care and treatment. It usually follows a period of hospitalization and is not chargeable as leave.

Emergency Leave

In the event of a personal emergency—the death of your father, for example, or the serious illness of your spouse—you will normally be granted emergency leave to take care of important personal matters. Since most family emergencies are highly time-dependent, the command will take swift, sensitive action on emergency leave requests.

How Leave Is Counted

The day you depart on leave, whatever the hour, is counted as a day of duty as long as it coincides with the granting of liberty. If you check in to your unit by the time liberty expires on your day of return, it will be counted as a day of duty and will not be charged as leave. If you return after liberty expires, it will be charged as another day of leave.

Compensatory Absence

Compensatory absence is a form of liberty granted to personnel at isolated units where normal liberty cannot be granted. It is granted insofar as possible for rest and rehabilitation, usually on the basis of seven days' absence for fourteen days of duty. There are limits on how much can be accrued.

At isolated overseas units, credit is earned at the rate of two and one-half days per month.

RETIREMENT

Retirement means that you are released from active duty but continue to draw a certain fixed pay for the rest of your life. The exact amount of this retired pay depends on many factors. Length of service is the greatest governing factor in most retirements. Degree of disability is the controlling factor in the case of disability retirements. In time of war or national emergency, retired personnel may be called to active duty.

There are several classes of retirement, including compulsory retirement at the age of sixty-two, voluntary retirement after twenty or more years' service, involuntary retirement after twenty or more years' service, and retirement for disabilities incident to service. Retired people are entitled to medical treatment in government facilities and are entitled to use commissary and exchange facilities at any military base.

PERIOD OF OBLIGATED SERVICE

All personnel, upon original enlistment in the Coast Guard, incur a statutory obligation as described in the Universal Military Training and Service Act. Various enlistment programs require different amounts of active duty, active reserve, or inactive reserve obligations. Regardless of the type of program, all personnel enlisting on or after 1 September 1984 incurred an eight-year period of obligated service.

Figure 13-1.
Master Chief Vincent W. Patton III served as the eighth master chief petty officer of
the U.S. Coast Guard from 1998 to 2002.

CHAPTER 13

Enlisted Rates and Ratings

A rating is the name given to an occupation in the Coast Guard that requires basic aptitudes related to the specialty plus training, experience, knowledge, and skills. A person who has a rating is generally a petty officer.

All ratings are equally important. Therefore, no one rating takes precedence over another rating. Among enlisted people present and regularly assigned to the same activity, or among those present in any gathering, the person with the longest period of continuous service in the highest pay grade shall be considered senior regardless of rating.

A rate is a pay grade that reflects a level of aptitude, training, experience, knowledge, skill, and responsibility. The order of pay grades is as follows:

Rates	Pay Grade
Seaman recruit	E-1
Seaman apprentice	E-2
Seaman	E-3
Petty officer third class (PO3)	E-4
Petty officer second class (PO2)	E-5
Petty officer first class (PO1)	E-6
Chief petty officer (CPO)	E-7
Senior chief petty officer (SCPO)	E-8
Master chief petty officer (MCPO)	E-9

DUTIES OF NONRATED PERSONNEL (E-1, E-2, E-3)

Seaman Recruit (SR). Recruits undergo formal basic training or boot camp at Coast Guard Training Center Cape May.

Seaman Apprentice (SA). Apprentices perform the simple duties of seamen, firemen, and airmen while under close supervision.

Seaman (SN). Seamen are nonrated personnel in training for a rating in deck, ordnance, electronics, administrative, and clerical or in miscellaneous groups other than engineering. They perform general deck and other detail duties and maintain deck equipment, compartments, lines, rigging, and decks. They perform as members of gun crews, small-boat crews, and security and fire watches.

DUTIES OF PETTY OFFICER RATINGS (E-4 THROUGH E-9)

The following is a brief description of duties and responsibilities of ratings in the Coast Guard.

Aviation Ratings

Aviation Maintenance Technician (AMT). AMTs inspect, service, maintain, troubleshoot, and repair aircraft engines, auxiliary power units, propellers, rotor systems, power train systems, and associated airframe and systems–specific electrical components. They service, maintain, and repair

Figure 13-2. Aviation maintenance technician (AMT).

Figure 13-3. Aviation survival technician (AST).

aircraft fuselages, wings, rotor blades, and fixed and movable flight control surfaces; they also bleed aircraft air, hydraulic, and fuel systems. AMTs also fill aircrew positions such as flight engineer, flight mechanic, loadmaster, drop master, sensor-systems operator, and basic air crewman. Training begins with five months of intensive instruction at ATTC.

Aviation Survival Technician (AST). ASTs function operationally as helicopter rescue swimmers and emergency medical technicians. ASTs may find themselves being deployed into a myriad of challenging rescues ranging from hurricanes and cliff rescues to emergency medical evacuations from ships at sea. ASTs also provide all survival training to aviators, such as swim tests, survival lectures, and shallow-water egress training. Other aircrew positions include HC-130H drop master, loadmaster, sensor-systems operator, and basic air crewman.

In addition, ASTs perform ground handling and servicing of aircraft and conduct routine aircraft inspections and aviation administrative duties. ASTs inspect, service, maintain, troubleshoot, and repair cargo aerial delivery systems, drag-parachute systems, aircraft oxygen systems, helicopter flotation systems, dewatering pumps, survival equipment for air sea rescue kits, and special purpose protective clothing. ASTs also store aviation ordnance and pyrotechnic devices.

Training consists of four months of intensive instruction at ATTC to include three weeks of emergency medical technician training at a training center in Petaluma, California.

Figure 13-4. Avionics electrical technician (AET).

Avionics Electrical Technician (AET). AETs inspect, service, maintain, troubleshoot, and repair avionics systems that perform communications, navigation, collision avoidance, target acquisition, and automatic flight control functions. In addition, they inspect, service, maintain, troubleshoot, and repair aircraft batteries, AC and DC power generation, and conversion and distribution systems as well as the electrical control and indication functions of all airframe systems, including hydraulics, flight control, landing gear, fuel, environmental control, power plant, drive train, anti-ice, and fire detection. AETs perform ground handling and servicing of aircraft and conduct routine aircraft inspections and aviation administrative duties. They also fill aircrew positions such as navigator, flight mechanic, radio operator, sensor-systems operator, and basic air crewman.

Qualifications include proficiency in solving practical mathematical problems and a high degree of electrical and mechanical aptitude. School courses in algebra, trigonometry, physics, electricity, and mechanics are extremely useful. Practical experience in the electrical trade is also helpful. Training begins with twenty-six weeks of intensive instruction at ATTC.

Deck and Ordnance Ratings

Boatswain's Mate (BM). Boatswain's mates, the most versatile members of the Coast Guard's operational team, are masters of seamanship. BMs are capable of performing almost any task in connection with deck maintenance, small-boat operations, navigation, and supervising all personnel assigned to a ship's deck force. BMs operate hoists, cranes, and winches to load cargo or set gangplanks; stand watch for security, navigation, or communications; and have a general knowledge of ropes and cables, including different uses, stresses, strains, and proper stowing. BMs must have leadership ability, physical strength, good hearing, normal color vision, and a high degree of manual dexterity. School courses taken in algebra, geometry, and shop are helpful. Any experience handling small boats is extremely valuable. Training for boatswain's mate is

Figure 13-5. Boatswain's mate (BM).

accomplished through fourteen weeks of intensive training at Yorktown, Virginia, or with on-the-job training through a striker program. Once this training is completed, BMs may go on to other advance training such as coxswain, heavy weather coxswain, aids-to-navigation basic and advanced, buoy deck supervisor, and law enforcement including fisheries, among others.

Gunner's Mate (GM). Gunner's mates are one of the oldest ratings in the Coast Guard. Carrying on a rating that was first formally established in 1797, GMs work with everything from pistols, rifles, and machine guns to 76 mm weapons systems. GMs are responsible for training personnel in the proper handling of weapons, ammunition, and pyrotechnics. Additionally, GMs receive intensive training in and develop skills in electronics, mechanical systems, and hydraulics, and perform maintenance on all ordnance/gunnery equipment: mechanical, electrical, and hydraulic.

GMs acquire skills in such a wide range of specialties, paving a firm path to a variety of great future careers. To be a GM, you should have an interest in all aspects of small-arms weaponry (marksmanship training, function, and usage) and the mechanical operation and electronic function of weapons systems. An aptitude in mechanics, basic electrical theory, mathematics, and attention to detail will help. Training for the

Figure 13-6. Gunner's mate (GM).

GM rating is through formal instruction located in Yorktown, Virginia. A gunner's mate requires skills in electronics and mechanical systems along with hydraulics. GM "A" School is currently thirteen weeks of formal training. After "A" school, most graduates immediately attend equipment- or system-specific "C" schools lasting from five days to fourteen weeks.

Intelligence Specialist (IS). The intelligence specialist performs a wide range of duties associated with the collection, analysis, processing, and dissemination of intelligence in support of Coast Guard operational missions. The duties performed by an IS include identifying and producing intelligence from raw information; assembling and analyzing multisource operational intelligence; collecting and analyzing communication signals using sophisticated computer technology; providing input to and receiving data from multiple computerized intelligence systems; preparing and presenting intelligence briefings; preparing planning materials for operational missions, conducting mission debriefings, analyzing results, and preparing reports; preparing graphics, overlays, and photo/map composites; plotting imagery data using maps and charts; and maintaining intelligence databases, libraries, and files. ISs start their careers with approximately fourteen weeks of specialized training at IS "A" School in Yorktown, Virginia.

Maritime Enforcement Specialist (ME). Protecting America's ports, waterways, and interests at home and abroad, MEs are trained in maritime law

Figure 13-7. Intelligence specialist (IS).

enforcement, antiterrorism, force protection, and physical security. MEs are a cadre of professionals well grounded in knowledge and skills pertaining to law enforcement and security duties. As such, members of this rating can be expected to be assigned challenging duties including traditional maritime law enforcement, antiterrorism force protection, and port security and safety as well as providing unit-level training in these fields. Required qualifications include the following:

- Must be eligible for a "secret" clearance.
- Must have normal color vision.
- Must have no domestic violence convictions or restraining orders prohibiting them from legally carrying a firearm.
- Must be in compliance with the Coast Guard weight and body fat standards.
- Must pass either the Boarding Officer and Boat Crew Physical Fitness test or the Deployable Specialized Forces Tier II Physical Fitness Test.

Operations Specialist (OS). Operations specialists perform a central role in the execution of nearly all Coast Guard operations. Operations specialists are tactical command-and-control experts, coordinating responses to a wide variety of Coast Guard missions, including search and rescue, maritime law enforcement, marine environmental protection, homeland security, and national defense. Operations specialists operate state-of-the-art communications systems, tactical tracking and identification systems, shipboard navigation systems, and advanced operational planning applications. OSs should have the ability to work in a stressful and high-paced environment, the aptitude for working with computer-based applications, and exceptional attention to detail. OSs must have normal color vision,

Figure 13-8. Maritime enforcement specialist (ME).

Figure 13-9. Operations specialist (OS).

normal hearing, be U.S. citizens, and become eligible to access classified information.

Hull and Engineering Ratings

Damage Controlman (DC). DCs assigned to cutters are responsible for watertight integrity, emergency equipment associated with firefighting and flooding, plumbing repairs, welding fabrication and repairs, and chemical, biological, and nuclear-warfare detection and decontamination. DCs are stationed throughout the Coast Guard, including Alaska, Hawaii, Puerto Rico, and Guam. Afloat assignments for DCs include all major cutters, buoy tenders, and river tenders. On board cutters, DCs are assigned to the engineering department, where they qualify to stand engineering watches. Shoreside assignments for DCs include base support units, air stations, sectors, tactical law enforcement units, and small-boat stations. Training to become a damage controlman can be accomplished by on-the-job training or by attending DC "A" School in Yorktown, Virginia. In the fifteen weeks of DC "A" School, students receive classroom instruction and lab time in each of the following areas: welding; oxy-fuel gas cutting; firefighting; carpentry; plumbing; watertight closure maintenance; chemical, biological, and radiological warfare defense; and shipboard damage control.

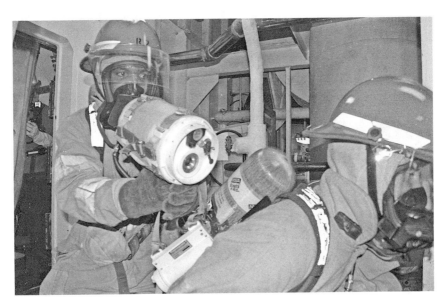

Figure 13-10. Damage controlman (DC).

Advanced training in welding, firefighting, and shipboard damage control procedures are available for DCs assigned to cutters.

Machinery Technician (MK). As one of the Coast Guard's largest ratings, machinery technician offers opportunities for assignment at every Coast Guard cutter, boat, and shore station. MKs are trained not only as technicians but also as managers and leaders—acquiring a breadth of knowledge in all areas of machinery operation and maintenance, from internal combustion engines (gas/diesel gas turbines) to environmental support systems (heating/ventilation/air conditioning), hydraulics, basic electricity, and areas of hazardous material recovery and control. In addition, MKs learn to work on the Coast Guard's computers and computer-based programs. Many MKs are also called on to act as federal law enforcement officers. MKs must have normal color vision and mechanical aptitude, with average or better ability to solve mathematical problems. Practical experience as a mechanic, machinist, or power plant operator is extremely valuable. School courses in mechanics, machine shop, electricity, and practical math are also desirable. MKs begin with thirteen weeks of training at MK "A" School in Yorktown, Virginia, or on-the-job training.

Electronics Technician (ET). ETs are responsible for the installation, maintenance, repair, and management of sophisticated electronic equipment,

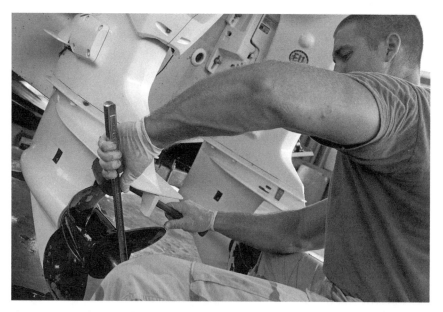

Figure 13-11. Machinery technician (MK).

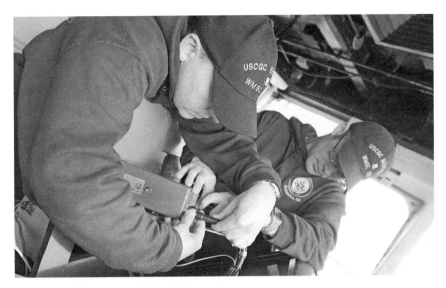

Figure 13-12. Electronics technician (ET).

including command-and-control systems, shipboard weapons, guidance and fire-control systems, communications receivers and transmitters, data- and voice-encryption equipment, navigation and search radar, tactical and electronic detection systems, electronic navigation equipment, and computers.

Being an ET requires a vast knowledge of electrical theory along with practical hands-on skills needed to repair and maintain command, control, and communication systems. Basic Electronics Technician School is one of the longest in the Coast Guard, at twenty-nine weeks, and is located in Petaluma, California (just one hour north of San Francisco). Students learn how to repair and maintain high-frequency single side-band trans- ceiver systems, antenna systems, very-high-frequency mobile transceivers, direction finders, global positioning system (GPS) receivers, small-boat radar, depth finders, and integrated control systems. ETs should have an interest in electronics and computer systems and an aptitude for detailed work and should be above average at solving mathematical problems. ETs must have normal color vision. Practical experience or prior training in electronic systems maintenance and repair is helpful but not required.

Information System Technician (IT). ITs are responsible for establishing and maintaining Coast Guard computer systems and analog and digital voice systems (telephones and voice mail), and for installing and maintaining the physical network infrastructure that ties the systems together. ITs at sea support tactical command, control, communications, and computer systems. Being an information system technician requires a vast knowledge of electrical theory along with practical hands-on skills needed to manage, repair, maintain, and install telephone systems and network cabling and computer systems. IT "A" School lasts for twenty-nine weeks and is located in Petaluma, California. Personnel learn how to install the Coast Guard's standard computer system image on server and client workstations, how to maintain network servers, how to install telephone and network copper

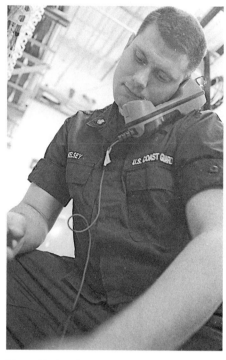

Figure 13-13. Information system technician (IT).

and fiber-optic cable, and how to perform moves, adds, and changes on private branch exchange and electronic key telephone systems.

ITs should have an interest in telephone and computer systems, an aptitude for detailed work, and be above average at solving mathematical problems. ITs must have normal color vision. Practical experience or prior training in telephone or computer systems maintenance and repair is helpful but not required.

Electrician's Mate (EM). EMs are responsible for installation, maintenance, repair, and management of sophisticated electrical and electronic equipment. These include

- Electrical power generation,
- Fractional and integral horsepower motor,
- Cutter propulsion plant control,
- Interior communication systems,
- Electronic navigation equipment, and
- Gyrocompass equipment.

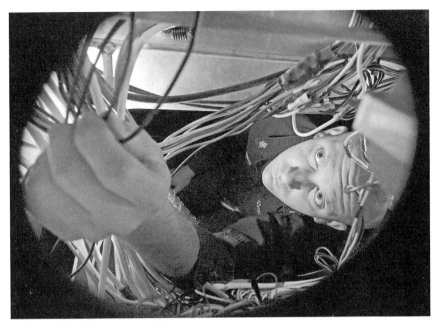

Figure 13-14. Electrician's mate (EM).

Afloat units are the EMs primary focus. All cutters 110 feet and larger as well as some smaller cutters have EMs working on board. These cutters are stationed throughout the world. On some cutters, the EM is the engineering petty officer (EPO) and, while acting in this capacity, is responsible for all engineering systems on board the cutter as well as for supervising the engineering department. EMs should have an interest in electrical, electronic, mechanical, and pneumatic systems; an aptitude for detailed work; and an above-average ability in solving mathematical problems. Since electricity itself is best never seen, the ability to visualize the theoretical working of the systems is a prime prerequisite. An EM must have normal color vision. Practical experience or prior training in electrical or electronic repair is helpful but not required.

Diver (DV). The Coast Guard announced the creation of the diver, or DV, rate and an associated chief warrant officer, or DIV, specialty on 31 January 2014. Coast Guard divers have a storied history that began in the 1940s with intelligence gathering and subsurface activities supporting the Office of Strategic Services, a predecessor of the Central Intelligence Agency. They were also assigned to the Navy Yard at Washington, D.C., to support salvage operations.

Figure 13-15. Diver (DV).

Today they sweep ports and waterways during coastal security missions; conduct salvage and recovery operations; inspect Coast Guard cutter hulls; survey coral reefs and environmental sensitive areas; repair, maintain, and place aids to navigation; conduct polar operations; and conduct joint operations with United States and international military divers.

Administrative, Support Services, and Scientific Ratings

Marine Science Technician (MST). As an MST, your job will change with the operational tempo set by the extensive number of missions you will respond to throughout your career in the U.S. Coast Guard. From protecting U.S. waters from aquatic nuisance species to supervising pollution and hazardous material responses, an MST plays the essential role of enforcing regulations for the safety of the marine environment and the security of the port.

As a domestic vessel examiner, an MST may conduct commercial fishing vessel examinations; inspect barges, cargo ships, passenger vessels, and liquefied gas–carrying vessels for compliance with U.S. laws; and conduct investigations on marine casualties, Merchant Mariner's license infringements, and pollution incidents.

As a port state control officer, an MST conducts vessel boardings to ensure compliance with applicable domestic laws and international treaties by checking structural and stability conditions; by verifying appropriate electrical, fire safety, lifesaving, mechanical, and navigation systems; and by examining living conditions for crew members on foreign-flagged vessels.

On the waterfront, an MST conducts harbor patrols and container inspections. An MST also assists with and conducts facility inspections to ensure that facilities have the proper equipment and capabilities to meet any environmental response.

As an MST, your ability to plan makes you an invaluable asset to the setup and exercise planning of any incident command system for response to anything from local port-security threats to natural disasters with nationwide impact.

In the maritime security role, an MST enforces security requirements at waterfront facilities and on domestic and foreign flag vessels. MSTs identify the actions required to respond to current and future security threats and concerns involving the maritime transportation system.

MSTs begin their career with ten weeks of intensive specialized training at MST "A" School in Yorktown, Virginia, which provides an introduction to the MST rating force and the foundational skills needed to investigate oil and hazardous material pollution incidents, supervise pollution cleanup operations, perform waterfront facility and security inspections,

Figure 13-16. Marine science technician (MST).

Figure 13-17. Yeoman (YN).

and conduct safety and security boardings on foreign-registered vessels coming into the territorial waters of the United States. MSTs must be people-oriented, possess strong attention-to-detail attributes, and be highly flexible and consistent. Training and experience in environmental sciences is helpful but not required.

Yeoman (YN). YNs are key problem-solvers, counselors, and sources of information to personnel on questions ranging from career moves, entitlements, and incentive programs to retirement options and veterans' benefits. They are the men and women who make things happen, the indispensable behind-the-scenes personnel who make any well-oiled organization hum. YNs can typically be found on large cutters and in most staff offices within the Coast Guard. Although most units that employ YNs are located in larger metropolitan areas, there are opportunities to serve at group offices in smaller localities.

Storekeeper (SK). SKs procure, store, preserve, package, and issue clothing, spare parts, provisions, technical items, and all other necessary supplies. They also keep inventories, prepare requisitions, and check incoming supplies. Storekeepers handle all logistical functions and are experts in the Coast Guard accounting system, preparing financial accounts and reports. They use all types of office equipment and computers extensively. They also operate all types of material handling equipment, including forklifts.

SK "A" School is an eight-week, performance-based course. Students perform actual SK tasks, with practical exercises in

- MILSTRIP (military standard requisitioning and issuing procedures), research Federal Logistics Data (FEDLOG);
- Create MILSTRIP messages using the Coast Guard Message System, or CGMS;
- Simplified acquisition procedures using Finance and Procurement Desktop;
- Fiscal procedures;
- Shipping and receiving;
- Inventory management and configuration management (Configuration Management Unit Level System, or CMPlus);
- Property management; and
- Transportation of freight.

Figure 13-18. Storekeeper (SK).

SKs should possess computer skills, organizational skills, and an aptitude for practical mathematics. Inventory and customer service experience is a plus although not required.

Public Affairs Specialist (PA). PAs are the Coast Guard's enlisted public communications experts. They write news releases and feature articles, shoot still and video imagery, serve as spokespersons, and maintain Web sites to raise public awareness of important Coast Guard issues and news stories. PAs typically focus their efforts in support of media relations and are stationed in major media markets throughout the United States. They work for public affairs officers on district and area commanders' support staffs or in small public affairs detachments located in major metropolitan areas. The ability to communicate with clear, accurate speech and writing is a must.

PAs need an excellent command of the English language and should be able to work effectively in a variety of media and situations. School courses

Figure 13-19. Public affairs specialist (PA).

in English, journalism, Web design, speech, photography, and typing are highly desirable. Experience with computers and video is also advantageous. Preferred traits are attentiveness, initiative, creativity, people skills, and poise under pressure. PAs receive entry-level training in an intensive twelve-week course held at the Defense Information School in Fort Meade, Maryland.

Culinary Specialist (CS). Culinary specialists receive top training in cooking skills, accounting, management, leadership, and the ability to organize and carry out many tasks. They also learn equipment use and safety, recipe conversions, basic food preparation skills and terminology, baking, sanitation, purchasing, storage, nutrition and wellness cooking, and dining-facility management. Being a CS provides excellent preparation and training in the fields of restaurant management, catering, cooking, or a variety of other jobs in the food service industries. Culinary specialists can be found in just about every duty station available throughout the United States and various locations overseas. They serve on every Coast Guard cutter, from harbor tugs to icebreakers.

To be a CS, you should have an interest in food preparation, an ability to understand and apply instruction and procedures for handling food, and a good mathematical background for recipe conversions. You should also have high standards of personal cleanliness, and an education in food

Figure 13-20. Culinary specialist (CS).

service and hospitality or experience in food service is helpful. A career as CS begins either with twelve weeks of specialized instruction at CS "A" School in Petaluma, California, or with on-the-job training.

Health Services Technician (HS). Health services technicians provide necessary routine as well as emergency health care services in large Coast Guard clinics or small sick bays, ashore or on cutters. Services could include providing direct medical care for personnel and families; assisting medical and dental officers; performing diagnostic testing, x-rays, and clinical lab tests; prescribing medications; administering immunizations; performing minor surgical procedures; and much more. The first duty station for HSs is usually at large medical clinics where professional supervision from highly qualified medical personnel helps to sharpen the skills needed to succeed.

If assigned to independent duty, an HS will provide for all of the crew's medical needs. HSs may also be involved in search-and-rescue or medical-evaluation missions. HSs should desire to help people needing medical and dental attention and have meticulous attention to detail and a pleasing personality and ability to work closely with others. Medical or dental experience is helpful.

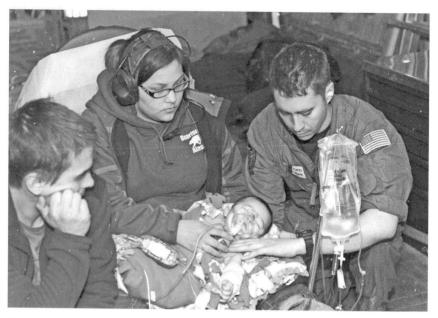

Figure 13-21. Health services technician (HS).

School courses in practical mathematics, hygiene, biology, physiology, and chemistry are an advantage. Nonrates begin with twenty-three weeks of HS "A" School at Petaluma, California, and experience intensive training in anatomy and physiology, patient examination, evaluation and treatment, and pharmacology.

Musician (MU). The U.S. Coast Guard Band recruits only the most highly skilled musicians, and the audition process is extremely competitive. Coast Guard Band auditions occur only in anticipation of projected vacancies. Openings are announced on the band's Web site, and notices are sent to college and university music departments. Auditions are held at the U.S. Coast Guard Academy in historic New London, Connecticut. All applicants must travel to New London, Connecticut, at their own expense.

The audition process is much like that of any major symphony orchestra, with a screened preliminary round to ensure anonymity, and with musical requirements that include performance of a solo, concert band and orchestral excerpts, and sight reading. Candidates are advanced to a final round based on their demonstrated musical ability; the winner of that round must complete an interview and reference check in addition to meeting Coast Guard requirements. The director makes the final decision

Figure 13-22. Musician (MU).

to award the position to the winner or winners, who then enlist in the U.S. Coast Guard for a period of four years at the rank of Musician First Class (E-6).

Reserve Ratings

Investigator (IV). Investigators provide support to Coast Guard law enforcement and intelligence missions. They conduct both criminal and personal background investigations, collect and analyze intelligence information, and provide personal protection services to high-ranking Coast Guard officials and other dignitaries.

Possible assignments include criminal investigations for crimes relating to Coast Guard missions, interagency law enforcement and liaison operations, investigations into felony violations of the Uniform Code of Military Justice, protective services operations, or law enforcement information collection.

Figure 14-1.
ADM Thomas Collins served as the twenty-second commandant of the
U.S. Coast Guard from 30 May 2002 to 25 May 2006.
His leadership was instrumental during the transfer of the Coast Guard to
the newly created Department of Homeland Security.

CHAPTER 14

Leadership and Moral Responsibility

A complex organization such as the Coast Guard doesn't just "run." It requires people who decide what must be done, how and when it will be done, and who will do it. These people may be petty officers, seamen, department heads, office chiefs, commanding officers, or admirals, but they all have one common responsibility—to lead their teams, often in challenging and sometimes extreme circumstances.

LEADERSHIP

Every Coast Guardsman must understand and employ the basic principles of leadership. These principles are an essential element of success when operating in the uncompromising environment of the sea.

The Coast Guard considers leadership the most important criterion for advancement in all pay grades. Leadership makes the difference between a good, well-functioning ship, unit, or organization and a poor one. A well-led unit is characterized by commitment to a unified vision; a strong sense of teamwork and pride; high morale; and motivated personnel. Successful leaders do two things well: they get the job done and at the same time develop their subordinates by allowing them to contribute, grow, and mature as military professionals.

Any time you attempt to influence the behavior of others to accomplish a task, you are leading. You may employ leadership as someone's supervisor, team leader, or commanding officer, or as a peer. In some cases, a more junior person, such as a small-boat coxswain, must exert leadership over more senior personnel, such as higher-ranking petty officers being transported to a boarding.

Leaders influence others to perform tasks by using two types of power. The first is position power, which is given to you by the organization based on the pay grade or job position you hold. Position power allows you to make people comply with your orders because you have the authority to do so. The second type of power is personal power, which is gained from other people's respect for you as a person and leader. Personal power encourages others to give you their commitment and trust because of your personal qualities and behavior toward them. Personal power encourages your followers to perform tasks not because they have to but because they want to.

You will learn about leadership from your own experiences (both as a follower and a leader), by observing other leaders (good, mediocre, and bad), and from formal classroom training. This formal training, provided by the Coast Guard Leadership Development Center, gives you the opportunity to reflect and answer the question, "What did I learn from my experiences, and how can I become a better leader?"

You will not become a good leader overnight. While for a few people leadership comes naturally, for most it takes time. The Coast Guard begins this developmental process by having you be a follower first. As you advance in your career, you will be put in positions of increasing responsibility and be given more leadership opportunities. Remember that today's master chiefs once were seaman recruits. Leadership and learning go hand in hand.

Leadership Style

During your service in the Coast Guard you will encounter a variety of leadership styles. Every person's style is unique. The majority of Coast Guard professionals exude a confident, communicative, and supportive attitude. Others are quieter or lead through consensus, while some are more energetic. You may on rare occasions encounter an officer, chief, or petty officer who lacks leadership skills and attempts to get the job done through bluster or intimidation. Fortunately, most people with such attitudes have truncated careers.

A fundamental rule of leadership is to be yourself. Do not try to adopt a persona that is foreign to you. Use the leadership principles you learn to build on your natural capabilities and attributes. If you are mild-mannered, there is no reason you need to yell to garner attention or get things done. That said, it is imperative that you observe the other leaders around you, recognize the things they do that are most effective (as well as those that are ineffective), and adopt the successful methods, as you are able, into your own leadership style. By doing this throughout your career, you will grow and improve your leadership abilities.

Leadership in Action

So what makes the best leaders? Strong leaders have positive, constructive attitudes. The power of a positive attitude is immeasurable. You have seen it throughout your life: the football team that bounces back after falling behind early in the game, the soldier who never gives up even when outnumbered by the enemy, or the coxswain who takes his boat into rough seas to rescue a family from a sinking boat. A positive "can do" attitude can lead a team to accomplish missions that might seem unlikely or sometimes extremely hard. Coupled with the positive attitude, the leader must also know the limits of her team and the resources at her disposal so as not to endanger a crew by attempting a physically impossible task.

Good leaders are good communicators. They do not need to be overly verbose or prize-winning authors, but they must be able to communicate without confusing their subordinates or leaving out key information. Clear and concise direction can mean the difference between an effective or ineffective operation. Be sure to communicate what needs to be done, and if necessary, how. A petty officer who directs a newly reported fireman apprentice to "take care of fueling the small boat" might want also to review safety procedures beforehand to ensure there are no gaps in the fireman apprentice's knowledge for how to do the job right.

Communication includes listening. Far too many people do not take the time to listen attentively to their bosses, peers, and subordinates. Encourage your team to speak their minds and contribute their ideas. By taking the time to hear what others are saying, you may be able to identify new solutions to problems, pick out the subordinate who is having a bad day, or potentially learn of a hazard that you did not know existed. Listen to your people.

Good leaders set standards and goals for their team, whether for a specific task—launching a helicopter, inspecting a cruise ship, loading supplies— or more generally for the operation of the unit. Subordinates are more motivated to give their best effort if they understand the goals, share a part in them, and are allowed to apply their talents fully to achieve them. On a broader note, every commanding officer will establish his or her command philosophy as well as a set of standing orders. These documents are used to provide clear direction as to what is expected from a crew and specific watch station personnel.

Subordinates also have a responsibility for assuring they understand the goals and direction provided by their leaders. If you are a deck watch officer on a cutter and find a section of the commanding officer's standing orders to be confusing, unclear, or ambiguous, you must ask for clarification.

The standards a leader sets should be as high as reasonably expected to be accomplished. Mediocrity has no place in the Coast Guard. A leader who sets high, clear, and well-defined standards will often surpass them by challenging his or her team to live up to their full potential. A leader who sets middling standards will be lucky to meet them. Every Coast Guardsman is a public servant with an ultimate goal of delivering the best performance possible for the nation. The difference between a unit that sets high versus low standards can sometimes be measured in lives lost.

After setting goals for performance, the leader must monitor or get feedback on that performance. This does not mean hovering over a subordinate's shoulder like a vulture or micromanaging another person's daily work, both of which can be annoying and counterproductive habits. It means that there must be some form of performance metric, and that metric is situational. If a chief boatswain's mate on a national security cutter has a well-trained deck force and experienced petty officers, the chief may need only to direct them to touch up the paint on a section of the cutter's hull and then see the results, in all their glory, the next day. However, if the deck force is new or key petty officers are not available to supervise the evolution, the chief may need to provide additional guidance and visit the work site a few times to ensure the task is being accomplished the right way.

Once a task is complete, a strong leader will provide immediate feedback to the participants and employ the "power of recognition" to acknowledge good work. Positive recognition is best given publicly and serves as a positive incentive to the personnel involved. The power of public recognition, even if nothing more than a "great job" or "thanks for your hard work," can go a very long way in motivating up-and-coming shipmates. Conversely, negative feedback is necessary sometimes to correct behavior or improve work performance. For a team, constructive criticism can be offered in a group setting, but for individual mistakes or problem, counseling is almost always best done in private.

Figure 14-2. Commodore Ellsworth P. Bertholf, commandant of the U.S. Revenue Cutter Service from 1911 to 1915 and commandant of the U.S. Coast Guard from 1915 to 1919.

A strong leader sets high standards for himself or herself and provides a clear example of top performance and behavior. The best leaders are impeccably honest, ethical, and moral. You can see it everywhere in society: a military officer or chief petty officer, a sports figure, a teacher, or a parent who through their own actions and personal example encourages you to perform to the fullest of your abilities. By setting a strong, positive example, you can help lift up others as a role model. This applies not only to your fellow Coast Guardsmen but to anyone outside your unit who may be observing your actions, such as the civilian population in the foreign port your cutter is visiting. By behaving responsibly, treating the people you encounter with courtesy and respect, and showing pride in your uniform and service, you bring credit not only to the Coast Guard but the United States.

On the opposite end of the spectrum, there may be nothing more destructive to unit morale than a leader who shuns the rules and regulations, looks sloppy in uniform, shows up late to work, or exhibits inappropriate behaviors. It is both insulting and completely improper for a senior to hold his or her subordinates to a higher standard than that leader is willing to meet. Do not be that person.

Mentorship is another key element of positive leadership. The good leader understands that every person in his or her team has strengths and weaknesses, individual training needs, a unique career focus, and sometimes special personal or family needs. Spend time to learn about your subordinates. Are they married? What "A" school are they interested in attending? Do they have any health issues? Offer them your perspective, share your past experiences, and give them your best advice on how to maximize their performance. The best thirty seconds you can spend each day is by asking each person that you directly supervise: "How are you this morning, and is there anything I can do for you today?"

As you advance in responsibility, you can develop your subordinates into leaders by letting them make decisions and learn from their mistakes as you coach them. Coaching involves spending time with your people daily, talking with them about your goals, listening to their ideas, and providing them with feedback on their efforts and performance. A positive, mentoring relationship can last a lifetime.

MORAL RESPONSIBILITY

The most accomplished leaders are servant leaders who put the well-being of their subordinates before their own. As military personnel and as public officials, we have a moral responsibility to take care of our sailors.

Your success as a leader can be measured in two ways. The first deals with the job. Was it done on time and with high-quality work? The second deals with your people. Did you effectively use their talents, and if they were given the choice, would they like to work for you again? This measure of leadership has to do with how the people were treated while they were performing the task. Servant leaders treat their subordinates with respect, compassion, and inclusion, even where monotonous, tough, repetitive, or dangerous work needs to be performed.

A historic example of servant leadership is the British naval tradition where the officers do not eat until all of their subordinates have been fed. More specific examples are the air station executive officer who spends extra time helping a deserving petty officer with her application to officer candidate school; the chief who quickly arranges unscheduled leave for a seaman whose young child is ill and needs comfort; the admiral who gives up a weekend to visit and compliment the crew of a cutter returning from a significant drug seizure; the acquisition specialist who spends his lunchtime training an office mate; or the sector chaplain who gives up his vacation to tend to a petty officer who has suffered a death in the family.

A simple way to think about servant leadership is how would I like to be treated, helped, or mentored? This does not diminish the essential need for setting high standards and for a leader to hold his people accountable to those standards. It means that high standards, top performance, and acting as a servant leader are completely compatible concepts.

One of the hardest aspects of leadership is dealing with others who refuse to live up to the high standards of the service, have lousy attitudes, or simply do not contribute to the mission. Being a servant leader does not mean that you give them a pass or just do their jobs for them. A good leader confronts those who are not contributing, helps those who are struggling, and ensures they know what is expected of them. The simple way to think of a strong servant leader is someone who stands up for her people but also stands up to them if warranted.

As to diversity, our nation's laws require equal treatment for all people regardless of race, ethnicity, religion, and other attributes. The military is the ultimate meritocracy: those who perform well do well, and those who are slackers do not. It is your performance, attitude, and loyalty that count—never your skin color, gender, accent, or place of birth. The Coast Guard has an increasingly diverse workforce, but every member of it must share the core values of honor, respect, and devotion to duty that make the service the finest military organization in the world. Across our workforce, and throughout our careers, we have an unwavering obligation to treat our fellow Coast Guardsman with honor and respect.

COAST GUARD LEADERSHIP DEVELOPMENT FRAMEWORK

The Coast Guard has developed a formal Leadership Development Framework that outlines twenty-eight leadership competencies. Leadership competencies are the knowledge, skills, and expertise the Coast Guard expects of its leaders. While there is some overlap in these competencies, they generally fall within four broad categories: leading self, leading others, leading performance and change, and leading the Coast Guard. Together, these four leadership categories and their elements are instrumental to career success. Developing them in all Coast Guard personnel will result in the continuous improvement necessary for us to remain always ready—Semper Paratus.

Leading Self

Fundamental to successful development as a leader is an understanding of self and one's own abilities. This includes understanding one's personality, values, and preferences while simultaneously recognizing one's potential as a Coast Guard member.

Personal conduct, health and well-being, character, technical proficiency, lifelong learning, followership, and organizational commitment are elements to consider when setting short- and long-term goals focused upon the leadership development of "self."

Leading Others

Leadership involves working with and influencing others to achieve common goals and to foster a positive workplace climate. A member of the Coast Guard interacts with others in many ways, whether as a supervisor, mentor, manager, team member, team leader, peer, or worker. Positive professional relationships provide a foundation for the success of our service. Showing respect for others, using effective communications, influencing others, working in teams, and taking care of one's people are elements to consider when evaluating one's capacity for leading others. Developing these qualities will increase the capacity to serve.

Leading Performance and Change

The Coast Guard and its members constantly face challenges in mission operations. To meet these challenges, leaders must apply performance competencies to their daily duties. Performance competencies include developing a vision, managing conflict, quality and daily management of projects, appraising performance, problem solving, creativity, innovation, decision making, and customer focus. Having these competencies enables each leader—and the service—to perform to the utmost in any situation.

Leading the Coast Guard

As leaders gain experience in the Coast Guard, they must understand how it fits into a broader structure of department, branch, government, and the nation. At a local level, leaders develop partnerships with public and private sector organizations in order to accomplish their missions. The Coast Guard "plugs in" via its key systems: money, people, and technology. A leader must thoroughly understand these systems and how they interact with similar systems outside the Coast Guard. An awareness of the Coast Guard's value to the nation becomes more important as one gets more senior. Leaders must develop coalitions and partnerships with allies inside and outside the Coast Guard. COMDTINST M5351.3 contains more information about the other components of the Coast Guard Leadership Development Framework.

Additionally, the commandant established unit-level and member-level leadership initiatives. The Unit Leadership Development program emphasizes teamwork and leader development within every unit, which contributes to an organizational climate that perpetuates the development of strong leaders. The Individual Development Plan encourages personal, professional, and intellectual development throughout the course of each member's career. The development of Coast Guard leaders at all levels is a command priority. Every commander, commanding officer, officer-in-charge, chiefs' mess, and supervisor is responsible for providing and supporting leadership development activities within their area of responsibility. Leadership development is an essential element of our unwavering commitment to mission excellence for serving and protecting the public trust.

Figure 15-1.
A military court room. You may never see the inside of one as long as
you have the "right attitude."

CHAPTER 15

Discipline

"Discipline" is a word with more than one meaning. "Discipline" can mean punishment. "Discipline" also refers to the mental, physical, emotional, and spiritual fortitude a person applies to getting the job done. Coast Guard personnel must have intense discipline when operating in dangerous seas, flying a helicopter at night, responding to a major oil spill, or working around the clock.

Well-disciplined people have the right attitude, are willing to follow orders, and, if necessary, put their own life on the line because they believe in what they are doing; have respect for their leaders; and know they are working for a greater cause. They behave in a military manner; wear the right uniforms; take pride in their division, their ship, and their station; and are ready to fight bravely in defense of their country.

When personal discipline fails or is not applied, mistakes often result. Despite mentoring, counseling, or frank sessions behind closed doors, there are occasions when rules are intentionally broken or duties are not fulfilled. In other cases, a Coast Guardsman may have personality flaws that lead to lying, stealing, physical assault, or worse. This is where the other form of discipline—punishment—is necessary.

Achieving unit discipline and cohesion is an all-hands evolution. Most units have masters-at-arms (MAA) who are assigned to maintain discipline and are the assistants to the executive officer. Large cutters or shore stations may have a chief master-at-arms (CMAA) with several assistants. When in port, a shore patrol may be assigned to monitor activity and respond to problems ashore. In addition to these formal structures, it is every Coast Guardsman's duty to help maintain the discipline of their unit, by behaving properly, encouraging others to do so, and reporting significant violations.

It is important to understand that, as a general principle, the minimum amount of counseling, extra instruction, or punishment should be applied that is necessary to correct behavior. For example, a seaman apprentice who shows up late for duty but otherwise has an unblemished record typically will be counseled by his supervisor and told not to do it again. A next step for a second occurrence could be additional counseling from more senior personnel, extra instruction, or a Page Seven entry into the person's service record. Should he continue to arrive late, despite the counseling, the seaman apprentice may receive nonjudicial punishment (NJP) from the cutter's commanding officer. If the person continued to flout the rules, a court-martial may be in order. Fortunately, this scenario is extremely unlikely, with most improper behaviors corrected through initial counseling. However, more serious infractions such as theft or assault may head straight to NJP or courts-martial.

UNIFORM CODE OF MILITARY JUSTICE

Discipline in all the military services is based on the Uniform Code of Military Justice (UCMJ), which became law on 5 May 1950. The UCMJ is the basis for the *Manual for Courts-Martial*. Before 1950 the individual services had different disciplinary systems. The only variations that now exist among the services are those that result from regulations specific to the department under which the service operates. The basic principles of the UCMJ are the same for all.

The manual recognizes the military authority of persons of one service over those of another; it even specifies that personnel can be tried by courts-martial composed of members of other services. Thus, Coast Guard personnel are subject to the orders of the Navy shore patrol and the Armed Forces Police. Likewise, Coast Guard shore patrols enforce the UCMJ upon members of the other armed services.

Necessity for the Disciplinary System

All of the armed services, including the Coast Guard, were established to accomplish objectives determined by Congress. The missions of the five services, both in war and peace, involve personal risk. A well-trained, closely knit, disciplined group of people can carry out a mission with a minimum loss of life, but the leaders who direct their operations must have the necessary authority.

Value of the Disciplinary System

The courts and laws of the nation exist to protect all people against criminal acts, and to prevent invasions of, or loss of, the rights of individuals. The disciplinary system of the Coast Guard protects all members of the service. The regulations of the Coast Guard and the UCMJ protect loyal, law-abiding military personnel from being victimized by criminal or lawless actions.

The disciplinary system also protects conscientious, hard-working members of the service from being unfairly burdened with people who refuse to do their share of the work. If such people get out of doing assigned work, those with a sense of responsibility must do it. The system sets up penalties for nonperformance of assigned duties.

It also prohibits compulsory self-incrimination (you cannot be forced to testify against yourself—Article 31) and cruel or unusual punishment (Article 55), specifies that parts of the UCMJ that must be explained to you (Article 137), and provides a way to make complaints if you are treated unfairly (Article 138). The UCMJ works both ways.

Protection of the Accused

Like the laws of our country, the laws of military justice protect those accused of offenses, even convicted lawbreakers. Accused personnel have the right to be informed about the charges against them, to reply to the charges, and to confront and question their accusers. They have the right to legal counsel of their own choosing, if possible under the existing military conditions. There are limitations on the questioning of prisoners and rules requiring humane treatment. Trials must be conducted fairly. Punishments are limited in accordance with the seriousness of the offense; cruel and inhuman punishments are prohibited. Finally, there is an elaborate system of reviews to ensure that all these provisions for protecting the rights of the accused have been observed and that the court's decision is just.

Mistaken Loyalty

A mistaken concept of loyalty to shipmates sometimes causes otherwise well-intentioned persons to refrain from reporting crimes. Tale-bearers who constantly run to officers with reports of minor derelictions of duty by other crew members are a nuisance, but you should be able to distinguish between tale-bearing and giving information about a serious offense.

If an act or neglect constituting a legal offense has been committed—or you have serious reason to believe one has been committed—it is your duty to report it to the proper authority. If you do not, you too are punishable under the UCMJ. In permitting an offender to escape punishment, you weaken the disciplinary system that protects you and your shipmates. You owe loyalty to those who keep the laws, not to those who break them.

Offenses under Military Law

The disciplinary laws of the Coast Guard are stated in the UCMJ, in *United States Coast Guard Regulations*, and in *Coast Guard Reserve Regulations*. Articles 77 to 134 of the UCMJ cover punitive or punishable offenses.

It is your responsibility to acquaint yourself with these laws. The oath you took on entering service obligates you to observe them. As in civilian life, ignorance of the law is no excuse for breaking it.

The UCMJ was explained to you when you entered the Coast Guard. Article 137 states that members shall have articles of the UCMJ explained to them upon enlistment, after six months of active duty, and upon each reenlistment. Copies of it are posted where you can read it for yourself. You should note especially the listing of offenses and their definitions. To assist you, certain legal terms are discussed here.

Differences between Military and Civilian Offenses

In general, acts that would constitute offenses against civilian laws are also offenses under military law, but some acts are military offenses that would not be cause for arrest in civilian life. Civilians who do not show up for work or who walk off the job will probably lose their jobs, but that is all that will happen. In a military organization, the same act would be unauthorized absence and perhaps neglect of duty or disobedience to an order. Civilians who quarrel violently with the boss will not be arrested as long as they do not harm the boss or threaten harm. In the service, a person could be charged with insubordination and possibly other offenses. Coming to work late, refusal to obey orders, drinking on the job—all of these would probably be cause for discipline by a civilian employer but not cause for arrest.

Yet in a military service each can send a person to nonjudicial punishment or a court-martial. The nature of military duty and the conditions of military life account for the difference. Your military duties are necessary for the Coast Guard to carry out its mission—the defense of the nation, the enforcement of national and international law, and the safety of life at sea. You cannot quit your job, be absent or late without permission, or refuse to do your work without endangering the mission. People who rebel against lawful authority and stir up trouble are a menace to the teamwork that is necessary for their own safety and that of their shipmates as well as for the accomplishment of the unit's assigned mission.

Principals and Accessories

A person who commits a crime or who aids in its commission is a principal. If a person advises, orders, or persuades someone to commit an offense, and that person does commit the crime, both are principals. After a crime has been committed, anyone who knows the person who committed it and who helps the offender to escape trial or punishment is an accessory after the fact. Principals and accessories are both punished according to the table of maximum punishment.

Lesser Included Offenses

If, during a trial, a person is charged with desertion and it develops that the defendant was absent without leave (AWOL) but did not intend to abandon the service permanently, the court may convict that person of unauthorized absence and punish accordingly. This is a lesser included offense within the more serious one of desertion originally charged. Punishment for a lesser included offense may occur under other charges. Housebreaking, for example, might be a lesser offense under a charge of burglary.

Attempt. An attempt to commit an offense is a violation even though the crime is not consummated. An attempt to commit a particular crime is always a potential lesser included offense when a perpetrator is tried for the crime.

Conspiracy. When two or more persons agree to violate the UCMJ by concerted action and any one of them performs any action seeking to accomplish the agreed objective, all of the parties to the agreement are guilty of conspiracy.

Offenses Related to War

Certain offenses become much more serious if committed when the United States is at war. Others are particularly related to wartime situations. Among those that become more serious are desertion and assaulting or disobeying an officer. For each of the offenses listed below, the death penalty may be imposed in time of war:

- Misbehavior before the enemy
- Subordinate compelling surrender
- Forcing a safeguard
- Improper handling of captured or abandoned property
- Aiding the enemy
- Misconduct as a prisoner of war
- Spying

General Article

Article 134, the last of the punitive articles, makes it an offense to commit acts or to neglect duties that result in creating disorders that are to the prejudice of good order and discipline or that bring discredit upon the Coast Guard. Such offenses are subject to trial by court-martial.

Liability under Civil Laws

What about civil laws? For example, suppose you do not pay your debts or that you run off with another person's spouse. Can you be given military punishment for such nonmilitary offenses? Yes, you can. In the first place, of course, you are liable to civil lawsuits and the actions of the civil courts and police. But military courts can also try you for conduct casting discredit on the Coast Guard and can award you various types of punishment.

Suppose you are arrested for speeding, for drunkenness, or for being criminally involved with drugs, and are held in jail until you are AWOL. You not only will have to pay your fine and suffer your imprisonment but you may also have to appear before a captain's mast and perhaps a court-martial for your AWOL time and for any discredit you have cast upon the Coast Guard. If you refuse to pay just debts, you will be warned by your commanding officer. You may lose your advancement opportunity because the Coast Guard does not promote people who are not honest. Finally, if you continue to run up unpaid debts, you could be administratively discharged or even receive an undesirable or bad conduct discharge. (On the other hand, if you are being gouged by unscrupulous dealers, see your executive officer or legal officer. They can help you in many ways.)

Purpose of Punishment

The punishment used by the Coast Guard—such as extra instruction, loss of liberty, loss of pay, and confinement—has three purposes:

- To deter you from breaking the rules;
- To encourage you to do your duty; and
- To provide an example to others.

The smart person learns from the mistakes of others. When others are being punished for being late, out of uniform, or careless in their attitude, smart people make sure that they are on time, in proper uniform, and attentive to duty.

If you receive punishment in the Coast Guard, remember, you brought it on yourself by your behavior. Take it like an adult; do not grouse, and do not feel sorry for yourself. Learn your lesson. Avoid the habits, attitudes, and companions that led you wrong. Do not hold grudges. The people who punished you are simply doing their duty—which they would not have had to do if you had done yours.

NONJUDICIAL PUNISHMENT AND COURTS-MARTIAL

Pretrial Procedure

Any Coast Guard member who knows an offense has been committed can *prefer charges*. Ordinarily, the facts are reported to the supervisor, who relays them to the commanding officer or officer in charge, who then takes the necessary action. No physical contact is necessary to apprehend, but the person must clearly be notified of being taken into custody. *Arrest* is moral restraint limiting a person's liberty pending disposition of charges. If there is reason to believe that people will not observe their military obligation to obey the order of arrest, they may be placed in *confinement*. This is forcible detention, but is not for punishment, and they must be so advised.

Preliminary Inquiry

The commanding officer orders a preliminary investigation to develop information to permit intelligent disposition of the case. The accused are informed of the charge and are permitted, but not required, to make a statement.

The accused are advised that any statement made may be used against them. After the investigating officer has developed the facts, a full report is made to the commanding officer. If an offense has been committed, the commanding officer may sign the charge sheet. If the offense is within the jurisdiction of the commanding officer, a court-martial or other appropriate action may follow. If the offense requires a general court-martial, the full report is forwarded to the district commander or other officer exercising general court-martial supervision over the unit. If a general court-martial is to be held, it must be preceded by a formal investigation.

Nonjudicial Punishment (Mast)

If you break a rule or are negligent, careless, or unmilitary in your conduct, you may be "placed on report" by a petty officer or commissioned officer. This means that you must appear before the captain at a specified time for a hearing and possible punishment. That is, you must appear at captain's mast, formally known as nonjudicial punishment. Captain's mast gets its name from the old sailing days when the setting for this form of naval justice was on the weather deck, at the foot of the ship's mainmast.

The commanding officer is authorized to assign certain punishments for minor offenses. Cases are tried and punishments given at captain's mast. At mast, the commanding officer may give the following punishments, which are designed to change behavior:

Type of Punishment	CO (LCDR or Above)	CO (LT or Below)	Officer in Charge
Admonition/reprimand	Yes	Yes	No
Correctional custody (E-3 and below)	30 days	7 days	No
Extra duties (E-6 and below)	45 days	14 days	14 days
Restriction	60 days	14 days	14 days
Forfeiture of pay	Half of 1 month's pay per month for 2 months	7 days' pay	3 days' pay
Reduction to next inferior grade (E-6 and below)	Yes	Yes	No

Mast is not always held for disciplinary purposes. A *meritorious mast* is held by the captain to give awards or commendations to people who have earned them.

On many ships and stations, a certain time of the day is set aside by the executive officer to hear requests from members of the crew. This practice is called *request and complaint mast*.

TYPES OF COURTS-MARTIAL

Summary Court-Martial

A summary court-martial is convened by the commanding officer and may not try commissioned officers, warrant officers, or cadets. A commissioned officer in the grade of lieutenant or above will sit as the judge, jury, prosecutor, and defense. All Coast Guard members may refuse summary court-martial. If refused, the matter may be referred to special court-martial. Members should be afforded an opportunity to consult with a lawyer (Coast Guard or civilian) prior to accepting or declining summary court-martial but are not entitled to military representation at the proceeding. Civilian lawyers may be hired by the accused at no cost to the government.

Special Court-Martial

A special court-martial is convened by a commanding officer or higher and consists of at least three members and a military judge. If the accused is enlisted, one-third of the "jury" may be enlisted, upon request. The staff judge advocate or convening authority will appoint trial counsel (prosecutor) and defense counsel. Punishments may include:

- Up to one-year confinement
- Bad conduct discharge
- Forfeiture of up to two-thirds of pay per month for twelve months
- Reduction to pay grade E-1
- Fine

General Court-Martial

A general court-martial is convened by a flag officer and consists of a military judge and at least five members. If the accused is enlisted, one-third of the "jury" may be enlisted, upon request. The staff judge advocate or convening authority will appoint trial counsel (prosecutor) and defense counsel. Any punishment may be assigned depending on the offense, including the death penalty.

Figure 16-1.
Coast Guardsmen from Sector Southeastern New England in Woods Hole,
Massachusetts, in full ceremonial dress, stand in line with Navy,
Marine Corps, Army, and Air Force honor guards during Red Sox opening day
ceremonies at Fenway Park.

CHAPTER 16

Personal Standards

The Coast Guard will train you to perform your military and professional duties, but your ability to perform them quickly and efficiently will depend upon the state of your physical and mental health. Good health habits and a positive attitude will make your job easier and improve relations with people. The following pages offer some suggestions.

PERSONAL CARE

As a member of the Coast Guard, you will have the facilities and the opportunity to maintain a good standard of living and health. Even the smallest unit has facilities to provide nourishing meals, comfortable berthing, and all the amenities necessary to live in a comfortable and productive fashion. It is important that you take advantage of the healthful environment provided by the service and adopt good personal habits and standards.

Hygiene

Personal hygiene is simply keeping healthy by good cleanliness habits. You must take care of your grooming every single day. While on board ship, you will be taking sea showers to conserve water. These are shortened showers to save time and water and do not excuse you from maintaining proper grooming. There are many things you can do to avoid body odor, including taking showers daily and after working out at the gym, using deodorant daily, and changing clothing and underclothing daily.

Poor personal hygiene is not only unhealthy for you; you can also expose your shipmates to many types of health risks. Good cleanliness habits begin with washing your hands. Frequently washing your hands will help to prevent the spread of germs and disease. Wash your hands:

- Before and after each trip to the head,
- Before and after each meal (if you are not able to wash your hands in the head, use waterless hand sanitizer), and
- Any time you cough or sneeze in your hand.

Care of Teeth

The three most common dental diseases are tooth decay (caries), inflammation of the gum (gingivitis), and disease of the gum and bone surrounding the teeth (pyorrhea). They cause needless loss of teeth, for they all can be prevented or controlled. All Coast Guard personnel shall have an annual dental exam. This is essential for timely detection of tooth decay as well as any other type of mouth disease.

Cavities

There is no surefire way to prevent tooth decay, but it can be cut down greatly by regular, correct brushing and by reducing the amount of sweets you eat. At the first sign of tooth decay, see a dentist. Untreated cavities can have serious long-term health consequences.

Gingivitis

Normal and healthy gums are pale pink and firm in texture. If they are swollen or puffy, hang loosely about the teeth, and bleed easily, you have gingivitis.

Pyorrhea

If gingivitis goes untreated, you may notice pockets or crevices between the tooth and gum. This is an indication of pyorrhea (sometimes called periodontitis). More teeth are lost from these two diseases than from tooth decay. Most dental diseases result from poor mouth hygiene combined with the misuse of the toothbrush and lack of flossing.

Toothbrushes

Proper use of the toothbrush is most effective in control or prevention of dental diseases. Keep at least two brushes, so you can always use a dry one. When the bristles become soft or flattened, throw the brush away.

Brush your teeth regularly, after each meal if you can. Use dental floss at least once a day. The proper tooth brushing technique will clean the teeth and gums, stimulate circulation in the gum tissues, and toughen the gums by the friction of the bristles; proper tooth brushing does not injure the teeth or gums. Brush with the mouth barely open, the muscles relaxed, permitting easier access to hard-to-get-at areas.

Diet

The Coast Guard takes great care to serve good-tasting and well-balanced meals, but they will do you little good unless you observe certain rules in your diet. Do not concentrate too much on one food group, such as meat or potatoes or especially desserts! Do yourself, your strength and stamina, your general health, and your teeth and gums a favor: eat a balanced diet. The basic food groups are listed below. To determine the proper amount of each, visit https://www.choosemyplate.gov/.

Grains	Vegetables
• Eat at least 3 ounces of whole grain bread, cereal, crackers, rice, or pasta every day. • Look for "whole" before the grain name on the list of ingredients.	• Eat more dark green and orange veggies. • Eat more dry beans and peas.

Fruits	Milk
• Eat a variety of fresh, frozen, canned, or dried fruit. • Go easy on fruit juice.	• Select low-fat or fat-free milk. • If you do not or cannot drink milk, choose lactose-free products or other calcium sources.

Meat and Beans
• Choose low-fat or lean meats and poultry; bake, broil, or grill meat and poultry. • Vary your choices with more fish, beans, peas, nuts, and seeds.

You will complete an annual Personal Wellness Plan to measure your current health status and see recommendations for change (when needed) for each major health area.

Dysentery

Dysentery is common in the tropics and can occur in the United States. The major symptom is diarrhea although nausea, stomach cramps, and vomiting may also occur. Dysentery is most often spread by food handlers or through produce fertilized with human waste. Water and flies are also factors. Ashore, when in countries where dysentery is known to exist, avoid uncooked food, particularly lettuce, celery, cabbage, and radishes.

If you must eat ashore, eat in reputable restaurants and avoid establishments that appear to have less than the highest standards of cleanliness. In areas where dysentery might exist, water should be boiled for at least five minutes and then stowed in sterile containers for use. If in doubt, drink bottled water and order your beverages without ice.

Drug Abuse

Drug abuse undermines morale, mission performance, safety, and health and will not be tolerated within the Coast Guard. The word "drugs," as used here, refer to barbiturates, narcotics, various hallucinogenic compounds, and other substances taken for nonmedical purposes. Drugs are classified as hallucinogens, stimulants, depressants and tranquilizers, narcotics (opiates), and inhalants.

Hallucinogens include LSD, mescaline, psilocybin, DMT, and THC (the active ingredient in marijuana and hashish). Hallucinogens, or psychedelics, are drugs that affect a person's perceptions, sensations, thinking, self-awareness, and emotions.

The effects of psychedelics are unpredictable. The physical effects include dilated pupils, higher body temperature, increased heart rate and blood pressure, loss of appetite, sleeplessness, dry mouth, and tremors. Sensations may seem to "cross over"—sometimes resulting in trauma, personality change, or psychosis.

Marijuana is a controlled substance and has clinically the same effects as hallucinogens, though usually to a lesser degree. Regular use of marijuana does three things to your body:

- The tars in marijuana smoke are far more likely to cause cancer of the lungs than those in tobacco smoke.

- The THC crystals in the smoke affect the chromosomes in the female ova, distort the male sperm, and endanger a fetus.

- The THC crystals collect in the nerve endings in the brain, ultimately affecting memory, physical activity, and learning ability.

Several states have approved ballot initiatives to legalize the use and possession of small amounts of marijuana in those jurisdictions, and more states have legalized marijuana use for medical purposes. Regardless of these state measures, using, growing, selling, distributing, or possessing any amount of marijuana remains illegal under the UCMJ. Active-duty service members are subject to the UCMJ at all times, regardless of leave status or geographic location. Reservists on active duty are also subject to the UCMJ. A service member who uses marijuana purportedly in compliance with state or foreign law is nevertheless still in violation of Article

112a of the UCMJ and may face adverse administrative or disciplinary action to include separation from the service and/or trial by court-martial. The maximum punishment for a single specification of marijuana use is five years confinement and a dishonorable discharge.

Stimulants make the entire body "run fast." People abusing stimulants may appear to comprehend but display poor judgment; they might also seem extremely nervous and be hyperactive and fatigued at the same time. Abusers may experience damage to the liver and kidneys and hemorrhages in brain tissue. The two most prevalent stimulants are nicotine in tobacco products, and caffeine, the active ingredient in coffee, tea, and some bottled beverages.

Depressants (tranquilizers) have an effect resembling alcohol intoxication (alcohol is a sedative). A person may be confused and disorganized and exhibit a loss of judgment. Pupils become constricted; speech is slurred, thick, and incoherent at times. Some users and abusers develop allergic reactions that are sometimes fatal. A common occurrence among abusers is called synergism, which is combining a depressant drug with alcohol; the result is to multiply rather than add to the cumulative effects of the drugs. Synergism often results in a fatal overdose.

Among the most addictive drugs are the *opiates*. Included in this group are codeine, morphine, and heroin. All of the opiates and their synthetic derivatives are addictive. Opiates briefly stimulate the higher centers of the brain and then depress activity of the central nervous system. Users may appear drowsy; they are able to understand and answer questions but are mentally dull. Their pupils are constricted, with little reaction to light; eyelids are droopy. Chronic users become psychologically and physically dependent on opiates.

In recovery from addiction (abstinence), some bodily functions do not return to normal for as long as six months. Sudden withdrawal by heavily dependent users who are in poor health has occasionally been fatal.

Cocaine is a local anesthetic. It is also a stimulant of the central nervous system. Chronic cocaine use eventually leads to a depressed, low-energy state characterized by flattened emotions, lack of interest in sex, and listlessness. The speed and probability of getting addicted is different for different people but is definitely higher in people with drinking problems. *Crack* is cocaine that is sold in the form of small, cream-colored chunks resembling rock salt. It is smoked. This leads to a reaction in less than ten seconds, lasting from five to ten minutes. Cocaine damages nasal membranes, causes bronchitis, and affects cardiac irregularities, often resulting in heart attacks. It can be instantly fatal.

Inhalants are breathable chemicals that produce psychoactive (mind-altering) vapors. Nearly all of the abused inhalants produce effects similar to anesthetics. Initial effects include nausea, sneezing, coughing, nosebleeds, feeling and looking tired, bad breath, lack of coordination, and a loss of appetite. Solvents and aerosols decrease the heart and breathing rate and affect judgment. Sniffing highly concentrated amounts of solvents or aerosol spray may produce heart failure and instant death. Repeated sniffing of concentrated vapors can cause permanent damage to the nervous system.

Do not use illegal drugs or prescription drugs for nonmedical reasons. Not only is it dangerous and illegal, drug abuse can get you discharged. As a law enforcement agency, the Coast Guard works to stop drugs from being smuggled into the country. Since the money that buys drugs supports smuggling, the use of drugs by Coast Guardsmen is counter to the efforts of the Coast Guard. Drug use impairs the users' abilities. Every Guardsman has a critical job that requires absolute attention. If you stand an improper lookout or cannot do your job on a repair party, lives may be lost. If your senses are impaired, you endanger more lives than just your own; you are a threat to everyone on board your ship.

The Coast Guard's drug policy is clear: drug abuse undermines and jeopardizes morale, mission performance, and safety; it will not be tolerated. Service members who have had a drug incident are processed for discharge. To eliminate drug use, commands randomly select members to submit urine samples, which are tested for evidence of drug use. Positive results from a confirmed urinalysis test constitute a drug incident, and the member will be processed for discharge. There is no second chance.

Alcohol Abuse

Alcoholic beverages are an accepted part of American life, and "social drinking" is considered normal behavior. The Coast Guard is a representative of our society. A person does not have to be a "bum" in order to have an alcohol problem. A person has an alcohol problem when drinking creates problem situations; the Coast Guard has a problem when alcohol problems cost the service time lost through absence, sickness, increased administration, and decreased performance. A person drinking sensibly is usually not a problem. Think of the word "sense," and use good sense by keeping all your senses alert and limiting yourself accordingly. Do not drink on an empty stomach, or when you are fatigued, or when you are taking medication, or when you have to drive or operate machinery.

If a man or woman develops an alcohol problem, that person should be encouraged to get professional help. Alcohol abuse training is not intended for "down and outers" only but also for less affected people. A person need not "hit bottom" in order to get help with an alcohol problem. That is why the Coast Guard offers help to its members.

The Coast Guard considers alcohol abuse a serious threat to job performance and safety. If you have an alcohol incident, your command will normally provide counseling for you. If there is a second alcohol incident, you will normally be processed for discharge. An alcohol incident is defined as alcohol abuse that has an outward detrimental effect on the abuser or others, such as involvement in any injury, loss of duty status, inability at any required time to properly perform duties due to alcohol consumption, damage or loss of property, or violation of the UCMJ or federal, state, or local laws.

Sexually Transmitted Diseases (STD)

Sexually transmitted diseases were for many years "swept under the rug" and not mentioned in polite society. STDs are infectious diseases that can be transmitted only by contact between two people. Anyone who is sexually active can catch an STD. The frequency of contact and number of sexual partners determine a person's risk of contracting STDs. Prostitutes or anyone else who has frequent sexual contact with different partners are at high risk of contracting STDs and passing them on to others.

All STDs can be transmitted from an infected person, man or woman, to an uninfected person, man or woman, through sexual intercourse. *Syphilis* can also be transmitted by a kiss if an infected person has an open sore on the lips. A woman can transmit syphilis to her unborn child, or *gonorrhea* to her child at birth.

The incubation period, the time from contact until the first symptoms appear, varies: ten days to ten weeks for syphilis; two to ten days for gonorrhea and *chancroid*; longer times for others STDs. Some types of gonorrhea, particularly in women, have no symptoms. A person who has become infected can transmit the disease to another before signs of infection appear. There is no way to tell that a person is not infected. Latent syphilis, the state in which clinical signs and symptoms of infection are absent, may appear as early as four years or as late as twenty years after infection. Among the infinite variety of results to be expected are destructive ulcers, disease of the heart or blood vessels, blindness, and insanity. Other kinds of STDs have other kinds of results, none of them pleasant.

STD control is a worldwide effort, but it all depends upon the individual. Medical department personnel emphasize that individual abstinence from sexual intercourse is the one sure way to avoid STDs. Medical department personnel can also instruct in the effective use of condoms as a preventive measure.

There is no military punishment for contracting STDs, but it is misconduct to conceal an infection. It is important that you conduct your social life so as to avoid contracting STDs, and important to report for medical attention promptly at the first sign of infection. This is necessary for two reasons: to commence treatment to cure the infection, and to make certain that the person from whom the infection was received is identified and treated. Remember—STDs do not just happen. They can usually be cured, but prevention is a lot less trouble and pain.

Acquired Immune Deficiency Syndrome (AIDS)

AIDS is a deadly disease that breaks down the body's resistance. The cause of the disease is the human immunodeficiency virus (HIV), which is transmitted through exposure to the blood, blood products, or semen of a previously exposed person. Means of exposure include sex, blood transfusions, and hypodermic needles. Not everyone who is exposed to the virus develops immunological deficiency or clinical illness, but the virus remains in the body for life. An individual who is exposed but who does not develop any illness can still transmit the virus to others. Once an individual develops immunological deficiencies, that person is prone to opportunistic illness, such as pneumonia and cancer. Victims die from these secondary illnesses rather than from the virus itself.

Exposure to the virus can be prevented. The Coast Guard recognizes the severity of this illness and has initiated a testing program to detect exposure to the HIV virus. Applicants for regular or reserve service, officer or enlisted, or for the Coast Guard Academy who test HIV positive will be rejected from Coast Guard service. Members of the Coast Guard and Coast Guard Reserve who test positive will be considered not fit for duty and will be evaluated by a medical review board. Members who test positive but do not demonstrate immunological deficiencies or clinical illnesses may be retained on active duty but will not be assigned to certain duty, including sea duty, isolated duty, aircrew duty, or small-boat crew duty. The Coast Guard constantly reviews its policy on AIDS as more is learned about the disease.

HUMAN RELATIONS

In the Coast Guard you will interact with people from various socio-economic, racial, ethnic, and religious backgrounds. Your day-to-day interactions are crucial to maintaining a healthy and professional environment. The first requirement in creating this environment is to have respect, tolerance, and consideration for others. The diversity that each person brings to the Coast Guard should be valued and enhances the organization. Every member plays an integral part in the accomplishment of Coast Guard missions.

Civil Rights

Civil rights are rights each person has just by virtue of being a citizen or a member of a civil society. These rights include fair treatment and equal opportunity regardless of race, color, religion, national origin, or sex. Treating members unfairly based on these factors can lead to illegal forms of discrimination that cannot be tolerated in the Coast Guard.

Discrimination

Illegal forms of discrimination include any actions, omissions, or use of language that deprives an individual or group of their rights because of race, color, religion, national origin, or sex. It does not matter whether the discriminatory action taken is intentional or not; discrimination is unproductive behavior that negatively affects the working and living environment. All members must have the opportunity to work to their full potential without feeling that they are being treated differently because of these nonmerit factors.

Sexual Harassment

Sexual harassment is a form of sex discrimination that is characterized by sexually oriented communication, comments, gestures, or physical contact that is unsolicited and unwelcomed. It includes offers or threats to influence or alter conditions of service in order to secure sexual favors.

Examples of sexual harassment are repeated requests for dates or promises of good duty assignments by a superior if the subordinate will go along with the sexual requests. Both men and women can harass or be victims of harassment. Acts of sexual harassment are inappropriate, diminish self-esteem, and undermine the professionalism of the Coast Guard. Put simply, sexual harassment is illegal behavior and will not be tolerated.

Positive human relations are the key to the growth and development of each individual in the Coast Guard. You have the responsibility to ensure that your conduct supports Coast Guard policy and fosters a positive human relations environment. You must ensure that you treat your shipmates and coworkers with the same respect and dignity that you expect from them.

Fraternization

Fraternization is defined as an inappropriate relationship between junior and senior personnel. This relationship could be between an officer and an enlisted, senior and junior enlisted, or with civilian employees. Even though you may feel your actions do not constitute fraternizing, others may. It is this perception by others that weighs heavily in determining whether an inappropriate relationship exists. It is important that you ensure that neither you nor your shipmates engage in any practices that might be considered fraternizing. Not only can it lead to disciplinary action and loss of personal credibility but it is also a significant detriment to military good order and discipline.

Relations with the Public

As a member of the U.S. Coast Guard, you are evaluated by your supervisors on how well you "represent the Coast Guard." You are an ambassador of the service, both on the job and off duty. Most often you will be in direct, personal contact with the public, but you may also have the opportunity to tell the Coast Guard story through the media. In either case, your unit commander establishes policy on how to deal with the public. A good guideline when talking to both the press and the public is never to say anything you would not like to read on the front page of the next day's newspaper. The same goes with your social media activities.

Foreigners

As a U.S. citizen, you regard people from other countries as foreigners. Bear in mind that when you are a visitor in any country in the world, it is you who are the foreigner. The customs of other people may seem strange to you, but it is their way of doing things, and they like it. Do not make fun of anyone or anything in another country; at best it is impolite, and at worst such thoughtlessness can result in serious trouble.

V

NAUTICAL AND
MILITARY SKILLS

Figure 17-1.
USCGC barque *Eagle* (WIX-327), designated "America's Tall Ship,"
is a three-masted, square-rigged sailing vessel.

CHAPTER 17

Seamanship

S eamanship is the oldest of all technical seagoing skills. Long before navigation, gunnery, and steam power were developed, sailors learned the elements of seamanship. Knotting, splicing, rigging, boat handling, anchoring, and mooring began thousands of years ago. Along with other aspects of going to sea, today's members of the Coast Guard also must learn to be good seamen. They must begin with the basic subjects in this chapter.

MARLINSPIKE SEAMANSHIP

Marlinspike seamanship is one of the oldest and most basic of all seagoing skills. It includes knot tying and the handling of ropes and lines.

Ropemaking

Ropemaking is essentially a series of three twisting operations. Fiber is first twisted from left to right to spin the yarn. The yarn is then twisted from right to left to form the strand. Strands are then twisted from left to right to lay the rope. This is the standard procedure, and the result is known as "right-laid" rope. When the process is reversed, a "left-laid" rope is produced (see figure 17-2). Opposite twists give rope stability. Rope must be kept twisted in the same way in which it was made if the rope is to stay in good condition.

Size of fiber line. The length of fiber line is given in fathoms. One fathom equals six feet. The size is specified by the circumference in inches. Thus a one-hundred-fathom coil of eight-inch nylon would be a piece of line six hundred feet long and two and a half inches in diameter.

The smaller sizes of cordage are known as *small stuff*—¾-inch, 1-inch, 1⅛-inch, 1¼-inch, and 1½-inch line—are frequently called 6-, 9-, 12-,

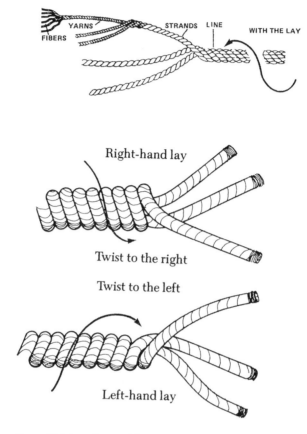

Figure 17-2. Lay of natural line.

15-, and 21-thread stuff. Sizes larger than 21-thread are termed *line*, and those five inches or more in circumference are called *hawsers*. Other terms for small stuff, depending on specific usage, are *ratline stuff*, *seizing stuff*, and *marline*.

Rope and Line

After rope has been acquired but before it has been made up for a specific use, it is called *line*. When placed into use, each *line* takes its name from the purpose it serves. Some examples are *mooring lines*, *towlines*, *manropes*, *heaving lines*, *seizing stuff*, and *sea painters*.

Synthetic Fiber Line

Synthetic fiber line is made from inorganic material (man-made). The four most common types of synthetic line are nylon, polypropylene, polyethylene, and polyester. The types most commonly used in the Coast Guard are nylon and polyethylene.

Nylon, which has replaced manila for the most part, is almost three times as strong as manila of the same size. Nylon comes in twisted, braided, and plaited construction. It can be used for almost any purpose except for lashings, high lines, and other purposes where its stretch and slippery surface are dangerous.

Nylon differs from other fiber ropes in that it stretches (elongates) under load, yet recovers to its normal size when tension is removed (elasticity). Depending on the type of nylon line used, it can be stretched up to 40 percent of its original length before it reaches a critical point.

Elongation may at times be a disadvantage, but you can halve it by doubling the lines. Doubling the line will reduce elongation, but it can cause an excessive strain on the deck fittings. Nylon rope can stand repeated stretching with no serious effect, but as it becomes older, the critical point is reduced.

Polyethylene is only half as strong as nylon, size for size. It is lighter than nylon, will float in water, and is usually brightly colored. It is used mainly for heaving lines and trail lines on life-ring buoys.

Dacron has characteristics similar to nylon but it will not stretch quite as much. It is becoming very popular since it wears well and handles quite well.

Cotton line is used for such things as the taffrail log, lead lines, heaving lines, and signal halyards. These lines are usually braided instead of laid.

Wire Rope

Wire rope is made in much the same way as fiber rope. It is used extensively for heavy lifts because it is much stronger than natural or synthetic fiber line. A four-inch manila hawser has a breaking point of 15,000 pounds whereas a one-inch wire, which is approximately the same size, has a breaking point of 46,000 pounds. *Always wear heavy gloves when handling wire rope.*

Line Handling

For convenience, the parts of a line are named in accordance with the way the line is laying or hanging. In taking turns around a bitt, take enough so you can hold the line. It is much easier to throw one turn off than to try an extra turn when you find that you cannot hold the line. Always stay out of the bight of a line.

Learning to handle line well is important since rescue operations, logistics, small-boat, and aids-to-navigation work all involve the use of line or rope in some critical part of the operation.

Opening New Coil of Line

When opening a new coil of fiber line, look inside the center tunnel of the coil and locate the line end. Then position the coil so that the inside line end is at the bottom of the center tunnel. Start uncoiling the line by drawing the *inside end up through the top of the tunnel* (see figure 17-3).

Once a line has been removed from the manufacturer's coil, it may be made up for storage or for ready use either by winding on a reel or in one of the following three ways:

Coiling down. This simply means laying it down in circles, roughly one on top of the other. Remember, right-laid line is *always* coiled down right-handed, or clockwise. When a line has been coiled down, one end is ready to run off. This is the end that went down *last* and that is now on *top*. If you try to walk away with the bottom end, a foul-up will result. If for some reason the bottom end must go out first, turn your entire coil upside down to free it for running.

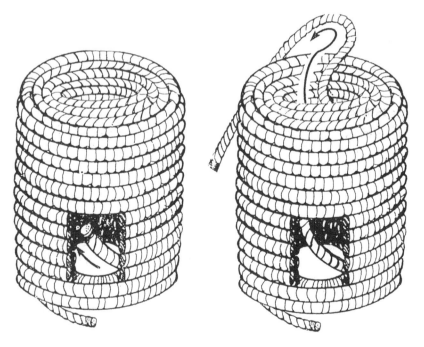

Figure 17-3. Uncoiling new line.

Faking down. This means laying it down in the same manner, except that it is laid out in long, flat bights, one forward of the other, instead of in round coils. With a long line, this saves the space a large coil might occupy, and faking down a heavy line is much easier than coiling it down.

Flemishing down. This means to coil it down first and then wind it tight from the bottom end so that it forms a close mat. Slack ends of boat painters, boat falls, or any other short lines about the decks that are not in continuous use should be flemished down for neatness.

Securing ends. Never leave the end of a line dangling loose without a *whipping.* It will begin to unlay of its own accord. Use tape to whip nylon line ends. A temporary plain whipping can be put on with anything, even rope yarn. Lay the end of the whipping along the line and bind it down with a couple of turns. Then lay the other end on the opposite way, bind it down with a couple of turns from the bight of the whipping, and pull the end tight (see figure 17-5).

Care of Line

Lines can be weakened to the breaking point by neglect or mistreatment. Here are a few pointers on keeping them in good condition:

- Always make up line neatly and in the correct direction, otherwise it will kink. This also prevents fouling.
- Work kinks and turns out of a line. Avoid removing kinks by putting a strain on the line; the kink may disappear, but the line may have been seriously weakened.
- Always make sure that the line is dry before you stow it. Wet line can develop mildew and rot.
- Avoid dragging line over sharp objects. Some of the fibers will surely be damaged.
- Do not allow dirt or sand to work into the strands; it cuts the fibers.
- Do not use frozen line. Line that freezes after becoming wet is easily broken and cannot be trusted. Thaw it out and dry it thoroughly before using it again.

Handling Nylon Line

When new nylon hawsers are used and strained, sharp, cracking noises will be heard. This is normal and does not mean the nylon will part unless the line is stretched too much (more than one-third of its length). When

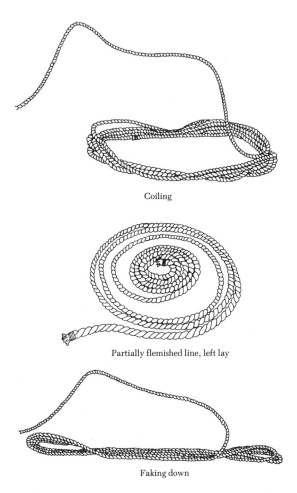

Coiling

Partially flemished line, left lay

Faking down

Figure 17-4. Three essentials in line handling: coiling, flemishing, and faking.

wet nylon is strained, it gives off water vapor that looks like steam. This is a normal, not dangerous, effect. If nylon line is parted by too much strain, it will first have stretched nearly one-half its length. When this stretched line parts, the ends whip back with great force and can easily injure any bystander. Stand clear!

Splicing

Splicing is a method of permanently joining the ends of two lines or of bending a line back on itself so as to form a permanent loop or eye. If properly done, splicing does not weaken the line, and a splice between two

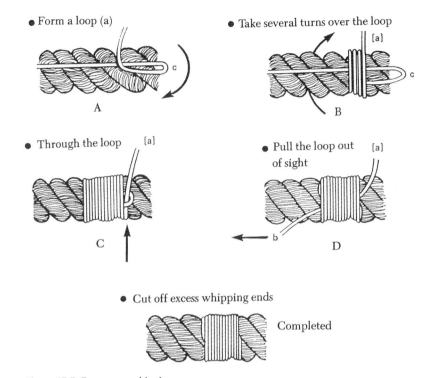

- Form a loop (a)

A

- Take several turns over the loop

[a]

B

- Through the loop   [a]

C

- Pull the loop out   [a]
  of sight

D

- Cut off excess whipping ends

Completed

Figure 17-5. Temporary whipping.

lines runs over a sheave or through a pulley much easier than a knot does. There are various forms of splices: long splice, short splice, and eye splice.

MOORING

Mooring and unmooring a ship to a pier, buoy, or another ship are the most basic jobs of the deck department. These involve skillful use of mooring lines (called line handling), capstans, and such fittings as cleats, bitts, bollards, and chocks. Quick, efficient line handling, when coming alongside or getting under way, is one of the marks of a smart ship and a well-trained deck force.

Mooring Lines

Ships are moored to piers by a system of mooring lines that vary according to the size and character of the ship. In all of the systems, the lines are classified in accordance with their employment, such as *breast lines* or *spring lines*.

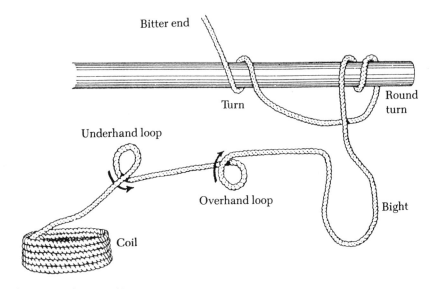

Figure 17-6. Elements of knots, bends, and hitches.

A breast line leads nearly perpendicular to the keel of the ship and controls the distance of that part of the ship from the pier. A spring line leads obliquely but nearly parallel to the keel and controls the fore-and-aft position of the ship with respect to her berth. "Springing" is a term applied to the use of spring lines to move the ship in toward the pier by surging forward or aft against a line that leads obliquely to the pier.

Deck Fittings

A *cleat* consists of a pair of projecting horns used for belaying a line. *Bitts* are cylindrical shapes of cast iron or steel, arranged in pairs on deck, forward and aft of each chock. A *chock* is a fitting through which mooring lines are led; the lines run from bitts on deck through chocks to bollards on the dock. Chocks are of three types: open, closed, and roller. A *bollard* looks somewhat like half a bitt but larger; it is on the pier, and the bight of a line is placed over it.

Mooring may often involve putting out fenders, handling camels, and placing rat guards. *Fenders* are shock absorbers of various types placed between ships or between a ship and a pier; they are dropped over the side and tended from the deck. *Camels* are floats used to keep a ship away from a pier or wharf so that overhanging structures will not strike objects on the

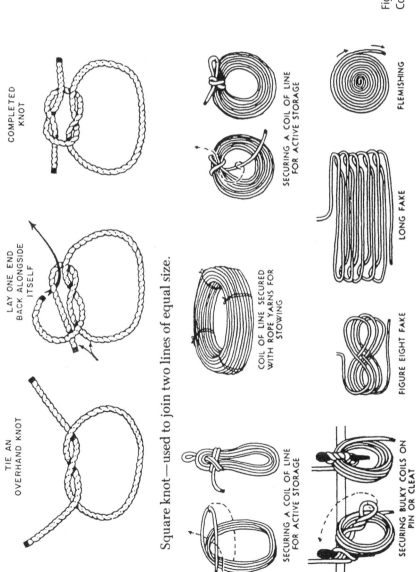

COMPLETED
KNOT

LAY ONE END
BACK ALONGSIDE
ITSELF

TIE AN
OVERHAND KNOT

Square knot—used to join two lines of equal size.

SECURING A COIL OF LINE
FOR ACTIVE STORAGE

COIL OF LINE SECURED
WITH ROPE YARNS FOR
STOWING

FLEMISHING

LONG FAKE

FIGURE EIGHT FAKE

SECURING A COIL OF LINE
FOR ACTIVE STORAGE

SECURING BULKY COILS ON
PIN OR CLEAT

Figure 17-7.
Coils, fakes, and flemishes.

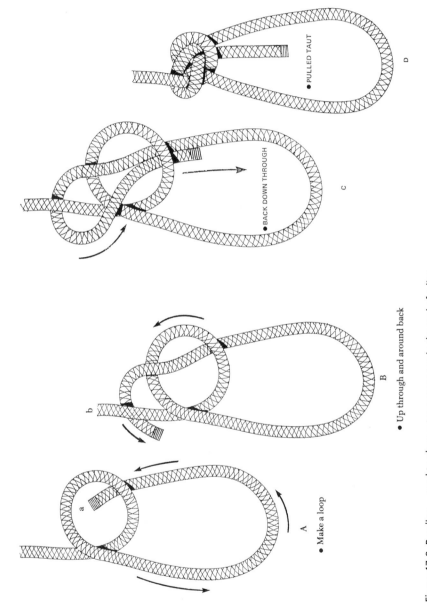

Figure 17-8. Bowline—used to place a temporary eye in the end of a line.

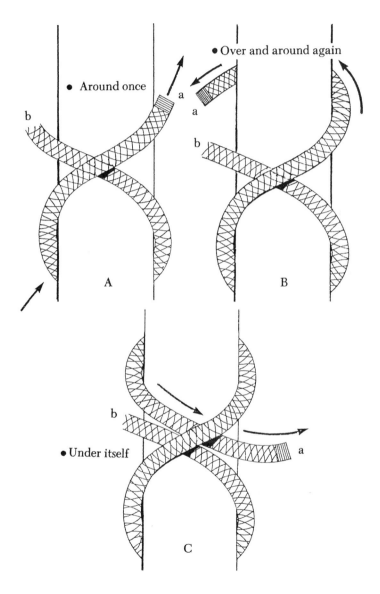

Figure 17-9. Clove hitch—used to make fast (tie) a line to a piling or railing.

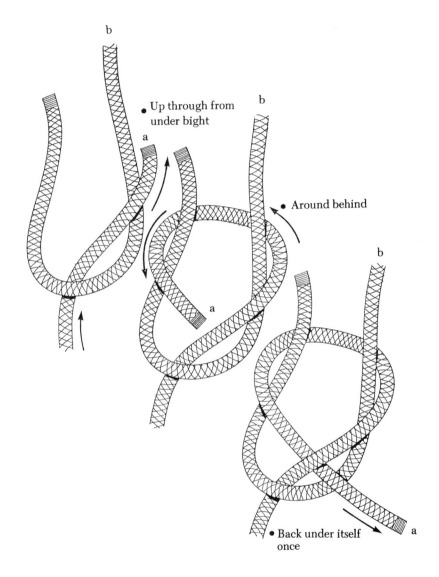

Figure 17-10. Single becket/sheet bend—used to join two lines of unequal circumference together.

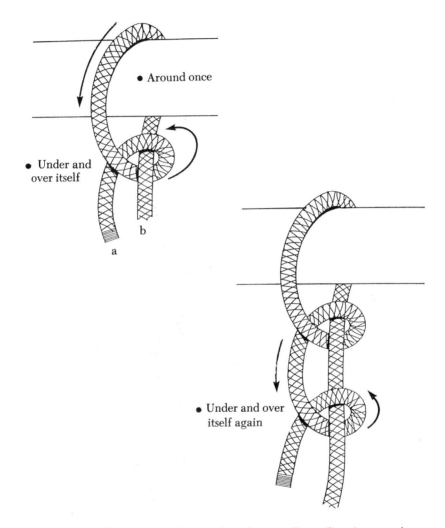

Figure 17-11. Two half hitches—used to make fast a line to a piling, railing, ring, or anchor.

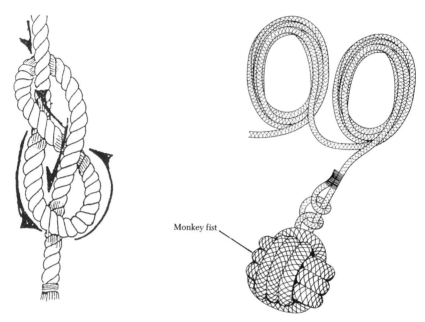

Figure 17-12. Figure eight—used as a stop knot for block and tackle.

Figure 17-13. Heaving line.

pier. *Rat guards* are circular metal disks lashed onto mooring lines to keep rats from coming on board.

In mooring, the messenger (a light line) is first sent over to the pier by heaving line, and then is hauled in with the attached mooring line by a person on the pier.

A *heaving line* is a light line with a monkey fist, plastic ball, or kapok bag on one end. After the ship is secured, the mooring lines are normally doubled up; a second line is passed to the pier or other ship, giving two parts of line each taking an equal strain, instead of only one part. The size of mooring line used depends on the type of line and type of ship.

Commands to Line Handlers

Commands to line handlers are listed here, with the exact meaning following each command.

"Stand by your lines."—Man the lines, be ready to cast off or let go.

"Let go," or "Let go all lines."—Slack off smartly to permit those tending lines on the pier or another ship to cast off.

"Send the lines over."—Pass the lines to the pier, place the eye over the appropriate bollard, but take no strain.

"Take to the capstan."—Lead the end of the designated line to the capstan, take the slack out of the line, but take no strain.

"Heave around on ————."—Apply tension on the line with the capstan.

"Avast heaving."—Stop the capstan.

"Hold what you've got."—Hold the line as it is.

"Hold."—Do not allow any more line to go out. Caution: this risks parting the line.

"Check."—Hold heavy tension on line, but render it (let it slip) as necessary to prevent parting the line.

"Surge."—Hold moderate tension on the line, but render it enough to permit movement of the ship (used when moving along the pier to adjust position).

"Double up."—Pass an additional bight on all mooring lines from the ship to the shore so that there are three parts of each mooring line fastened to the pier instead of only one part of each line.

"Single up."—Take in all bights and extra lines so that there remains only a single part of each of the normal mooring lines.

"Take in all lines."—Used when secured with your own lines. Have the ends of all lines cast off from the pier and brought on board.

"Cast off all lines."—Used when secured with another ship's lines in a nest. Cast off the ends of the lines and allow the other ship to retrieve her lines.

"Shift."—Used when moving a line along a pier. Followed by a designation of which line and where it is to go: "Shift number three from the bollard to the cleat."

If auxiliary deck machinery is to be used to haul in on a line, the command is given, "Take one (number one) to the winch (capstan)." This may be followed by, "Heave around on one (number one)" and then, "Avast heaving on one (number one)."

Figure 18-1.
Navigation is a skill that can be learned only through study and practice.

CHAPTER 18

Navigation

T here are three kinds of navigation: marine navigation, air navigation, and space navigation. Only marine navigation, which includes four methods of determining position (piloting, dead reckoning, electronic navigation, and celestial navigation) is discussed here. Even though you are not a navigator, you should understand the elements of navigation.

All navigational methods depend on exact measurements of distance, speed, direction, and time. Marine navigation sometimes requires the measurements of the depth of water, called *soundings*. The result in any method is a position or location, usually called a "fix."

The location anywhere on earth is determined by latitude—the distance north or south of the equator—and longitude—the distance east or west of the prime meridian, which runs from the North Pole to the South Pole through Greenwich, England. Latitude is measured in degrees north or south of the equator, with 0 degrees at the equator and 90 degrees at each pole. Longitude is measured in degrees from Greenwich, 180 degrees east and 180 degrees west. The place where 180 degrees east and 180 degrees west meet, halfway around the world from Greenwich, is called the International Date Line. The location of a ship at sea is established on a chart.

NAVIGATION AIDS

Charts

Charts, which show ocean areas and shorelines, and *maps*, which show landmasses, are marked off in parallels of latitude (degrees north or south) and meridians of longitude (degrees east or west). Each degree (°) is divided into sixty minutes (') or nautical miles. A nautical mile measured along the equator or a meridian is 6,076.11549 feet or roughly 2,000 yards. Any position at sea or place ashore is stated in degrees and minutes north or south and east or west; Cape May, New Jersey is 38.92°30'N and 74°89'W; the island of Funafuti in the South Pacific is 8°30'S and 178°30'E.

Distance

Distance at sea is measured in nautical miles. A nautical mile is one minute, or one-sixtieth of a degree. Speed is measured in knots, a seaman's term meaning nautical miles per hour. A ship makes 27 knots but never 27 knots per hour. In electronic navigation, distance measured by radar is called *range*.

Direction

Direction is determined by a compass, either magnetic or gyrocompass, which is described later. The four cardinal directions are north, east, south, and west. All directions are measured from north on a system of 360 degrees, in which east is 090 degrees, south is 180 degrees, west is 270 degrees, and north is either 360 or 000 degrees.

Time

There are two kinds of time predominantly used at sea: local apparent time as determined by the passage of the sun across the sky and Greenwich Mean Time (GMT), which is mean time (time based on the sun) at the prime meridian in Greenwich, England, from which all meridians of longitude east and west are measured. All standard time is also measured from that meridian (see "Zone Time," chapter 21). GMT is used for observations in celestial navigation and is shown by chronometers (highly accurate clocks).

The world is divided into twenty-four standard time zones, each covering fifteen degrees of longitude, or nine hundred miles at the equator. Each standard time zone bears a number, a plus (+) or minus (–) sign, and a letter. The number refers to the difference in time between that zone and the Greenwich zone. The sign tells whether the time is earlier (+) or later (–) than the Greenwich zone time; the sign shows how to find Greenwich Time from the standard time in any zone.

If a ship is in zone + 4, and the clock showed the time to be 1300 on board the ship, it would be 1300 + 4 or 1700 in the Greenwich zone. In radio traffic, when the time of origin of a dispatch is expressed in GMT, it is indicated by ZULU after the date-time group.

In recent years a new basis of timekeeping has been developed, required by the exacting needs of modern science, electronics, and navigation. Called Universal Coordinated Time (UTC), it is based on the frequency of vibration of a radioactive element, usually cesium. Because UTC is based upon an unchanging atomic time standard, and because GMT can be affected by small variations in the motion of the earth, there can be as much as a 0.9-second difference between the two. However, the two systems are kept in close agreement by the periodic insertion of a "leap second" by international time authorities so that the actual difference rarely approaches this maximum and can be disregarded for most normal navigational purposes. Most *radio time signals* broadcast by various maritime nations by which navigators check their chronometers use UTC.

Soundings

Soundings are made with an electronic device, usually an echo sounder. A ship is said to be "on soundings" when she is in water shallow enough that a lead line can be used to determine depth. Deep-sea soundings show when a ship crosses a submarine canyon, sea mount, or other bottom feature, and when a chart shows bottom contours, soundings may be used to establish the ship's position by an actual fix.

METHODS OF DETERMINING POSITION

Piloting

Piloting is the oldest method of navigation, used before man ventured out of sight of land and across the seas. It is a method of directing the movements of a ship by referring to landmarks, navigational aids such as lighthouses or lightships, or soundings. Piloting is generally used when entering or leaving port and in navigating along the coast, and may be used at sea when the bottom contour makes the establishment of a fix possible.

Navigational aids used in piloting include the *compass*, to determine the ship's heading; the *bearing circle*, to determine direction of objects on land, buoys, ships, and so forth; *charts*, which depict the outlines of the shore as well as the positions of landmarks and seamarks and the standard depths of water at many locations; *buoys*; *navigational lights*; the *echo sounder*, which determines the depth of water under the ship's keel by measuring the time it takes a sound signal to reach the bottom and return to the ship; and the *lead line*, which determines the depth of the water by actual measurement. *Coast Pilot* and *Sailing Directions* are books containing detailed information on coastal waters, harbor facilities, and so forth, for use in conjunction with the chart of the area. *Tide Tables* predict the times and heights of the tide, and *Tidal Current Tables* predict the times, direction of flow, and velocity of tidal currents in harbors, bays, and inlets.

In clear-weather piloting, the ship's position is usually determined by taking simultaneous gyrocompass bearings on three objects of fixed position. In addition, radar is used to obtain ranges and bearings to landmarks and seamarks.

A ship's position can be fixed by simultaneous visual bearings of two known objects as well as by a radar bearing and *range* on a single object, both of which are *plotted* or drawn on the chart (see figure 18-2). In figure 18-2, the lighthouse bears 035° from the ship, so the navigator draws a *line of position* (LOP) in the reverse direction, or 215° (180° + 35°) from the lighthouse in the direction of the ship. A factory bears 121°; an LOP 301° is similarly drawn *from* the factory. Each line is labeled with the bearing, 035° and 121°, below the line as shown.

The intersection of these two lines represents the actual position of the ship on the chart. A position that has been accurately established is called a *fix* and is so labeled, together with the time that it was established. The symbol used to represent a fix is a small circle placed over the intersection of the LOPs. A line drawn from the fix in the direction in which the ship is steaming is called a *course line*. The direction, or *course*, is labeled above the line; speed in knots is labeled below the line.

The manner of obtaining a fix by radar bearing and range or distance is shown in figure 18-2. Radar gives a bearing of 112° on a prominent tower and a range of 3,080 yards. The navigator plots a line from the tower, uses dividers to measure 3,080 yards on the chart scale, then puts one leg of the dividers on the tower location on the chart and marks the bearing line with the other end. This establishes the fix by bearing and range. Good piloting permits almost continuous position fixing to a high order of accuracy but demands great care and judgment. These examples demonstrate general principles. It should be noted that two LOPs are known as an estimated position while three LOPs are necessary for a fix.

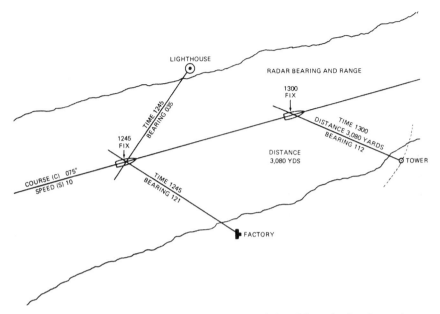

Figure 18-2. Fixes established by gyrocompass (1245 fix) and by radar bearing and range (1300 fix).

Dead Reckoning (DR)

This is a method in which the position of a ship is calculated by plotting the course steamed and distance covered by the ship from the last well-determined position. DR gets its name from "deduced (abbreviated "de'd") reckoning," used in the old days when navigators "deduced" a new position mathematically rather than determining it graphically, as is the modern practice. Today the navigator determines the DR position on a chart by plotting courses and distances run from the last known position.

When celestial navigation or piloting is used, DR provides a check and helps uncover errors. If other means of navigation fail, the navigator must rely solely upon DR.

Figure 18-3 shows the DR plot on the chart. The 1200 fix is plotted and labeled. A course line is drawn from the fix on the ship's course of 073°. Course is labeled above the line and speed of 15 knots below the line. At 15 knots, in one hour the ship will cover one-quarter degree, or 15 minutes on the chart. To determine the 1300 position, the navigator uses dividers to measure 15 minutes of latitude on the vertical latitude scale printed on each side of the chart. (One degree of latitude equals sixty nautical miles;

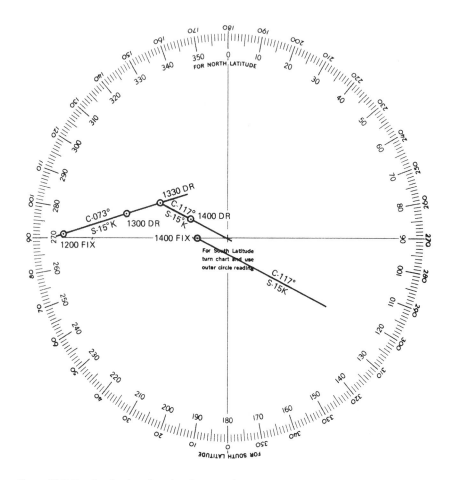

Figure 18-3. Dead reckoning plot, showing 1200 fix, 1300, 1330, and 1400 DR position. At 1400 a new fix is taken and ship's position on the chart is changed accordingly.

one minute of latitude equals one mile.) This distance is marked off from the fix along the course line, and the resulting spot is labeled "1330 DR," as shown.

The captain orders the officer of the deck to put the ship on a new course, 117° at 1330. Using his dividers, the navigator marks a spot seven and one-half miles from the 1300 DR position along the direction in which the ship is steaming, labels it "1330 DR," and draws in a new course line in the direction 117°. When properly maintained, the DR plot permits the ship's approximate position to be quickly determined.

Electronic Navigation

The electronic navigation division will be further developed during the coming years. As far as basic techniques are concerned, it is analogous in many respects to piloting, differing primarily in the methods by which the data are collected. *Radar, satellite,* and *LORAN* are examples of electronic navigation equipment. Increasing use is being made of satellite-based electronic navigation systems for mid-ocean navigation. The global positioning system (GPS) consists of satellites in synchronous orbits capable of furnishing continuous fix information worldwide.

Celestial Navigation

While GPS has become the predominant means of navigation in the open ocean, knowledge of celestial navigation is still necessary. Celestial navigation is the determination of position by the observation of celestial bodies (sun, moon, planets, and stars) and is still a widely used offshore navigation method. Observations, called *sights,* are made with a sextant and involve measuring the altitude above the horizon of navigational stars or other bodies. When taking a sight, the navigator usually calls it "shooting a star."

Many of the navigational aids used in piloting are used in celestial navigation, but additional equipment, such as chronometers and sextants, is required. A DR plot is always maintained. On some ships this is done automatically by a DR tracer.

NAVIGATIONAL INSTRUMENTS AND EQUIPMENT

Sextant. The sextant is a precision instrument that can measure angles in degrees, minutes, and seconds. Through a system of mirrors, the image of a star is brought down to the horizon; then very exact scales allow the navigator to read the angle between the actual star and the horizon. This angle is called the *altitude,* which is the basis of all celestial navigation.

In establishing a position by star sight, several observations are taken. Each one is *reduced* or worked out by means of the *Nautical Almanac* and the reduction tables; the results produce a single LOP, a line passing through the ship's position. The ship's location is represented by the point at which the various lines of position intersect on the chart. This is the ship's location at the time of the observation and is marked "2000 posit," "0530 posit," and so forth.

Stadimeter. The stadimeter measures the distance of an object of known height, such as a masthead light, between heights of 50 to 200 feet at distances of 200 to 10,000 yards. Like a sextant, the stadimeter operates on the principle of measuring an angle. The height of an object is set on a scale, and the reflecting image is brought into coincidence with the actual direct image and the distance is read off another scale.

Azimuth circle and bearing circle. An azimuth circle is a metal ring that fits over a compass bowl and measures bearings of objects on the surface of the earth and *azimuths* (or bearings) of celestial bodies.

Plotting equipment. Position plotting on a chart is usually done with a universal drafting machine, also called a parallel motion protractor, which is clamped to a chart table and allows both distance and bearing to be plotted at once. Sometimes a simple plastic protractor and straight edge are used.

Chronometer. A chronometer is a highly accurate clock mounted in a brass case, which is supported in gimbals in a wooden case, so as to eliminate much of the ship's motion. Chronometers are kept in a cabinet in the chartroom, usually on the centerline of the ship, where they are protected against shock and temperature changes. Chronometers are set to show GMT; they are wound every day at exactly the same time, and this fact is always reported to the captain with the noon reports. Once a chronometer is started, it is never allowed to stop, and it is not reset while on board ship. A record is kept of whether it is running fast or slow, but a good chronometer will never deviate more than a hundredth of a second from its average daily rate. Chronometers are checked against radio time signals, which are broadcast all over the world. To do this, a navigator uses a comparing watch, or hack watch, and calls the process "getting a time tick." The exact GMT time, as determined by radio, is *never* used to change the chronometer but only to show whether it is running fast or slow.

Exact time, as indicated by the chronometer, is necessary because all celestial navigation is based on tables using GMT. The time of a celestial observation, anywhere in the world must be converted to GMT before the navigator works out the position. Electronic chronometers are becoming the standard in use now.

COMPASSES, COURSES, AND BEARINGS

Compass

A *compass* is a device for determining direction on the surface of the earth. There are two main types of compasses: *magnetic* and *gyroscopic*. In each is a *compass card* from which the directions can be read.

Figure 18-4. Crew member checks heading on compass on board the Coast Guard barque *Eagle*.

Magnetic Compass

Magnetic compasses have the following general characteristics. The magnetized compass needles align themselves with the earth's magnetic field and are fastened to either a disk or a cylinder marked with the cardinal points of the compass: north, east, south, and west. North, on a magnetic compass, points to the magnetic North Pole, which is several hundred miles from the geographic North Pole.

The card and needles are supported on a pivot. No matter how the ship, aircraft, or boat swings, the card is free to rotate until it has realigned itself to magnetic north. The moving parts are contained in a bowl or housing provided with a window through which the compass card may be seen. Ship's compasses usually have a flat glass top for all-around visibility and for taking bearings.

The *lubber's line*, a mark in the window of the compass or on the compass bowl, indicates the fore-and-aft line of the ship or boat. The compass direction under the lubber's line tells the ship's heading. A *deviation card* is attached to the binnacle (the stand in which the compass is housed) or near it. The card gives the deviation for various headings in this form:

000° (360°)	14°W	120°	15°E	240°	4°E
015°	10°W	135°	16°E	255°	1°W
030°	5°W	150°	12°E	270°	7°W
045°	1°W	165°	12°E	285°	12°W
060°	2°E	180°	13°E	300°	15°W
075°	5°E	195°	14°E	315°	19°W
090°	7°E	210°	12°E	330°	19°W
105°	9°E	225°	9°E	345°	17°W

Compass error. The magnetic North Pole and the true, or geographic, North Pole are not at the same location, so the magnetic compass does not point directly north in most places. Usually there is a difference of several degrees, known as *compass error*, which is made up of variation and deviation. The *variation* of a compass is caused by changes in the magnetic field of the earth from place to place; hence, the needle follows the magnetic field and "varies" from the real north. The *deviation* of a compass is caused by the presence of metal objects around the compass, such as the iron and steel of the ship and her equipment. Even a knife in the pocket of the helmsman can cause the compass to deviate. Never bring a metal object near a compass.

Variation. The difference between geographic north and magnetic north is called *variation* (figure 18-5). Variation for any given locality, together with the amount of yearly increase or decrease, is shown on the *compass rose* of the chart for that particular locality. Variation remains the same for any heading of the ship at a given locality.

Deviation. An error of deviation is caused by the magnetic effect of any metal near the compass. It is different for different headings of the ship. Every so often, the navigator goes through an operation called "swinging ship." The ship steams in a complete circle from 0° to 360°, and the amount of her compass deviation is noted at every 15°. The results are compiled in a deviation table that is kept near the compass. There is a similar table near the magnetic compass in every aircraft. Compass error for any compass is the final effect of the variation of the locality and the deviation of the ship's heading. In some cases these must be added; in other cases one is subtracted from the other, as explained below.

True course. The heading of the ship or boat in degrees measured clockwise from true north is its true course.

Magnetic course. The heading of the ship or boat in degrees measured clockwise from magnetic north is its magnetic course.

Compass course. The reading of a particular magnetic compass, that is, the course the compass actually indicates, is called the compass course.

Correcting for Compass Error

Variation and deviation combined give what is known as *magnetic compass error*. The course on which the ship is to head is the *true course*, worked out from the chart, on which only true courses and bearings are given. Given the true course, it is necessary to find the *compass course* that you must steer to make good the true course. Do this by applying variation and deviation

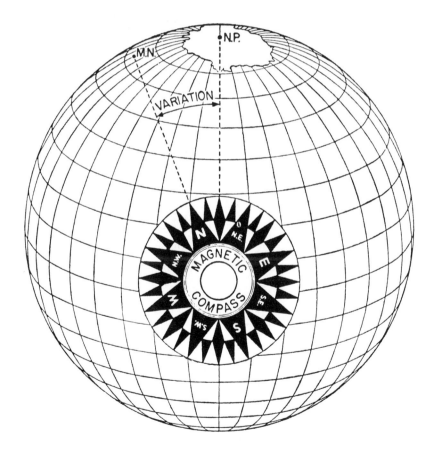

Figure 18-5. Diagram showing how variation affects the compass. The magnetic needle points to the magnetic pole (MN) instead of the geographic North Pole (NP).

to the true course. This is done by a simple rule, remembered as a nonsense statement: *Can Dead Men Vote Twice?*

In order to change from true to compass course, or vice versa, set up the columns as follows:

Can	Compass
Dead	Deviation
Men	Magnetic
Vote	Variation
Twice	True

Going up, or changing from true to compass, is called *uncorrecting*. Coming down, or changing from compass to true, is called *correcting*. Just remember this rule: When *correcting*, *add* easterly and *subtract* westerly error. When *uncorrecting*, *subtract* easterly and *add* westerly error. All compass errors are either easterly or westerly. (There are no northerly or southerly errors.) To correct a compass course of 270° to the true course, first correct for deviation, and then correct for variation. An example is given below. The deviation table described earlier shows that the deviation for 270° is 7° west. Assume that the chart shows the variation to be 12° east. Make a table as follows:

Compass	270°	Compass	270°
Deviation	7°W	Deviation	7°W
Magnetic		Magnetic	263°
Variation	12°E	Variation	12°E
True		True	275°
Total Error		Total Error	5°E

To find true course, the 7°W deviation is subtracted from the compass course of 270° (column two), which gives a magnetic course of 263°. The variation, 12°E, is then added to the magnetic course giving the true course of 275°. The total compass error is 5°E, which is the difference between 7°W and 12°E.

But how do you decide whether to add or subtract the deviation or variation? Remember: when *correcting*—going from compass course to true course—*add* easterly errors and *subtract* westerly errors. Note that *true* differs from *magnetic* by the amount of *variation*, and that *magnetic* differs from *compass* by the amount of *deviation*.

Uncorrecting

The process of finding the *compass course* from the *true course* is called *uncorrecting*. Suppose that the given true course is 180° and variation is 10°W.

Apply variation	Compass	
	Deviation	
	Magnetic	190°
	Variation	10°W
	True	180°
	Total Error	
to true course	Compass	176°
	Deviation	14°E
	Magnetic	190°
	Variation	10°W
	True	180°
	Total Error	4°E

This is *uncorrecting*, so reverse the rule and *add* westerly variation, giving a *magnetic course* of 190°.

Refer to the deviation table previously discussed. Take the deviation nearest the heading you are on. In this case the nearest deviation is 14°E, or that shown for 195°. Remember: when *uncorrecting*—going from *true to compass*—*add* westerly errors and *subtract* easterly errors.

Gyrocompass (Gyro)

The gyrocompass is essentially a heavy flywheel driven at high speed by an electric motor and mounted on gimbals so that it is free to move in all directions. It is usually located in a well-protected place below deck. *Repeaters*—compass cards electrically connected to the gyrocompass and placed on the bridge and in other parts of the ship—show the same readings as the master gyrocompass.

Gyro error. The gyrocompass is not affected by either variation or deviation. The motion of the earth will cause the rotor to move so that its axis lies in a north–south direction. For mechanical reasons and because of the vibrations of the ship, even the best gyrocompass will occasionally vary from true north. This gyro error is rarely more than a few degrees, and normally it is constant over a long period of time and on any heading of the ship.

Correcting for gyro error. Gyro error is determined by taking an azimuth, or bearing, on a celestial body where the exact bearing can be determined. This error is applied every time the compass is used, and the rule for the magnetic compass is followed: when correcting, add easterly errors and subtract westerly errors.

Bearings

Bearings are lines drawn, pointed, or sighted from one object to another. For accurate navigation, a system of true and relative bearings has been worked out so that all directions at sea are given in bearings that are measured in degrees (see figure 18-6).

Relative bearings are based on a circle drawn around the ship itself, with the bow as 000°, the starboard beam as 090°, the stern as 180°, and the port beam as 270°. Thus, if a ship is on a course true north (000°), another ship sighted dead ahead would bear 000° *true* and 000° *relative*. But if the ship were on a course true east and sighted a ship dead ahead, it would bear 090° *true* but would still be dead ahead or 000° *relative*. Relative bearings are used wherever there is no compass. Lookouts cannot have accurate compasses at hand, nor can they be expected to know the course of the ship and the true directions that lie about it. They need a way to point out where objects lie, and this method must be fast, accurate, and unmistakable. By using the Navy system of relative bearings measured in degrees from the bow of the ship, a crew member can soon learn to report objects in such a way that anyone can locate them immediately.

True bearings can be obtained directly from a gyro repeater, from a magnetic compass situated to make bearings on outside objects possible and with the appropriate corrections applied, or by calculation from a relative bearing.

Pelorus. A pelorus is a flat, nonmagnetic metal ring mounted on a vertical stand about five feet high (see figure 18-7). The inner edge of the ring is graduated in degrees from 0° at the ship's head clockwise through 360°. This ring encloses a gyro repeater. Upon the ring is mounted a pair of sighting vanes. These vanes are sighted through on an object much like the sights on a rifle.

Gyro bearings. Since the gyrocompass, and therefore the gyro repeater in the pelorus, are already closely lined up with the true geographic directions, taking a true bearing over the gyro card is the easiest and most common method. Merely line the vane sights of the bearing circle on the

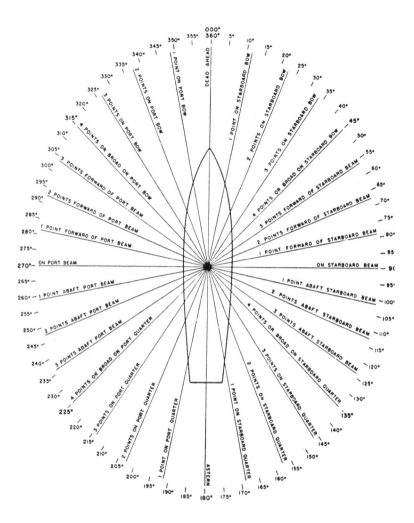

Figure 18-6. Relative bearings from a ship. The old system of giving bearings by the thirty "points" of the compass is indicated but no longer used.

object, steady the compass bowl in its gimbals (rings that compensate for the roll of the ship) until the leveler bubble shows that the vanes are level, and then read off the bearing in degrees on the compass card.

TIDES

Tides are very important in many shiphandling operations. In some harbors, deep-draft ships may be able to enter only at high tide. Large ships

Figure 18-7. A pelorus.

are usually launched, or dry-docked, at high tide. Ships going alongside piers in channels subject to strong tidal currents will usually wait for slack water, when the tide is neither ebbing nor flooding. Every Coast Guard person concerned with the handling of a vessel, from the largest cutter to a motor whaleboat, must understand what causes tides and the meaning of the names given to different tidal conditions.

The term "tide" describes the regular rise and fall of the water level along a seacoast or in an ocean port. Gravitational attraction of the moon is the primary cause of tides (see figure 18-8). It exerts a very considerable

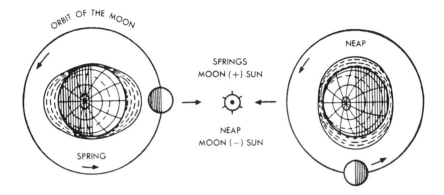

Figure 18-8. Relations of positions of the sun and moon to height of tides.

pull on the sea, piling the water up on that part of the earth nearest the moon. There is an almost equal *bulge* of water on the opposite side of the earth because the centrifugal force of the earth piles the water up where the pull of the moon is weakest.

Because the moon orbits the earth every twenty-four hours and fifty minutes, there are two low and two high tides at any place during that period. The low and high tides each are twelve hours and twenty-five minutes apart. The sun also affects the tide, but it is so much farther away than the moon that its pull is not nearly so great. A tide rising or moving from low to high water is said to be *flooding*; when the tide is falling, after high tide, it is said to be *ebbing*.

The difference in water depth between a high and the next low tide is considerable in many harbors; areas that are safe for a powerboat at high tide may be completely dry at low water.

CURRENTS

In most harbors and inlets, the tides are the chief causes of currents; however, if the port is on a large river, the river's flow may have a marked effect on the tidal currents. The flow of such a river will prolong the duration of the ebb current compared to that of the flood, and the velocity of the ebb current will be considerably greater than that of the flood.

Where the currents are chiefly caused by the rise and fall of the tide, their direction and speed are largely governed by the shape of the shorelines and the contour of the bottom. Where there is a long *reach*, or straight section of the waterway, the current will tend to flow most rapidly in the center and considerably slower in the shoal water near each shore. If your boat goes with the current, you should generally want to stay near the center of the waterway; if the boat goes against the current, you should stay as near shore as the prevailing water depth will allow.

In many wide inlets, near the time of *slack water,* when the current is at the end of the ebb or flood, the current may actually reverse itself in part of the inlet; while the ebb is still moving out in the main channel, a gentle flood current may start near one shore. This condition, where it exists, can be very helpful to a small-boat operator.

Where there is a bend in the channel, the current will flow strongest on the outside of the bend. This effect is very marked, particularly with a strong current. In some areas, a strong current can create areas of very rough water, called *tide rips.* These are usually shown on charts and should be avoided.

Every vessel, regardless of size, must make some allowance for the set and drift of the current, or this will affect the course to steer. One more thing to bear in mind about currents is that only on the seacoast does the time of the turn of the current agree with the turn of the tide; that is, only on the seacoast does the current *flood* until the time of high water and *ebb* until the time of low water. Owing to the effect of the shape of the land on the water flow, at many ports there may be a very considerable difference between the time of high or low water and the time that the current starts to ebb or flood.

A strong wind, blowing for a considerable period of time, will increase the speed of the current if they are both moving in the same direction. A wind will also prolong somewhat the duration of that current. Similarly, a strong wind may blow so much water out of a harbor that the water level will be considerably lower than its predicted depth for that time.

EFFECT OF WIND AND CURRENTS ON A LIFE RAFT

For the greatest part, the movement of a raft will be governed by prevailing winds and currents. These, of course, cannot be altered, but they sometimes can be intelligently used if you know the direction in which you want to go.

Wind and current do not necessarily move in the same direction in a given area. One may be favorable; the other unfavorable. The lower the raft rides in the water and the lower its occupants remain, the greater will be the effect of current. This effect can be increased by the use of a sea anchor or drag (or drogue) if the current is setting toward land. On the other hand, if the wind should be favorable, the raft should be lightened as much as practicable. Survivors should sit erect to offer wind resistance. Any sort of makeshift sail would be of help. To use wind or current advantageously, two things are required: Knowledge of the direction you wish to go, and knowledge of the direction of wind and current.

NAVIGATION AIDS IN THE HEAVENS

The Sun

The sun rises in the east and sets in the west. If you are north of latitude 23°27′N, the sun will invariably pass to the south of you on its daily trip across the sky. Latitude 23°27′N is an imaginary line passing approximately through Marcus Island, Taiwan, the tip of Lower California, along the northern shore of Cuba, and through the northern part of the Arabian Sea (figure 18-9). The sun follows the upper path at about 21 June each year before starting slowly south.

If you are south of latitude 23°27′S, the sun will always pass to the north of you on its daily trip from east to west. Latitude 23°2′S passes roughly through Noumea, Rio de Janeiro, and the southern part of Madagascar. These imaginary lines represent the Tropic of Cancer (north) and the Tropic of Capricorn (south).

Planets

Planets such as Mars, Jupiter, and Venus resemble stars—except that they do not twinkle. Planets are known as wanderers, since they "move about" among the stars, and are not much help in determining direction. If possible, learn something about the planets and how and where they usually appear so that when a planet appears in some constellation, you will not think that you have mistaken the order of the stars and therefore of the constellation.

Stars

Like the sun, the stars move across the sky from east to west. Their position relative to one another remains fixed. Once you learn the relationship of stars and groups of stars to one another, you can find them more easily.

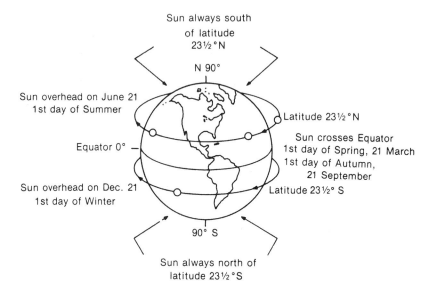

Figure 18-9. Diagram showing the sun's apparent path around the earth.

Upon locating one or more stars or constellations, you can use them as markers that tell you where to look for others.

You will not find the same stars in the same part of the sky every night. This is because the sun moves westward around the earth at a slightly greater speed than do the stars. Consequently, stars that may be just appearing over the horizon at midnight in one month may be high in the sky at midnight another month. Or they may not appear at all. The latter happens when they travel across the sky within a few hours of the sun, which, of course would be during daylight—for the stars travel across the sky in daylight just as they do at night.

Groups of stars are known as constellations. Some of the most prominent and easily identifiable constellations and single stars are described in the following pages. Train yourself to recognize them.

Orion

The constellation of Orion (figure 18-10) always rises true east, sets true west, and follows a path directly over the equator. Notice how the three stars forming the belt are almost parallel to the equator. The two brightest stars of the constellation, Betelgeuse and Rigel, lie approximately the same distance north and south of the line, respectively. Orion can be seen from any position on earth.

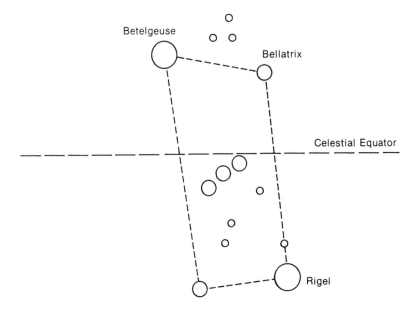

Figure 18-10. The constellation Orion is the biggest and brightest in the Northern Hemisphere; it includes several navigational stars.

On a clear night around Christmas, Orion is visible for almost the entire night, rising around 1800 and setting about 0600. Each month thereafter, it is visible for shorter periods of time, appearing at dusk in a progressively higher position until March when it first appears near its zenith (highest position in relation to you). From then until June it is in a lower position to the west of you. In June Orion is not visible at all, as it makes its passage during daylight hours. Then in July it once more makes its appearance in the east, around 0400, and each month thereafter appears about two hours earlier.

The Big Dipper (Ursa Major)
The Dipper is a distinctive constellation containing seven stars. If you are in northern latitudes, it will be the most important and useful constellation for navigational purposes.

Polaris (the North Star)
As the Big Dipper is the most important constellation in northern latitudes, Polaris is the most important star. Since it is almost directly over the North Pole, for practical purposes it can be considered to be due north of you wherever you may be.

Polaris is not very bright and is sometimes hard to see when the sky is hazy. Without the Dipper it would be very hard to find. But the two "pointers" of the Dipper, whatever its position, "point" to Polaris. *The number of degrees Polaris is above your horizon will always be nearly equal to your latitude.* This means that if Polaris is 30° above the horizon, you are in latitude 30°N; if it is 50° above the horizon, you are in 50°N. It is directly overhead at the North Pole.

To estimate the angular distance of Polaris above the horizon, you must first estimate the point in the sky that is exactly overhead (the zenith). From the horizon to that point is 90°. Halfway from the horizon to the zenith is 45°; one-sixth (or one-third of halfway) is 15°, and so on.

Southern Cross

In the Southern Hemisphere, Polaris is not visible. There, the Southern Cross is the most distinctive constellation. As you go south, the Southern Cross appears shortly before Polaris drops from sight astern. An imaginary line through the long axis of the Southern Cross points toward the South Pole. The Southern Cross should not be confused with a larger cross nearby, known as the False Cross. The latter, though the stars are more widely spaced, is less bright. It has a star in the center, making five stars in all; the Southern Cross has only four, two of which are among the brightest stars in the sky.

There is no star at the South Pole to correspond to Polaris at the North Pole. In fact, the point where such a star would be, if one existed, lies in a region devoid of stars. This point is so dark in comparison with the rest of the sky that it is known as the Coal Sack.

To find the South Pole, extend an imaginary line southward along the long axis of the Cross. Join the two bright stars to the east of the Cross with an imaginary line. Bisect this line with one at right angles. Where this east line intersects the one through the Cross is (approximately) the point above the South Pole. This point can be used to estimate latitude in the same manner as Polaris is used in northern latitudes—by its height above the horizon.

ORIENTATION

It is possible to determine north, east, south, and west by day or night. If you can determine any of the cardinal directions, you can easily determine the others.

Well out to sea, the prevailing winds from about latitude 8°–10°N to about latitude 40°N are from the northeast blowing toward the southwest. They are called the northeast trades and will carry you in a southwesterly direction. North of 40°N and south of 40°S the winds are usually from the west and will tend to carry you eastward.

These are only general rules. There are many local and some seasonal exceptions. Learn as much as you can by observation and questions about the winds and sea currents in the areas in which you operate. Learn to pick out the stars that have been mentioned and many more. Learn where to look for them in the sky. This is how the first navigators found their way, and, without navigational instruments, it is still the best method.

NAVIGATIONAL PUBLICATIONS

The principles of all methods of navigation can be learned, but much detailed information necessary in working navigational problems must be obtained from complicated charts and tables. One of these is the *Nautical Almanac*, which enables the navigator to determine the exact location in the sky of the planets, sun, moon, and major stars for any second of GMT. *Tide Tables* show the predicted times and heights of high and low tides. This publication also shows the exact time of sunrise and sunset. *Tidal Current Tables* list the predicted times of slack water and the predicted times and velocities of maximum flood and ebb currents as well as the flood and ebb current directions.

Figure 19-1.
The wooden deck of USCGC *Blackthorn* (WLB-391) as shown under way
with buoys on board.

CHAPTER 19

Rules of the Road and Buoyage Systems

Just as an automobile driver must know traffic signals and the laws governing speed, parking, and passing on curves, so people who handle ships and boats must know the signals, lights, laws, buoyage systems, and other matters governing traffic on the high seas, their connecting waters, and in crowded harbors. The rules are not just for big ships; they apply to boats and ships of all sizes.

SEAGOING TRAFFIC RULES

The nautical traffic laws are embodied in a set of rules termed collectively the Rules of the Road. On the high seas and in the navigable inland waters of countries without special local rules of their own, ships are governed by a set of laws established by agreement among maritime nations known as the International Rules. In adjacent rivers and harbors of the United States, its territories, and possessions, however, ships are subject to local regulations known as Inland Rules.

Content of the Rules

The Rules of the Road are easy to think of in terms of automobile operation, but whereas cars and trucks are mostly limited to roadways, ships and boats travel open waterways and oceans. Roadways can have lanes marked with paint, have signs for directions and warnings, be controlled by signal lights, and be illuminated by street lights. Except in harbors with marked channels for deep-draft vessels, navigating a ship or boat on open water is a lot like driving a car around a large, mostly empty parking lot.

The nautical Rules of the Road, to be useful, must provide for both day and night situations and for when vessels can see each other and, because of poor visibility, when they cannot. The rules should let one vessel know something about the other, what it is doing, where it is going, its size, whether or not it can maneuver, and, on congested waters, its intentions. Information like this may not be important to a driver on a roadway, but in a large, nearly empty parking lot that information might be very useful.

The Rules of the Road apply to a vessel when it is "under way," meaning it is not at anchor, made fast to the shore, or aground, and they apply when it is at anchor or aground. They do not apply when it is moored, or made fast, to the shore.

There are rules that are mostly common sense, such as keeping a lookout, being cautious, going at a safe speed, using the equipment you have to minimize risk, avoiding dangerous situations before they happen, considering the need for other vessels to move in narrow waterways because of their draft and not blocking them or embarrassing them, keeping to the right in narrow channels, and obeying special navigation rules and regulations. There are rules that govern power-driven vessels, sailing vessels, and other vessels that have limited maneuverability when they are in sight of one another. Certain rules describe which vessels have the right-of-way. The vessel with the right-of-way is called the *stand-on vessel*, and the one that has to keep out of the way of the other is called the *give-way vessel*. Then there are rules for vessels in restricted visibility, rules for lights and shapes, rules for sound and light signals, and rules for distress signals.

The basic rule of safety for ships and boats is practically the same as that for automobiles: safety first, keep to the right, and proceed cautiously when in doubt. And while doing this, report immediately to the officer in charge of the ship or boat all lights, sounds, and signals of any sort.

Application and Scope of the Rules

It is important that persons who navigate vessels have a full and complete understanding of the Rules of the Road. But you should also realize the purposes of the rules and, being aware of their importance, take an intelligent and active interest in your duties as lookout, crew member, or passenger in a boat. You *must* realize that all lights and sounds at sea have meaning, and that no light or sound signal should ever be overlooked. This chapter gives an idea of the importance of the rules and some details as to their application to small boats. A section on buoys is included. Although not actually part of the Rules of the Road, the buoyage system (the traffic lights and signals of harbors and channels) also governs water traffic.

Navigation Lights

At the same time that the new mariner learns the meaning of the words "port" and "starboard," the colors red and green pop up. Red for port and green for starboard are two of the four colors used in navigation lights. Sometimes navigation lights are called *running lights*. The other two colors are white and yellow. White navigation lights on a vessel are used along with red and green navigation lights to let an observer know in which direction a vessel is going when only its lights can be seen. The yellow light is used in special circumstances such as towing situations.

Red port and green starboard lights are used to mark the sides of the vessel, and the white lights are usually located on the vessel's centerline. Navigation lights are not used by the navigator to see by, as are the headlights of a car, but rather to allow navigators and lookouts on other vessels to see the vessel. The red and green lights are called *side lights*. A white light at or near the stern is called a *stern light* and one, and sometimes two, white lights on the vessel's superstructure, or on masts, are called *masthead lights*.

These navigation lights are allowed to show only in certain sectors around the vessel. Using the 360-degree system, each of the two side lights are positioned to show from right ahead to 22.5 degrees abaft the beam on each side for a total of 112.5 (90° + 22.5°) degrees on the horizon on each side. The stern light is positioned to show from right aft to 67.5 degrees on each side of the vessel for a total of 135 (67.5° + 67.5°) degrees on the horizon. Masthead lights are positioned to show from right ahead to 22.5 degrees abaft the beam on both sides or 225 (22.5° + 90.0° + 90.0° + 22.5°) degrees on the horizon. Masthead lights combined with the stern light show at least one white light all around the vessel (masthead 225° + stern 135° = 360°). When two masthead lights are required, such as on vessels 50 or more meters in length, or when they are optimally installed on vessels of less than 50 meters, the aftermost masthead light is higher than the forward one.

Side lights let you know which side of another vessel you see so you know in which general direction the other vessel is heading. If you do not see the side lights, then you know you are looking at the stern of the other vessel. When you see both side lights, the other vessel is headed right at you. The use of two masthead lights on larger vessels lets you see changes of heading on the part of the other vessel. If the two white lights get closer together, the ship is changing its heading toward you. If the white lights separate, it is heading away.

A single white light seen on another vessel that is under way is usually the vessel's stern light. That means that you are astern of the other vessel. When a side light is visible, that means that you are nearly abeam of the other vessel or forward of it. The masthead lights and one or both side lights on another vessel should be visible at the same time, and the other vessel's stern light should never be visible to you at the same time as one of its side lights. A lot of other kinds of lights are shown by vessels at night. Some are used for towing, some for anchoring, and some for a special purpose or situation of the vessel displaying the lights.

Sound Signals

There are two reasons for using sound signals. One is to let a vessel, which is aware of the presence of another, announce the direction of maneuver or intended maneuver. On the inland waterways of the United States and under overtaking situations in international narrow channels and fairways, agreement with the other vessel's intentions is also signaled by sound. Warning signals are used when needed in connection with maneuvering signals. The other reason for sound signals is so that one vessel can let other vessels know of its presence when visibility is restricted. The sound signals are made by blasts of the ship's whistle. A short blast is about one second in duration; a prolonged blast is four to six seconds in duration.

Maneuvering and Warning Signals

On the high seas under the International Rules of the Road, one short blast means, "I'm altering my course to starboard" and two short blasts means, "I'm altering my course to port." On inland waters, one short blast means that the vessel intends to leave the other on its port side; two short blasts means an intent to leave the other vessel on the starboard side. The direction of the maneuver is the same for both U.S. inland and international waters, but on U.S. inland waters, the vessel being signaled must repeat the signal when in agreement or sound the danger signal of five or more short and rapid blasts. Three short blasts, both inland and international, means a vessel is operating astern propulsion. The danger signal for both rules is the same: five or more short and rapid blasts.

On international waters in narrow channels or fairways, overtaking signals that include intent signals and agreement signals are exchanged. Intent to overtake on the starboard side is signaled by two prolonged blasts followed by one short blast; on the port side, the signal is two prolonged and two short. The agreement signal is one prolonged, one short, one prolonged and one short (or "Charlie" in Morse code).

On U.S. inland waters, one prolonged blast is sounded when other vessels may not be observed, such as at a bend of a channel, or behind other obstructions, and when leaving docks.

Lights may be used in conjunction with whistle signals to show the sound signal. The light is flashed for the same duration as the whistle is sounded. The lights used are all-around (360°) lights; International Rules specify the color white of five miles' minimum visibility, and Inland Rules specify white or yellow of two miles' visibility.

On U.S. waters when agreement is reached by bridge-to-bridge communications, a vessel in a meeting, crossing, or overtaking situation is not obliged to sound whistle signals but may do so. If agreement is not reached, whistle signals shall be exchanged.

Sound Signals in Restricted Visibility

Sound signals, often called "fog signals," are sounded during the day or night whenever visibility is restricted. A power-driven vessel making way through the water shall sound, at intervals of not more than two minutes, one prolonged blast. If the vessel is under way but stopped and not making way through the water, it shall sound, not more than two minutes apart, two prolonged blasts with a two-second interval between the blasts. These signals are required on both international and U.S. inland waters.

There are additional special signals for vessels at anchor, vessels not under command, fishing vessels, pilot vessels, sailing vessels, towing and towed vessels, and vessels aground. There are some exceptions for vessels of less than twelve meters. Basically, the "fog signals" for both U.S. inland and international waters are the same.

Right-of-Way

A vessel having the right-of-way is the *stand-on* vessel. Instead of doing whatever it wishes, the stand-on vessel has a number of obligations. Likewise, the other vessel—the *give-way* vessel—has its own obligations.

Deciding which vessel is stand-on and which is give-way depends upon a number of factors. Manner of propulsion, or lack of it, is one factor;

whichever vessel is to the port of the other is another factor; and sailing vessels approaching one another are governed by the position of the wind. Moreover, narrow-channel situations, Vessel Traffic Regulation situations, and overtaking situations create their own requirements. But both the U.S. Inland Rules and International Rules are alike with few exceptions on the matter of determining stand-on and give-way vessels.

The rules involving two sailing vessels are somewhat complicated and may require a detailed explanation. Except where a power-driven vessel is confined to a narrow channel, is being controlled by a traffic service, or is being overtaken, a sailing vessel is the stand-on vessel in a power-driven–sailing vessel encounter. The power-driven vessel is thus the give-way vessel. The give-way vessel must keep out of the way of the stand-on vessel and, as far as possible, take early and substantial action to keep well clear.

In any overtaking situation, the overtaking vessel must keep out of the way of the overtaken vessel; therefore, the overtaken vessel is the stand-on vessel, and the overtaking is the give-way vessel. Overtaking means when one vessel is coming up on another vessel so that at night the vessel coming up would only see the other's stern light. That would mean it was coming from a direction more than 22.5 degrees abaft the overtaken vessel's beam.

In a head-on situation where two power-driven vessels are meeting on reciprocal or nearly reciprocal courses, each must alter course to starboard

Figure 19-2. Buoy tenders keep our waterways safe.

so that each shall pass on the port side of the other. There is no give-way or stand-on vessel in this situation.

Whenever two power-driven vessels are crossing so as to involve risk of collision, the vessel that has the other to starboard becomes the give-way vessel and, hence, shall keep out of the way. If the circumstances of the case allow, the give-way vessel must avoid crossing ahead of the other vessel. On the Great Lakes, U.S. Western Rivers, or other specified waters, a vessel crossing a river shall keep out of the way of a power-driven vessel ascending (going upstream) or descending (going downstream) the river.

Stand-on vessels must keep their course and speed. Otherwise, the give-way vessel's maneuver to keep clear could result in a collision. A stand-on vessel can take action to avoid a collision by its maneuver alone as soon as it becomes apparent that the give-way vessel is not taking appropriate avoiding action. If it is necessary for the stand-on vessel to take this action, it shall not alter course to port for a vessel on its port side. The stand-on vessel shall take such action to avoid a collision when it is so close that action by the give-way vessel alone would not avoid a collision.

Except in vessel-traffic controlled situations, narrow-channel situations, and overtaking situations, a power-driven vessel shall keep out of the way of a vessel not under command, a vessel restricted in its ability to maneuver, a vessel engaged in fishing, and a sailing vessel. Sailing vessels under way shall keep out of the way of a vessel not under command, a vessel restricted in its ability to maneuver, and a vessel engaged in fishing. A vessel engaged in fishing when under way shall so far as possible keep out of the way of a vessel not under command and a vessel restricted in its ability to maneuver. Seaplanes on the water shall generally keep out of the way of all vessels and avoid impeding their navigation. When a risk of collision exists between a seaplane and a vessel, the seaplane shall comply with the Rules of the Road.

Rule of Good Seamanship

Nothing in the rules shall exonerate any vessel or the owner, master, or crew thereof from the consequences of any neglect to comply with the rules or of the neglect of any precaution that may be required by the ordinary practice of seamen or by the special circumstances of the case.

General Prudential Rule

In construing and complying with the rules, due regard shall be paid to all dangers of navigation and collision and to any special circumstances, including the limitations of the vessels involved, which may make a departure from the rules necessary to avoid immediate danger.

Submarine Distress Signals

A yellow smoke bomb fired into the air indicates a submarine is coming to periscope depth from below periscope depth. A red smoke bomb fired into the air indicates the submarine is in trouble and will surface immediately if able to do so. Any colored smoke bombs fired into the air at short intervals indicate that a submarine is in trouble and requires assistance.

AIDS TO NAVIGATION IN U.S. WATERS

There are two systems of navigational aids used throughout the world. They are known as IALA A and IALA B. The United States uses IALA B as shown in appendix D and elsewhere in this chapter. The most basic difference between the systems is that the use of red and green is reversed.

Buoys Used in U.S. Waters

Buoys are plastic, foam, or metal floats of various types. They direct vessel traffic in the same way that road markers direct automobiles on highways. There are seven types of buoys.

Nun buoy. A conical-shaped buoy, used principally to mark the right-hand side of a channel, that is, the right-hand side when facing inland. When speaking of right hand or left hand with regard to buoys, it is understood that the ship or boat in question is approaching land from seaward.

Can buoy. A cylindrical buoy, used principally to mark the left-hand side of the same channel. The nun and can buoys are the only buoys whose shapes have special significance. When a ship is going toward land, the nun buoy should be to starboard, the can buoy to port.

Bell buoy. A float with a flat top surmounted by a framework in which a bell is fixed. The motion of the sea causes the bell to ring.

Gong buoy. Similar to a bell buoy but has four gongs, each of a different tone.

Whistle buoy. Usually a conical-shaped buoy bearing a whistle. The motion of the sea causes the whistle to sound.

Lighted buoy. A buoy that contains batteries. It is surmounted by a framework supporting a light. Shapes of lighted buoys are not uniform.

Combination buoy. A buoy in which light and sound signals are combined, such as a lighted bell buoy, lighted gong buoy, or lighted whistle buoy. Shapes of combination buoys follow no set pattern.

Purposes of Buoys

Buoys of many shapes and colors are used to mark the sides of channels, junctions of one channel with another, obstructions, anchorages (including quarantine anchorages), areas where fish nets and traps are permitted, and areas where dredging operations are being conducted.

All types of buoys are used for all the purposes listed, except that nun buoys are not found on the left-hand side of a channel and can buoys are not found on the right-hand side of a channel as viewed from seaward.

At night or in fog, the nun and can buoys are difficult to see; therefore, other types of buoys may replace them at important points.

Colored Buoys

Buoys are painted in various colors and combinations of colors according to location and purpose as follows:

- *Red*: Right-hand side of a channel (as seen from seaward).
- *Green*: Left-hand side of the same channel.
- *Red and green horizontal bands*: Obstructions in the channel, or junction where two channels meet. If the top band is red, the better channel is to the left of the buoy, which usually is a nun buoy. If the top band is black, the better channel is to the right of the buoy, which usually is a can buoy. Such marking always means that there is a channel on either side of the buoy. A junction where a secondary channel leads off from the main channel at a right angle usually is not marked by a junction buoy.
- *Red and white vertical stripes*: Middle of the channel. Buoy may be passed on either side.
- *White*: Information and regulatory marks.
- *Yellow*: Special marks.

Numbered Buoys

In addition to being colored, some buoys are numbered, lettered, or marked with combinations of numbers and letters. Red buoys bear even numbers, green buoys odd numbers, with numbers running from seaward toward the shore. If there are more buoys on one side of the channel than on the other, some numbers are omitted so as to keep these buoys numbered in approximate sequence. Numbers are not given to banded or striped buoys, but letters are assigned to some of them for identification purposes.

"Red, Right, Returning"

"Red, right, returning" is a widely used phrase that helps in remembering the significance of the color on channel markers. The *red* buoys should be on the *right* when a ship is *returning* from the sea.

Lighted Buoys

Lighted buoys are used in important locations, particularly where they must be located and identified at night. If the lights are off longer than they are on—that is, if the period of darkness is longer than the period of light—they are called *flashing lights*. If the lights are on more than they are off, they are called *occulting lights*.

Any regularly flashing or occulting lighted buoy on which the light goes on every two or three seconds marks the side of a channel. If it goes on and off regularly but at a much faster rate, it marks a point on the side of a channel where special caution is required, such as a turn. Colors used for such buoy lights are red or white on the right side of the channel, and green or white on the left side of the channel. Note that white may be used for either side, but a red light is used only for the right side and a green light only for the left side.

Lights that flash rapidly a number of times, pause, and then repeat the flashes are called *interrupted quick-flashing lights*. They are used on lighted obstruction and lighted junction buoys. A white light means there is no preferred channel. If the light is red, the preferred channel will be followed by keeping the buoy to starboard; if green, keep the buoy to port (as viewed from seaward).

Lights that produce groups of short-long flashes, each group appearing at regular intervals, are called *short-long flashing lights*. They are always white and always represent black-and-white vertically striped mid-channel buoys.

Fixed Aids

Fixed aids called *daymarks* are often positioned along U.S. inland waterways in place of buoys. Sometimes these aids are on pilings over water, sometimes they are on shore. Fixed aids are occasionally placed in line with other fixed aids to form a "range" to mark the center of a channel. Aids, buoys and fixed, use the same system of numbers, lights, and colors.

Temporary Markers

Temporary channel markers usually consist of floats carrying lights, pennants, or lights and pennants. Red pennants mark the right side of the channel and green pennants mark the left side. Red and green vertically striped pennants mark obstructions and channel junctions, and red and white vertically striped pennants mark the mid-channel (also called the fairway). If lighted, the right side of the channel will be marked by red lights; the left, by white lights; obstructions, by blue-over-red combination lights; and fairways, by green lights.

Mariners must keep on the alert and not rely solely on these sea markers. Buoys have a way of shifting about; lights burn out, or the blinking mechanism fails; whistles, bells, and gongs fail to sound loudly when the sea is calm. Good navigation practices require the use of "backup" systems.

Storm Warnings

In U.S. waters, information regarding the weather and the approach of storms is furnished by the National Weather Service by means of bulletins, radio broadcasts, and, in certain seaport towns, by storm warnings signaled by a system of flags, pennants, or lights (see appendix E). The night equivalent of these signals consists of lights in a vertical line.

Small Craft

When conditions dangerous to the operation of small craft are forecast for the area (winds up to 33 knots), a red pennant will be displayed by day, and a red light over a white light will be displayed at night.

Gale

Two red pennants or a white light over a red light indicate that winds ranging from 34 to 48 knots are forecast.

Whole Gale

A square red flag with a black center or two red lights indicate that winds from 48 to 63 knots are forecast.

Hurricane

Two red flags with black centers or three lights (red, white, red) indicate that winds of 64 knots and above are forecast.

Figure 20-1.
National security cutters carry the MK110 57 mm gun.

CHAPTER 20

Weapons

The Coast Guard employs various small arms and weapon systems to fulfill its many law enforcement and military missions. This chapter offers an introduction to weapons in general use by the Coast Guard.

The Coast Guard classifies small-arms weapons as any weapon that is .50 caliber or smaller. A caliber is a unit of measure of the diameter of a weapon's bore. The standard personal defense weapon in Coast Guard use is the .40-caliber S&W Sig Sauer P229R DAK handgun. Other Coast Guard small arms in use are the M16A2 rifle, the Remington M870 riot shotgun, the M240 machine gun, and the M2 .50-caliber machine gun. Weapons larger than a .50-caliber are considered to be a weapon system.

FOUR WEAPONS SAFETY RULES

Safety is the most important aspect of weapons handling and training. Safety rules and procedures must be followed at all times! The following are the four weapons safety rules that apply at all times to any weapon, whether on the range, a boarding, a ship, at a station, or even at home.

1. Treat every weapon as if it were loaded, regardless of perceived or actual condition. This rule is intended to prevent unintentional injury or damage due to careless handling of weapons or perceived low risk. Treat every weapon with the respect due a loaded weapon.

2. Keep your weapon pointed in a safe direction at all times. NEVER point a weapon at anything you do not intend to shoot. This rule is to reinforce the importance of muzzle control and muzzle awareness.

Figure 20-2. Training is vital for all personnel using weapons.

When at the range, the safe direction is down range; in the line of duty, the safe direction is any direction that does not cross the path of another person or vessel. ALWAYS handle a weapon so that you control the direction of the muzzle, even if you stumble.

3. Keep your weapon on safe until aimed in on target and the decision to shoot has been made. (The Sig P229R DAK is the only exception to this rule.) This rule is to enforce the use of the weapon's own safety feature.

4. Keep your finger outside the trigger guard, indexed along the receiver, until the decision to shoot has been made. This rule is intended to minimize accidental discharge due to an individual being startled, bracing in a fall, or experiencing a rush of adrenaline.

COAST GUARD SMALL ARMS

As previously mentioned, the Coast Guard deems all weapons that use .50-caliber ammunition and smaller as small-arms weapons.

P229R DAK Personal Defense Weapon. The P229R DAK is a short-recoil, semiautomatic, magazine-fed, double-action pistol with an intermediate reset point, chambered for the .40-caliber cartridge. Each pull of the trigger cocks and releases the hammer to fire the pistol from first shot to last.

M16A2 Rifle. The M16A2 rifle is a lightweight, air-cooled, gas-operated, magazine-fed, shoulder-fired weapon capable of firing all NATO 5.56 mm ammunition with a burst-control device that limits the number of rounds fired in automatic mode to a three-round burst. It also contains a muzzle compensator designed to reduce position disclosure and improve controllability and accuracy.

Remington M870 12 GA Shotgun. The M870 12 GA shotgun is a manually operated, pump-action, single-shot, shoulder-fired weapon. The

Figure 20-3. Safety is a paramount concern.

shotgun is chambered for standard 2¾-inch and 3-inch shotgun shells and is equipped with a four-round magazine tube. The M870 shotgun is also equipped with an advanced combat optical gun reflex sight that is to be used with both eyes open, allowing for fast and rapid target acquisition in all light conditions.

COAST GUARD SMALL-BOAT WEAPONS

M240B Machine Gun. The M240B machine gun is a belt-fed, air-cooled, gas-operated, fully automatic machine gun. The weapon fires from the open-bolt position. Ammunition is fed into the weapon from a 100-round bandolier containing a disintegrating metallic link belt. The weapon features a fixed headspace, which permits rapid changing of barrels; each weapon is issued with two barrels.

M2HB .50-caliber Machine Gun. The Browning M2HB .50-caliber machine gun is a belt-fed, recoil-operated, air-cooled, crew-operated machine gun. The weapon is capable of single shot as well as automatic fire. Maximum surface of the barrel and receiver are exposed to permit air-cooling.

CUTTER WEAPONS SYSTEMS

MK75 MOD 0 GWS. The MK75 is a fully automatic, remotely controlled, dual-purpose, lightweight weapon that stows, aims, and fires 76 mm/.62-caliber ammunition. The gun mount has a high rate of fire and large ready-use stowage. The mount is capable of firing on short notice in an unmanned condition. The GWS can be used to engage air, surface, and ashore targets.

28 mm MK38 MOD 1 MGS. The MK38's primary use is for unit offense and defense, primarily against surface targets in low- to mid-intensity environments. It may also be used to engage slow, low-flying aircraft such as helicopters. The gun system uses standard steel-cased ammunition. The system is designed for durability, accuracy, and safety.

M61A1 Close-in Weapon System (CIWS). The CIWS is a six-barrel, electrically fired, automatic gun that uses an ammunition feeder, which handles the M7 disintegrating loading links to fire 3,000 to 4,500 rounds per minute. The CIWS is used for automated fast-response defense against incoming aircraft and missiles.

57 mm MK110 MOD 0. The MK110 57 mm gun is a multi-purpose, medium-caliber gun used for the ship's self-defense against ships, attack aircraft, and antiship missiles. Based on the Bofors 57 MK3, the MK110 can fire salvos at up to 220 rounds per minute and has a range of nine miles. Current mounting for the weapon include the national security cutters. To increase lethality and flexibility, the ammunition comes equipped with a smart programmable fuse with six modes: contact, delay, time, and three proximity fuse modes.

Figure 20-4. Rounds shot from the close-in weapon system (CIWS) are captured in midair. The system is able to search, detect, and track hostile targets.

AIRCRAFT WEAPONS

M107 Barrett Rifle. The M107 is a recoil-operated, semiautomatic, anti-matériel rifle used to disable the engines of go-fast boats carrying illegal drugs.

M240B/G Machine Gun. As previously mentioned in the small-boat weapons section, the M240B machine gun is a belt-fed, air-cooled, gas-operated, fully automatic machine gun that fires the 7.62x51 mm NATO cartridge. The weapon fires from the open-bolt position. Ammunition is fed into the weapon from a one-hundred-round bandolier containing a disintegrating metallic link belt. The weapon features a fixed headspace, which permits rapid changing of barrels; each weapon is issued with two barrels. The M240B is used for warning shots and for self-protection.

M14 Rifle. The M14 is a gas-operated, semiautomatic rifle with a twenty-round detachable box magazine. The M14 is an American selective fire automatic rifle firing the 7.62x51 mm NATO cartridge. The M14 is used for self-defense and against lightly armored targets.

Figure 21-1.
Coast Guard members work on the electronic and communications systems on board
a forty-seven-foot motor lifeboat at Station Marblehead, Ohio.

CHAPTER 21

Communications

Communication is defined as an exchange of information. This includes operations specialists sending and receiving messages and bridge boatswain's mates making flag hoists, but it also includes people talking to people. Messages, signals, conversations, and written reports are all communication. The bow lookout using a sound-powered telephone to report to the bridge is communicating just as the OOD, using bridge-to-bridge communication to advise another ship of a course change, is communicating. The lookout is using internal communication, and the OOD is using external communication; all communication can be classified as one or the other.

INTERNAL COMMUNICATION

Internal communications are those that take place within a ship- or a shore-based unit. They can include everything from "passing the word" over the 1MC system to using "squawk boxes," the sound-powered telephone system, regular dial phones, bell and buzzer systems, boat gongs, or messengers. Internal communication also includes printed or written material such as the plan of the day, visual display systems such as the rudder-angle indicator and engine-order telegraph on the bridge, the plot in the Combat Information Center, and even the "On Board–Ashore" display board for officers at the quarterdeck. Everyone must keep tuned in to the internal communication system at all times.

Passing the Word

Before public address systems were developed, information or orders for the crew were passed by a boatswain's mate, who first sounded the appropriate call on the pipe and then repeated "the word," fore and aft throughout the ship. Now, on the 1MC system, the word is broadcast all over the ship at once. An announcement is usually preceded with "Now hear this," or "Now hear there," unless a boatswain's call is used. When a boatswain's call (or pipe) is used, "All hands" is piped before any word concerning drills and emergencies, and "Attention" before routine words.

Sound-Powered Telephones

These phones are used on all ships. They do not require outside electrical power; the user's voice acts on a carbon-filled cell and diaphragm to generate enough current to power the circuit. All handset phones look much like ordinary desk telephones and are used for routine ship's business; headset phones are somewhat more complicated and are used for drills and exercises. The handset phone has a button on the bar between receiver and mouthpiece that must be depressed to talk or listen. On the headset phone, this button is in the top of the mouthpiece.

Testing

Always test the phone immediately; press the mouthpiece button and transmit (for example: "Bridge, fantail, phone check"). If the phone does not work, immediately notify the person in charge of your station; use a spare phone if one is available. Never turn an earphone away from your head; it will pick up and transmit all noise in the area and drown out the circuit. If the mouthpiece is damaged, you can talk through one earphone and listen through the other; if both earphones become damaged, you can talk and listen through the mouthpiece.

Securing

Remove the jack from the jack box by holding it in one hand and backing off the collar with the other. When the collar is free, pull the jack out. Do not pull on the lead cord or drop or bang the jack or you will have a bad connection the next time you use the phone. Replace the jack box cover finger-tight; otherwise dust, dirt, and salt-air corrosion will soon short-circuit the jack box.

To make up the phone, remove the headband and drape it over the yoke, then coil the lead cord beginning from the headset, making ten-inch clockwise loops in one hand. Next remove the headpiece from the yoke and hold it with the lead cord; hold the mouthpiece in the same hand. Unhook the neck strap from the breastplate, fold the yoke flat (being careful not to bend the mouthpiece cord), and wrap the strap around the coiled cord and the headband, leaving a short end that hooks to the breastplate again.

The phone should now be in a compact package, ready to hang up or stow in the storage box provided at the station. Be careful not to crowd or jam any cords.

Caution—remember that the jack must always be unplugged when the phone is not in use, otherwise the phone will transmit any noises through the circuit. Do not lay the phone on deck for the same reason; besides, someone could step on it.

Telephone Circuits

Sound-powered phones are hooked up in circuits according to their use. Most ships will have these circuits:

JA	Captain's battle control	1JS	Sonar control
JC	Ordnance control	1JV	Maneuvering, docking,
JF	Flag officer		catapult control
1JG	Aircraft control	JW	Ship control, navigation
JL	Battle lookouts	JX	Radio and signals
2JC	Dual-purpose battery control	JZ	Damage control

Every outlet on a circuit will have a number. Some circuits may have auxiliary circuits; these have the same letter designation, prefixed by an X. All circuits feed into main switchboards where more than one can be tied together, or certain stations on a circuit can be cut out.

Telephone Technique

The way you talk in ordinary conversation is not the way to use a telephone. The person on the other end of the line cannot see you, they may not know you, and they may be unfamiliar with the things you say. Telephone talkers must speak clearly, be specific, and act businesslike; they are not on the circuit for social chitchat. Follow these suggestions:

- Use a strong, calm voice. Speak slowly, pronounce words carefully.
- Do not mumble, run things together, or talk with gum or a cigarette in your mouth.
- Use standard terms and phraseology. Avoid slang words or phrases.
- When transmitting numbers and letters, use approved communications procedure. The expression "Item 5C" may sound like

"Item 9D," but "FIFE Charlie" will not be mistaken for "Niner Delta."

Circuit Discipline

Circuits are like a conference call—everyone can talk, and listen, at the same time. To prevent such a confusing state of affairs, strict circuit discipline must be maintained.

- Send only official messages.
- Keep the button in the off position except when actually talking.
- Do not leave your station or engage in other work or pastimes without permission.
- Use only standard phraseology.
- Never show anger, impatience, or excitement.

Each phone talker is a key link in the ship's interior communication chain. Unauthorized talking means that the chain is weakened. Do not do it; do not permit it.

Circuit Testing

To find out if stations on the circuit are manned and ready, the control-station talker says: "All stations, Control, report when manned and ready." Each talker then acknowledges in the assigned order (or sequence). Here is how it would go on a gun circuit:

> Gun one: "One, manned and ready."
> Gun two: "Two, manned but not ready" (for example).
> Gun three: "Three, manned and ready."

Normally each station answers in order but does not wait more than a few seconds if the station ahead of it does not acknowledge. If you are on gun three and gun two does not answer, acknowledge for your gun. Gun two can then come in at the end. The test is not complete until each station has answered and any equipment faults have been checked.

Message Form

Most messages have three parts: the name of the station being called, the name of the station calling, and the information to be sent. This order must always be followed. Call the station the message is for, identify your own station, and then transmit the message. Remember this order: to, from, about. If you are on the anchor detail and want to call the bridge, the message is "Bridge (to), forecastle (from): anchor secured (about)." (Forecastle is pronounced "FOKE-sul.")

Messages are acknowledged when understood by repeating back the entire message. This lets the sender know the message has been received by the station called and that it is understood. If you don't understand, then transmit "say again."

Sending a Message

First name the station being called. Next name the station doing the calling. Then, the message: "Bridge, forecastle: anchor ready for letting go."

Receiving a Message

First, identify station calling; then your station; then acknowledge the message. If your station is the forecastle, and the bridge has just ordered the anchor "let go," acknowledge with, "Bridge, forecastle: Let go aye, aye."

Sometimes there are three or four steps involved, for example: "Forecastle, bridge: how many lines are to the pier?" If you don't know, you say, "Bridge, forecastle: wait." After getting the desired information, call the bridge: "Bridge, forecastle: five lines to the pier." The bridge will acknowledge: "Forecastle, bridge: understand, five lines to the pier."

Requesting Repeats

If an incoming message is not clear, the receiving station says, "Say again." After the message is repeated and understood, the receiving station acknowledges by repeating back the message.

Spelling Words

Difficult or little-known words are spelled out, using the phonetic alphabet: "Stand by to receive officer from NASA. I spell NOVEMBER ALFA SIERRA ALFA—NASA."

Securing the Phones

Never secure until you have permission from the control station:

> Forecastle: "Bridge, forecastle: permission to secure."
> Bridge: "Forecastle, bridge: wait."

The bridge talker learns that the forecastle may secure and responds:

> Bridge: "Forecastle, bridge: you may secure."
> Forecastle: "Bridge, forecastle: securing."

Remember to make the phones up properly and stow them before leaving your station.

Squawk Boxes

Intercommunication voice (MC) units, or "squawk boxes," are installed in most important stations of most ships. They are normally used only by officers and should be limited to emergency use if paralleled by sound-powered telephones. No matter who uses the MC system, circuit discipline and standard phone talker procedure must be observed. A complete list of MC circuits follows.

Circuit Announcing System

One-way system		*Two-way system*	
1MC	Battle and general	19MC	Ready room
2MC	Engineers	21MC	Captain's command
4MC	Damage control	22MC	Radio room
6MC	Boat control	23MC	Distribution control
11MC	Turret	24MC	Flag officer's command
16MC	Turret	25MC	Wardroom
17MC	Antiaircraft	26MC	Machinery control
18MC	Bridge	27MC	Sonar control
		29MC	Sonar information
		31MC	Escape trunk

EXTERNAL COMMUNICATIONS

External communications are all those between a ship and another ship, aircraft, or command whether it is afloat or ashore. They may be

- Visual—flag hoist, semaphore, signal searchlight, blinker;
- Sound—whistles, bells, foghorns, or even a gun (for distress signal);
- Physical delivery—guard mail and United States mail; and
- Electronic—e-mail, voice radio, facsimile (fax), text messages, telephone when it is possible to hook into shore circuits, and even handheld radios.

With ships and aircraft operating twenty-four hours a day around the world, there must be rapid, reliable communication, not only for tactical and strategic control in wartime but also for administrative and logistic purposes at all times. No ship or aircraft is ever out of touch with its base of operations or its tactical, type, or administrative commander.

Visual Signals

Visual signals began centuries ago, when flags and pennants were given special meanings. Columbus hoisted a lantern in the rigging to let his ships know land had been sighted. Even with dozens of radios on a ship, visual signals are still used extensively, especially in fleet operations, for two reasons—speed and security. Radio messages may be intercepted by anyone with a receiver, perhaps hundreds of miles away. Visual signals are limited to line-of-sight transmissions—the receiver and sender have to be where they can see each other—and do not require complicated equipment. The three main systems of visual signals are flashing light, semaphore, and flag hoist.

Flashing Light Signaling

This system uses short and long flashes of light to spell out dot-and-dash messages. The transmitting signalman sends one word at a time, with a slight pause between each letter. The receiving signalman flashes a dash after each word, meaning it was received and the receiver is ready for another.

Directional method. In this system the sender aims the light directly at the receiver. Only people near the receiver can see the light. The signalman controls the length of the flash by a shutter on the face of the light. Other types of directional gear are the blinker tube (or blinker gun) and the multipurpose lamp—both battery-operated with trigger switches to control the light flashes.

Nondirectional method. Another term for this type of signaling is "all-around." Most of it is done by yardarm blinkers, lights mounted high on a mast and controlled by a signal key on the signal bridge. This method is best for sending a message to several ships at once.

Nancy. Whereas other signal systems use "white" light, Nancy uses invisible infrared light that can be seen only by those who have a special Nancy receiver that gathers the infrared rays and converts them to visible light. Nancy has a range of from 10,000 to 15,000 yards and can be used only at night. It is a very secure method of communication.

Semaphore

This requires the least equipment of all—two small handheld flags. A good signalman can send or receive twenty-five five-letter groups a minute. Only thirty flag positions need to be learned.

AND
ANSWERING
SIGN

Figure 21-2. The semaphore alphabet is easily learned and can be used by anyone.

Flag Hoist

This is the most rapid system of visual signaling but, like semaphore, it can be used only in the daytime. Usually it is used for tactical orders; the meanings of each signal must be looked up in a signal book. There is a signal flag for each letter of the alphabet, one for each numeral from 0 through 9, and others with special uses. A complete set of signal flags will have sixty-nine flags and pennants (see appendix C); with them, thousands of signals can be sent.

Most ships carry only two or three complete sets of flags, and substitutes are used when particular flags are already flying: the first substitute repeats the first flag or pennant in the same hoist, the second substitute repeats the second flag or pennant, and so on.

The following five flags, and their meanings, should be known to every crew member, whether they stand bridge watch or not:

- BRAVO—ship is handling explosives or fuel oil
- OSCAR—man overboard
- PAPA—all hands return to ship
- QUEBEC—all boats return to ship
- FIVE FLAG—breakdown

Absence Indicators

When the commanding officer, or any flag officer, is absent from the command, an "absentee pennant" is flown as follows:

- *First substitute (starboard yardarm).* The admiral or unit commander, whose personal flag or pennant is flying, is absent.
- *Second substitute (port yardarm).* The chief of staff is absent.
- *Third substitute (port yardarm).* The commanding officer is absent. (If the commanding officer is to be gone more than seventy-two hours, the pennant shows the absence of the executive officer.)
- *Fourth substitute (starboard yardarm).* The civil or military official whose flag is flying is absent.

Other Visual Signaling Systems

Other visual systems involve special methods for special occasions. Here are some of them:

- *Speed indicators.* These are flags and pennants, or red or white lights in combinations of flashes, used to show a ship's speed.

- *Pyrotechnics.* These are colored smoke and flare signals, usually used for distress or emergency purposes.
- *Panels.* Large strips of colored cloth laid out in different designs on the ground or the deck of a ship to signal aircraft.

Procedure Signs

Procedure signs (prosigns) are a form of communication shorthand. They provide, in brief form, orders, requests, instructions, and other types of information used often in communication. Shown below are some of the more common prosigns that may be sent by radiotelegraph (also called continuous wave radiotelegraph, or CW), semaphore, or flashing light.

Prosign	Meaning
AA	Unknown station
AR	Out (End of transmission)
AS	Wait
B	More to follow
BT	Break
DE	This is (from)
EEEEEEEE	Error
F	Do not answer
HM HM HM	Emergency silence
IMI AB	Repeat all before (word)
IMI AA	Repeat all after (word)
IMI WA	Repeat word after (word)
IX	Execute to follow
K	Over (invitation to transmit)
NR	Number
R	Received (also Routine)
T	Relay to (Unit)

The International Morse Code

All CW and flashing light signaling uses the International Morse Code, a system in which letters and numbers are formed by combinations of dots and dashes, or "dits" and "dahs." A skilled radioman or quartermaster sends code in evenly timed dots and dashes, in which a dot is one unit long, a dash is three units long; there is a one unit interval between dots and dashes in a letter, three units between letters, and seven units between words.

Letter	Code	Letter	Code
A	• —	N	— •
B	— • • •	O	— — —
C	— • — •	P	• — — •
D	— • •	Q	— — • —
E	•	R	• — •
F	• • — •	S	• • •
G	— — •	T	—
H	• • • •	U	• • —
I	• •	V	• • • —
J	• — — —	W	• — —
K	— • —	X	— • • —
L	• — • •	Y	— • — —
M	— —	Z	— — • •

Letter	Code	Letter	Code
1	• — — — —	6	— • • • •
2	• • — — —	7	— — • • •
3	• • • — —	8	— — — • •
4	• • • • —	9	— — — — •
5	• • • • •	10	— — — — —

Electronic Communication

Electronic communication can be made through various methods. Dedicated direct-wire circuits, such as district teletype circuits, go directly from the sending unit to the receiving unit. The signals on these circuits cannot be intercepted except by physically cutting into the circuit. Radio communication such as radiotelegraph (CW), radiotelephone (RT), and facsimile (fax) use the electromagnetic waves from antennas to broadcast signals through the atmosphere.

Radiotelephone (RT)

Radiotelephone, or voice radio, is one of the most useful military communication systems. Because of its directness, convenience, and ease of operation, voice radio is used almost exclusively between ships and aircraft for short-range tactical communication (up to twenty-five miles). RT transmissions on VHF (very high frequencies) and UHF (ultra high frequencies) are good for line-of-sight communication.

RT transmissions on MF (medium frequencies) and HF (high frequencies) are used for longer-range communication. Although most RT communication is made in the clear, equipment is installed at most Coast Guard units to allow for covered and secure voice transmissions. Since RT circuits are used by many people, strict circuit discipline is necessary at all times.

Transmitting Techniques

Because voice radio is used so widely, everyone must understand the basics of circuit discipline (discussed earlier) and transmitting techniques. Because voice radio can be heard by anyone with the proper equipment, the following are strictly prohibited:

- Violation of radio silence
- Unofficial conversation or procedure words
- Transmitting without permission
- Unauthorized plain language
- Profane, indecent, or obscene language

The following techniques will ensure good communication:

- Listen before transmitting.
- Do not break in on a message.
- Speak clearly, distinctly, and slowly, but transmit phrase by phrase, not word by word.
- Use standard pronunciation, use standard terminology, and always use prescribed procedures.

Procedure Words

Procedure words (prowords) are words and phrases used to speed up radio communication. They perform the same functions and are used in the same manner as prosigns. Many prosigns and prowords have the same meaning.

Proword	Meaning
Correction	An error has been made.
I say again	I am repeating transmission or portions indicated.
I spell	I shall spell the next word phonetically.
Out	End of transmission, no receipt required.
Over	Go ahead; or, this is end of transmission, reply is necessary.
Roger	I have received your last transmission or message.
Say again	Repeat your last message.
Wilco	I have received your message, understood it, and will comply. (Used by commanding officer only.)

Pronouncing Numbers—Always pronounce numbers as follows:

Number	Pronunciation	Number	Pronunciation
0	Ze-ro	44	Fow-er fow-er
1	Wun	90	Nin-er ze-ro
2	Too	136	Wun thu-ree six
3	Thu-ree	500	Fife ze-ro ze-ro
4	Fow-er	1478	Wun fow-er sev-en ait
5	Fife	7000	Sev-en tou-sand
6	Six	16000	Wun six tou-sand
7	Sev-en	16400	Wun six fow-er ze-ro ze-ro
8	Ait	812681	Ait wun too six ait wun
9	Nin-er		Decimal is pronounced "day-see-mal."

Ranges and distance given in miles are transmitted as cardinal (whole) numbers, such as "range two thousand." *Speeds* are given as individual numbers, such as "speed two zero." *Altitude* is given in feet (except for weapons orders, which are always in yards) and transmitted in cardinal numbers.

Pronouncing Letters—In all communication, a phonetic alphabet is used to avoid confusion between letters that sound alike, such as C, D, and T. Each letter is designated by a word, as follows:

Letter	Name	Spoken
A	ALFA	*AL*-FAH
B	BRAVO	*BRAH*-VOH

Letter	Name	Spoken
C	CHARLIE	*CHAR*-LEE
D	DELTA	*DEL*-TAH
E	ECHO	*ECK*-OH
F	FOXTROT	*FOKS*-TROT
G	GOLF	GOLF
H	HOTEL	HOH-*TEL*
I	INDIA	*IN*-DEE-AH
J	JULIETT	*JEW*-LEE-*ETT*
K	KILO	*KEY*-LOH
L	LIMA	*LEE*-MAH
M	MIKE	MIKE
N	NOVEMBER	NO-*VEM*-BER
O	OSCAR	*OSS*-CAH
P	PAPA	PAH-*PAH*
Q	QUEBEC	KAY-BECK
R	ROMEO	*ROW*-ME-OH
S	SIERRA	SEE-*AIR*-RAH
T	TANGO	*TANG*-GO
U	UNIFORM	*YOU*-NEE-FORM
V	VICTOR	*VIK*-TAH
W	WHISKEY	*WISS*-KEY
X	XRAY	*ECKS*-RAY
Y	YANKEE	*YANG*-KEY
Z	ZULU	*ZOO*-LOO

Standard Phraseology—Because of the specialized and technical nature of most operations, it is necessary that, in all communication, everyone "talk the same language." There is one correct, proper way to say things, and it must be used to avoid confusion, which in many high-speed operations can easily lead to danger and disaster.

Cryptography

Since all communication may be intercepted by an enemy, messages that contain vital information must be "disguised" so that they cannot be read. This is done by putting messages in codes or ciphers.

Cryptography is the science of enciphering and deciphering messages; *cryptanalysis* is the breaking of enemy ciphers or codes. All ciphers are designed to allow a large volume of traffic and at the same time to defeat enemy cryptanalysts.

Codes are word-for-word substitutions of letter or number groups with prearranged meanings for entire phrases (such as are found in signal books). They are used where speed is preferred to security and are changed frequently since they are subject to compromise.

Ciphers offer letter-for-letter substitutions and greater flexibility and security than codes and are easier to use in rapid communication. Most ciphers are changed periodically. Simple tactical codes and ciphers are changed frequently; those considered more secure may be used for months or years.

Sound Signals

Sound signal devices include the ship's bell, whistle, siren, gong, and sometimes underwater sound devices (sonar). The *bell* is sounded when the ship is at anchor in fog, mist, snow, or heavy rain. The *gong* is sounded under certain conditions when the ship is in international waters. The *whistle* or *siren* is sounded in fog and whenever required by Rules of the Road, either as a signal of intent or as a danger signal. Ships use whistles, sirens, and bells for transmitting emergency warning signals (such as air raid alerts) and, in wartime, for communication between ships in convoy.

On most small vessels, the voice tube is the primary means of interior communication. The voice tube requires only lung power; its effectiveness decreases according to its length and the number of bends in it. On large ships, voice tube communication is for short distances only, as between an open conning station and the pilothouse.

Physical Delivery

A vast amount of administrative detail concerning personnel, supplies, logistics, and operations is handled by official mail, which is carried through the U.S. postal system. Official mail between ships or stations in the same port is carried by *guard mail*. Guard mail petty officers are designated in writing. They carry registered-mail log books and must ensure that all mail is logged and receipted. Classified mail is carried by *officer couriers*. Personal mail is handled much as it is in a civilian community. Always have your correspondents use the correct zip code.

Zone Time

The solar day contains twenty-four hours, the time it takes the earth to turn once on its axis. The earth is divided into twenty-four different time zones so that at any moment of the day there are twenty-four different times around the world. The United States has four time zones—Eastern, Central, Mountain, and Pacific. When the time is 0800 in Boston, it is 0500 in San Francisco, 0600 in Denver, and 0700 in Chicago.

The standard time zones begin at Greenwich, England, the point from which all longitudes east and west are measured. All navigation and most communication are based on Greenwich Mean Time (GMT), the time at the 0, or prime, meridian at Greenwich. Time zones are numbered from 1 to 12, east and west of Greenwich; those west are + and those east are −. (The line between east and west is the 180th meridian, halfway around the world from Greenwich.)

If your ship is in Boston, Zone +5, and you need to know GMT at 0800 Eastern Standard Time, you add 8 and 5 and find that it is 1300 GMT. In Hawaii, which is five zones farther west, the local time is naturally five hours earlier than in Boston, or 0300. To check on this, note that Hawaii is in Zone +10, and 0300 plus 10 equals 1300 GMT.

The 180th meridian is also called the *International Date Line*. When it is Saturday east of the line, it is Sunday west of the line. As a ship steaming around the world crosses each time zone, it has its clocks *advanced* one hour if it is steaming *east*, and *set back* one hour if steaming *west*. To compensate for this gain or loss of twenty-four hours, the date is changed when crossing the date line. A ship steaming west and setting the clocks back an hour for each time zone it crosses will advance the calendar one day when it crosses the International Date Line; a ship going east will set the calendar back and have the same day twice.

This change does not have to be made the instant the ship crosses the date line; it may be delayed for a reasonable period. It would not be desirable to lose a Sunday, nor to have Sunday twice in a row. Usually, a ship will advance the clocks (shorten the day) during the midwatch (0000–0400), and slow down the clocks (lengthen the day) during the second dogwatch (1800–2000).

When a date-time group ends in ZULU, the time used is GMT. If your ship is operating out of Hawaii, which is in time zone WHISKEY (+10), you add 10 to the local zone time to find GMT.

Figure 22-1.
The twenty-five-foot Defender-class boat (RB-HS/RB-S).

CHAPTER 22

Security

Security that is used in most training manuals and publications refers to the safeguarding of classified information. The security of the United States in general, and of military operations in particular, depends greatly upon the success attained in the safeguarding of classified information. Security takes in more than just classified printed information (i.e., classified or registered publications, manuals, and charts). It may include such forms of communication as radio, visual signals, mail, and knowledge of ships' movements. Security, in the full sense of the word, has to do with the safeguarding of materials and the physical safety of ships and stations. It is important that everyone understands what classified information is, its importance, how to safeguard it, how to obtain clearance to work with or have access to it, and the penalties for security violations. Refer to COMDTINST M5500.11A for information.

SECURITY CLASSIFICATIONS

Information, oral or written, that is vital to the national security is classified. Three classifications are used to protect this information: *Top Secret*, *Secret*, and *Confidential*. The term "FOUO," for official use only, is assigned to sensitive information and is not considered a security classification.

Top Secret

Information of such importance to the national security of the United States that unauthorized disclosure or loss would result in *exceptionally grave damage* to the United States or its allies is classified as Top Secret. Examples of exceptionally grave damage include hostilities against the United States or its allies, disruption of foreign relations affecting the national security, and the compromise of vital national defense plans or complex cryptologic and communications systems.

Secret

Information of such importance that unauthorized disclosure could reasonably be expected to cause *serious damage* to national security is classified as Secret. Examples of serious damage include disruption of foreign relations, compromise of significant military plans or intelligence operations, and compromise of significant scientific or technological developments relating to national security.

Confidential

Information of such importance that unauthorized disclosure could reasonably be expected to cause *damage* to the national security is classified as Confidential. Examples of damage include the compromise of information that indicates the strength of ground, air, and naval forces in the United States and overseas; disclosure of technical information used for training, maintenance, and inspection of classified munitions of war; and performance characteristics, test data, design, and production data on munitions of war.

For Official Use Only (FOUO)

The term "for official use only" (FOUO) is assigned to official unclassified information that requires protection against uncontrolled release to the public. Examples of information that qualifies for the FOUO designation are the results of investigations, names and locations of informants, service-wide examination questions, bids on contracts, and plans to buy or lease real estate. It is important to remember that FOUO is not a classification but is only a designator.

DISCIPLINARY ACTIONS FOR SECURITY VIOLATIONS

Anyone responsible for the loss or unauthorized disclosure of classified material and anyone who violates security regulations will be subject to disciplinary action under the Uniform Code of Military Justice. Disciplinary action may include trial by court-martial.

SECURITY CLEARANCE

Before you can be granted access to classified information, you must have a valid security clearance. The clearance must be equal to, or higher than, the classification of the information to which you will have access.

Standards for Granting Clearances

All the standards required for the granting of a security clearance are listed in the *Military Personnel Security Program* (COMDTINST M5520.12). A security clearance is an administrative determination by competent authority that you are trustworthy and dependable and possess the integrity required for a position of trust. This determination is based upon a security investigation. Once this determination has been made, you may be granted access to classified information up to, and including, the level of your clearance.

Security Investigations

Various types of security investigations are used to grant security clearance. Personnel requiring access to top secret material must have a background investigation (BI) completed prior to access. Personnel requiring access to information classified secret and below must have a National Agency Check completed prior to access.

Background Investigation. All of the paperwork requirements for a BI can be found in the *Military Personnel Security Program* (COMDTINST M5520.12). To maintain continuous access to top secret material, BIs must be updated every five years. Neglecting to update the investigation will result in removal of access to top secret material until the investigation has been updated.

National Agency Check (NAC). Every person entering the Coast Guard has an NAC completed. This investigation allows commanding officers to grant access to material classified up to, and including, secret. To maintain continuous access to this material, the NAC must be updated every ten years. Neglecting to update the investigation will result in removal of access to all classified material until the investigation has been updated.

Clearance Requirements

A final Top Secret or Secret clearance can only be granted by the Coast Guard Security Center. This clearance may be granted after a BI has been completed along with a review of your personnel and health records. Interim clearances are normally granted only for short-term access to classified information during special operations or emergency situations. The granting of interim clearances will be kept to a minimum.

Security investigations remain valid and may serve as the basis for the issuance of future clearances unless derogatory information becomes available, indicating a need for further investigation; or the individual is assigned to a particularly sensitive billet requiring a greater clearance criterion than indicated by the foregoing; or continuous active service in the armed forces or civilian employment in the government service is broken by a period longer than one year.

Access

Access to classified information is granted based on a "need to know." If you need to have classified information in order to perform your assigned duties, you will be granted access by your commanding officer or officer in charge. Having access does not mean that you will be able to look at all the classified material held by the command. You will receive only the information you need to perform your job.

All classified material entrusted to you will be clearly marked with its appropriate classification. There is never an excuse for the careless handling of classified material. There may be times when you will be exposed to classified information that you do not have a "need to know." Do not discuss this information with anyone or attempt to pry into the details of this information.

THREATS TO SECURITY

The compromise of classified information, through loss or disclosure to unauthorized persons, presents a threat to national security. The seriousness of that threat must be determined and measures taken to negate or minimize the adverse effects of the compromise. Any individual becoming aware of a compromise or subjection to compromise of classified material must immediately notify the most readily available command. Take any measures necessary to keep unauthorized persons from viewing the material until someone is sent to take charge.

To keep yourself from getting involved in a compromise, remember the following: What you do, see, and hear at work stays at work. Bragging or talking freely about what you know and do may impress your family and friends, but it could jeopardize your safety and the safety of your shipmates. There was a saying during World War II, "Loose lips sink ships." Believe it or not, that saying is as applicable today as it was then.

Figure 22-2. Coast Guard Investigative Service seal.

COAST GUARD INVESTIGATIVE SERVICE

The Coast Guard Investigative Service is a diverse group of special agents, investigations assistants, and support personnel who assist Coast Guard commands in maintaining internal integrity, good order, and discipline. In addition, Coast Guard Investigative Service agents conduct external investigations, addressing issues such as drug and alien smuggling, environmental crimes, and crimes against the government generally. These men and women, both military and civilian, serve in much the same capacity as their counterparts in the Naval Criminal Investigative Service, the Air Force Office of Special Investigations, and the Army Criminal Investigation Command.

INTELLIGENCE COLLECTION METHODS

Unfriendly nations and drug smugglers are always interested in classified information on developments, weapons, techniques, materials, and ship and aircraft movements and their operating capabilities. The people who attempt to collect such information will not look or act like spies in TV thrillers. They are just like the people who lived next door when you were growing up. This ability to blend in with the normal populace allows them to be successful in their work. This brings us back to what was discussed in "Threats to Security," above. If you have access to and work with classified material, you should never discuss it with anyone other than those you know have access and a need to know. Intelligence agents collect many odd little bits of information. From these little bits, they can piece together a fairly complete picture. Do not make their work easier.

SAFEGUARDING CLASSIFIED INFORMATION

If you accidentally come across some classified matter—a letter, booklet, or a device—that has been left unguarded, misplaced, or not properly secured, do not read or examine it or try to decide what to do with it. Notify the nearest officer or petty officer in charge, and then stand by to keep unauthorized personnel away until a responsible person arrives to take charge.

RESTRICTED AREAS

Restricted areas are established at units to provide additional safeguards for matériel and equipment. These areas will be clearly marked with warning signs. Persons on watch in such areas should not hesitate to ask visitors for identification. No responsible person will object to being stopped politely, but firmly, pending identification.

Three levels of restricted areas meet the security requirements for the protection of the information stored in the space:

- *Exclusion Area:* Access to this area constitutes, for all practical purposes, access to classified matériel or equipment.
- *Limited Area:* A limited area is an area in which uncontrolled movement would permit access to classified matériel or equipment.
- *Controlled Area:* Controlled areas are the areas adjacent to or surrounding exclusion or limited areas.

A complete breakdown of each area can be found in chapter 12 of the *Security Manual* (COMDTINST M5500.11).

EXTERNAL SECURITY

Every person on board ship, whether on watch or not, must always be security-minded and on the alert for any sign of danger to the ship. A ship in port should be relatively safe, but it can be threatened by many things—for example, a hurricane, tidal wave, flooding, fire, explosion, sabotage from within the ship, saboteurs, sneak attack, civil disorder, and riot.

Threats to security may originate outside the ship. Strangers who approach the ship should be regarded with suspicion, even though they appear to be ordinary visitors. All individuals coming on board must be identified by the officer of the deck or designated representative, and packages, parcels, brief cases, tool boxes, and so forth should be inspected. Crew members standing gangway or quarterdeck watches assist the OOD in identifying approaching boats, screening visitors, and checking packages.

Sentries and Guards

Sentries and guards, posted for security purposes, are guided by written instructions and must know how to challenge boats in order to identify occupants before they come alongside. All sentries may be armed when the situation demands. An armed guard should be reasonably proficient in the use of the weapon; one who does not know the weapon is useless and a danger to ship and shipmates.

Sneak Attack and Sabotage

Particularly at night, ships moored or at anchor are vulnerable to sneak attack or sabotage. The ship could be approached by swimmers, small boat, midget submarine, or submarine. Boarders may pose as bumboat crews; saboteurs may mingle with the returning liberty party, pose as visitors, or sneak on board when ships are moored to a pier. Where such dangers of attack exist, the operations officer will organize special watches and issue instructions.

The Signal Bridge

In port, the bridge and signal watches perform the following functions with regard to security. They report to the OOD any boats approaching the ship or operating in the vicinity of the ship in a suspicious or aimless manner, and also report to the OOD any unusual disturbances or signs of distress in the harbor, on board other ships, or ashore.

INTERNAL SECURITY

The safety of a ship may be threatened from within. Sabotage is possible, particularly in times of great international tension. Abrasives in oil, nails driven into multiple conductor cables, or foreign objects placed in turbines or reduction gears can cause great damage. Fire or flooding, accidental or otherwise, is always a danger. All ships maintain watches to help maintain internal security.

Emission Control (EMCON)

Emission control is necessary in wartime because modern science enables an enemy to detect almost any electronic emission. When EMCON is set, cell phones, personal computers, and many other electronic devices on board ship may not be used.

Darken Ship

This order demands rigid observance by everyone going topside. On a dark night, the glow of a cigarette can be seen for miles; the loom of light from an improperly shielded hatchway will let a submarine make a successful periscope attack.

Quiet Ship

Observance of this order is most important. Banging or hammering can give away the position of an otherwise perfectly silent vessel.

Trash, Garbage, and Oil Slicks

Orders covering trash disposal and pumping bilges must be carefully observed. A ship littering the ocean with floating debris can be tracked down by an alert enemy. It is also violating both U.S. and international laws.

Keys

The commanding officer is custodian of all keys to all magazines but may designate others to have duplicate keys. Heads of departments are responsible for keys to locked spaces under their cognizance. Keys to other spaces are kept in the custody of designated officers or petty officers. Each department head maintains a key locker containing all the keys to spaces in his or her department. Keys to these key lockers are available to the OOD at all times.

Crew members must provide locks for their personal lockers and should carry their keys at all times. Any other keys with which they may be entrusted should never be carried off the ship.

SHIPYARD SECURITY

When a ship is in a shipyard, all workmen coming on board must be identified. Compartments containing classified matter must be secured, either by locking or by sentries. Fire watches are normally assigned each welder and burner who comes on board. Special precaution must be taken after each shift to inspect spaces for all fire hazards.

GENERAL SECURITY MEASURES

Security refers to all measures taken to protect a ship or station against damage by storm or fire and to guard against theft, sabotage, and other subversive activities. Security involves sentry duty, fire watches, guard duty, and barracks watches. Sentry duty is a formal military duty governed by specific orders. Guard duty may be the same as sentry duty, or, at times, guards may be permitted to relax military bearing so long as they are on the job and ready for action. A fire watch may cover an assigned area on foot, or with a vehicle, or simply be an assignment to a certain place for a certain period. A barracks watch may sometimes stand a sentry watch or sometimes merely be available to answer a phone, check people in and out, turn lights off and on, and preserve order and cleanliness.

SENTRY DUTY

The requirements for standing sentry duty are the same as for all watches: keep alert, attend to duty, report all violations, preserve order, and remain on watch until properly relieved. The rules or orders for sentries form the basic rules for all security watches. Special Orders cover duty as a sentry with regard to the specific watch in question. They will be passed on and explained to you by the petty officer of the watch or corporal of the guard.

VI

SAILORS AT SEA

AND ASHORE

Figure 23-1.
Duty on board a Coast Guard cutter is an exciting opportunity.

CHAPTER 23

Living On Board Ship

The Coast Guard is first and foremost a seagoing service; therefore, many of you will serve in the Coast Guard's fleet of cutters at some time in your careers. Duty on board a cutter is your opportunity to serve our country in a way rich in challenge, adventure, and tradition.

The sailor's life is unique and demanding, an experience for which you are actually beginning preparation as a recruit, cadet, or officer candidate. The training received in these roles and supplemented by the information in this and other chapters in this manual is intended to help smooth your transition from "landlubber" to "salt"; additional reference reading on much of the information discussed in this chapter is published in *United States Coast Guard Regulations*.

Welcome aboard!

SHIP'S ORIENTATION AND INDOCTRINATION PROGRAM

It has long been recognized that newly reporting enlisted personnel, particularly those who are reporting to their first ship, have a definite need for a complete formal introduction to their new and perhaps initially bewildering world. For this reason, each Coast Guard ship includes just such a program as a part of its training plan and assigns a senior enlisted member to act as your sponsor. Sponsors administer the ship's Orientation and Indoctrination Program to you and monitor your progress as you take your important place in the ship's crew.

Your sponsor will provide you with your personal qualification standards (PQS) and check-in sheet, which will document your completion of the various steps of your training. As well the technical aspects of a ship's operation, PQS and check-in sheets require new "hands" to learn very useful information on many subjects, including history and mission of the ship, ship's routine and regulations, personnel procedures, education services, career benefits, legal services, morale and religious services, the diversity program and human relations, drug and alcohol abuse, medical and dental services, safety, security, and vehicle regulations.

THE *ORGANIZATION AND REGULATIONS MANUAL* AND THE *SHIP-BOARD REGULATIONS MANUAL*

On board ship, order is understandably important, and because of the great need for efficiency in all operations, the Coast Guard's regulations prescribe a standard organization to ensure each ship is organized in a way efficient and meaningful to its mission. This organization is detailed for each type of Coast Guard ship in a publication called the *Organization and Regulations Manual*. The specific regulations that support the organization of Coast Guard ships of all types are detailed in another Coast Guard publication, the *Shipboard Regulations Manual*.

The *Organization and Regulations Manual* and the *Shipboard Regulations Manual* provide ship's personnel with a ready source of information about their duties and responsibilities for administering, operating, living in, and fighting the ship. Because commanding officers may modify these regulations to some extent to meet specific needs of the ship, you will be required to familiarize yourself with your ship's particular *Organization and Regulations Manual*. This will normally be accomplished in the initial stage of your orientation and indoctrination.

Pay careful attention as you read it, and ask your sponsor any questions you may have about the information it contains, for you will be required to sign a form stating that you have read and *understand* this very important publication.

SHIPBOARD ORGANIZATION

Commanding Officer

The officer ordered to command a ship is always called the "captain," regardless of actual rank. The *commanding officer* (CO) is responsible for the operation and efficiency of the ship and the conduct and performance of the crew. The CO enforces all regulations and directives from higher authority. All authority, command direction, and responsibility for the command rest with the CO. In practice, the CO delegates duties for carrying out functions of the ship to the executive officer, department heads, other officers, and certain crew members.

Executive Officer

The officer ordered as assistant to the commanding officer is the *executive officer* (XO). The XO is next in the chain of command below the captain. The XO is in charge of the personnel, routine, and discipline of the ship. Department heads may consult directly with the captain, but the XO must be informed of any decisions the captain makes.

Command Master Chief

The senior enlisted assigned to a unit is the *command master chief* (CMC). The CMC functions as an adviser to the commanding officer on matters relating to the welfare of enlisted personnel. While the CMC does not replace the chain of command, issues that cannot be resolved in the normal manner should be brought to the CMC's attention.

Departments and Divisions

Coast Guard ships are organized, under a commanding officer and an executive officer, into at least three departments: operations, engineering, and deck or weapons. Depending upon the size and type of ship, there may also be a supply department, medical department, and, if a helicopter detachment is assigned, an aviation department. These departments are subdivided into divisions; the smaller the department, the fewer the divisions. Each department is directed by a commissioned, warrant, or petty officer called the *department head*, who is assigned by the commanding officer. Similarly, each division is headed by a *division officer* or petty officer. In small ships, an officer may head two departments or be a division officer as well as a department head. The department head, assisted by the division officers, represents the commanding officer in all matters pertaining to the department and is responsible to the commanding officer for the general condition of the department. All members in a department make requests to the XO through the division officer and the department head.

Operations Department

The officer in charge of the operations department is the *operations officer*. As you might expect, the operations officer is responsible for the operation of the ship at its missions and for handling operational and combat intelligence. The operations officer may be assisted by division officers such as the *navigator, communications officer, electronics material officer,* and *combat information center officer.*

Navigator

The navigator is responsible to the CO for the safe navigation of the ship. The navigator reports first to the CO and then to the operations officer on all matters having to do with navigation. The navigator is responsible for the proper condition of all charts and instruments and the steering equipment of the ship.

Communications Officer

The communications officer is responsible for all exterior communications, including electronic and visual signals. The communications officer oversees the handling of all official messages to and from the ship and ensures proper communications security is employed by the ship's communicators.

Electronics Material Officer

The electronics material officer is responsible for the material readiness of all electronic equipment on board.

Combat Information Center (CIC) Officer

The Combat Information Center officer is responsible for the operation and readiness of the CIC. The CIC collects, displays, evaluates, and distributes tactical information such as surface or air radar contacts.

Engineering Department

The officer in charge of the ship's engineering plant is the *engineer officer.* The engineer is responsible for the propulsion plant, auxiliary machinery, all related systems, hull repairs, and damage control. The engineer officer may be assisted by division officers such as the *main propulsion assistant, auxiliary officer,* and *damage-control officer* or *assistant.*

Deck Department

The officer in charge of the deck department is the first lieutenant. The first lieutenant is responsible for all deck and ordnance equipment and for upkeep and cleanliness of parts of the ship not assigned to other departments. The first lieutenant handles ground tackle, mooring lines, rigging, towing gear, and all boats. The first lieutenant may be assisted by a *weapons officer*, who is responsible for armament and ammunition. On high endurance cutters, this function is performed by the *weapons or combat systems department*, of which the deck *division* is normally a part.

Officer of the Deck (Day)

There is much activity on board a ship even when it is quietly moored in a safe harbor. It is impossible for the captain and the executive officer to stay awake day and night supervising all operations of each watch. Hence, it is necessary that there be an officer to act as a central clearing station and keep the daily happenings organized. This officer must have the ability and authority to act immediately in an emergency.

When the ship is under way, it is even more important that all operations be under the control of an officer of the watch. The engine room watch cannot be expected to know whether the ship should suddenly be stopped, slowed, or speeded up. The lookouts, observing an object in the water, cannot run to the bridge and tell the helmsman which way to turn. There must be an officer constantly on watch to receive the incoming reports, act on some of them immediately, and decide what others should be referred to the captain, executive officer, or other officer. At sea, the officer so in charge of each watch is known as the *officer of the deck* (OOD); in port, this officer is identified as the *officer of the day*. At sea, the OOD takes station on the bridge. In port, the OOD is stationed on the quarterdeck or at the main gangway, depending on the construction of the ship and the decision of the commanding officer as to the limits of the quarterdeck.

The OOD is the officer on watch in charge of the ship. The OOD is responsible for the safety of the ship; every person on board who is subject to the captain's orders, except the executive officer, is likewise subject to the orders of the OOD. Only the captain and the executive officer are superior to the officer carrying out the duties of the OOD. All the ship's company must cooperate fully with the OOD in the performance of their duty, and the OOD's orders must be obeyed quickly and respectfully.

Engineering Officer of the Watch

The *engineering officer of the watch* (EOOW) is the officer on watch in charge of the ship's main propulsion plant and the associated auxiliaries. The EOOW ensures that the engineering log, the engineer's bell book, and the prescribed operating records are properly kept and that all orders received from the OOD are promptly and properly executed.

The EOOW may be directed in the duties of the watch by the engineering officer or the main propulsion assistant, either of whom may assume charge of the watch if they consider such action necessary. The EOOW reports to the OOD and to the engineering officer any defects of machinery, boiler, or auxiliaries that may affect the proper operation of the ship. The EOOW ensures that frequent inspections of engines, machinery, and boilers are made and that all safety precautions are observed.

COLLATERAL DUTIES

Some jobs on board ship are not assigned directly to the executive officer or to department heads. These are known as collateral duties and may be assigned to any individual on board ship, who then acts as an assistant to the executive officer. Collateral duties may include educational services, commissary and crew's mess, property accounting, and supply.

Boards and Committees

A board or committee is a group of ship's personnel organized under a leader to inspect, deliberate, and evaluate a function, situation, or problem. Officers, and in some cases enlisted personnel, are assigned to boards and committees by the commanding officer or executive officer. Some examples of such groups are the Morale Committee and the Awards Board.

Watch Organization and Log Keeping

A ship in commission *always* has personnel up, about, and actively engaged in the operation or oversight of the ship. These many different duties are organized into periods called *watches*. Watches and related log keeping are discussed in chapter 24.

SHIPBOARD BILLS

A commanding officer sets forth detailed written instructions for assigning personnel to duties or stations within the ship for the purpose of executing every specific operational job (*evolution*) or administrative function that carries out or supports the ship's missions. Each such instruction is called a "bill" and systematically describes the "who," "what," "where," "when," and "why" for every evolution and function performed by the ship. The *Organization and Regulations Manual* for each ship describes *administrative bills*, *operational bills*, *emergency bills*, and *special bills*.

Additionally, two documents devoted to organizing and assigning members of the ship's company that are common to all ships but not included in the *Organization and Regulations Manual* are the *battle bill* and the *Watch, Quarter, and Station Bill* (WQSB); these are developed individually by each command.

Administrative Bills

Administrative bills set forth procedure for the everyday administration of the ship and its company. Examples include

- personnel assignment bill;
- cleaning, preservation, and maintenance bill;
- damage-control inspection and maintenance bill;
- material conditions of readiness bill;
- berthing and locker bill;
- security bill;
- zone material inspection bill; and
- crew recall bill.

Operational Bills

Operational bills detail procedure and specific assignments of crew members for evolutions of a periodic nature. Such bills may include

- boat bill;
- cold weather bill;
- darken ship bill;
- aircraft ditch and rescue bill;
- drydocking and undrydocking bill;
- fuel transfer and loading bill;
- heavy-weather bill;
- helicopter operations bill;
- hurricane/typhoon preparedness bill;
- law enforcement bill;
- navigation bill and navigation standards;
- pollution control bill;
- search and rescue bill;

- replenishment at sea bill;
- rescue of survivors bill;
- rescue and assistance bill;
- ship's silencing bill;
- special sea and anchor detail bill;
- towing bill; and
- visit and search, boarding, and custody crew bill.

Emergency Bills

Emergency bills detail procedures and specific assignments of the ship's company to perform a critical evolution where there is danger of injury or loss of life, damage to or loss of important material, or damage to the ship itself. Examples include

- general emergency bill (e.g., collision, grounding, abandon ship);
- man overboard bill;
- chemical, biological, and radiological defense bill; and
- emergency destruction of classified material bill.

Battle Bill

The battle bill provides the information necessary for all hands to fight the ship efficiently. It includes

- officer assignments to battle stations bill;
- officer succession to command bill;
- enlisted assignments to battle stations bill;
- battle messing bill;
- casualty treatment and collection stations bill; and
- general degrees of readiness/personnel assignments bill.

Special Bills

Special bills detail procedures and specific assignments of crew members for evolutions of an irregular nature or for evolutions unique to special or unusual mission assignments for the ship, including

- antisneak/antiswimmer attack bill,
- civilian evacuation bill; and
- prisoners and prisoners of war bill.

Watch, Quarter, and Station Bill (WQSB)

The WQSB is actually an amplification of the bills discussed above, inasmuch as it displays in one place the duties of each crew member in the emergency and battle bills and for each watch condition. The WQSB also shows each member's duties in some or all of the ship's administrative and operational bills.

BILLETS AND BILLET NUMBERS

The ship's company is organized for bills by assigning each member to a *billet*—a set of duties. Because it would be impossible to place enough crew members on board ship so that each person could have only one duty supporting one bill, each crew member is assigned a group of duties that provide them a part in most, if not all, of the ship's bills. Every billet is identified by a *billet number*. This number is the same on each bill. A billet number consists of four digits or a letter followed by three digits. The first digit or letter may, depending upon the size of the ship's company, indicate the member's department or division, the second the section, and the third and fourth the member's seniority within the section. A member whose billet number is O-103 is in the operations department, and in the first section, and is the third most senior member in the section. The last digit may also indicate port or starboard watch, odd or even, when a two-section watch condition is in effect for the ship. When reporting on board, you will be provided with a billet slip by your division officer or department head. This slip will list your billet number and duties for the various ship's bills. The skills required for some or all of these duties or stations may be provided you within the scope of the ship's orientation and indoctrination program, or by means of the PQS system included in the ship's training program. In either case, it is ultimately your responsibility to learn and know your station and duties for each bill *thoroughly*.

SHIP'S BELL

The passage of time is marked by the striking of the ship's bell at the end of each half hour of the watch period. The bell is struck an increasing number of times so that it is struck once at the end of the first half hour, twice at the end of the first hour, and so on to eight times at the end of the fourth hour. After eight bells are struck, the sequence starts all over again, with an odd number of bells marking a half hour, and an even number marking an hour.

For the relation between time, bells, and watches, see the table below:

Midwatch		Morning Watch		Forenoon Watch		Afternoon Watch		Evening Watch		Night Watch	
Time	Bells	Time	Bells	Time	Bells	Time	Bells	Time	Bells	Time	Bells
0030	1	0430	1	0830	1	1230	1	1630	1	2030	1
0100	2	0500	2	0900	2	1300	2	1700	2	2100	2
0130	3	0530	3	0930	3	1330	3	1730	3	2130	3
0200	4	0600	4	1000	4	1400	4	1800	4	2200	4
0230	5	0630	5	1030	5	1430	5	1830	5	2230	5
0300	6	0700	6	1100	6	1500	6	1900	6	2300	6
0330	7	0730	7	1130	7	1530	7	1930	7	2330	7
0400	8	0800	8	1200	8	1600	8	2000	8	2400 (0000)	8

Standard Routine

As discussed earlier in this chapter, the *Organization and Regulations Manual* for each ship type provides each captain with a somewhat flexible standard for organizing the ship and its company. The routines provided as examples below may vary greatly from those published for many ships, based upon each commanding officer's desires, the specific ship's type and administration, and the local situation.

Time	Daily routine in port
0345	Relieve the midwatch.
0500	Call the galley force.
0530	Call the relief for the two-hour watches.
0545	Call the OOD, master-at-arms (MAA), messcooks, and other early risers.

At five minutes before sunrise—stand by lights.
At sunrise—secure aircraft warning, deck, and in-port standing lights.

0545	Relieve the watch.
0600	Reveille. Up all hands.
	Liberty expires for messcooks.

0615	Scrub down weather decks.
	Wipe down deckhouses with fresh water.
	Sweep down compartments.
	MAA reports "crew turned out" to the OOD.
0630	Morning meal.
0715	Secure from morning meal.
0745	Liberty expires.
0745	Relieve the watch.
0755	Quarters for muster.
	Report approach of eight o'clock to the captain.
	First call to colors.
0800	Colors (moored or anchored).
	Turn to.
0900	Sick call.
	OOD is relieved.
1000	Knock off ship's work.
	Coffee break for the crew.
	Up all late sleepers.
1015	Secure from coffee break.
	Turn to ship's work.
1045	Division officers inspect compartments.
1100	Inspection of messcooks and food handlers.
	XO's request and complaint mast.
	Secure the mess deck.
1115	Knock off all hands from work.
	Noon meal for the watch.
1130	Noon meal for crew.
1145	Relieve the watch.
1155	Wind ship's chronometers.
	Report chronometers wound to the captain; request permission to strike eight bells.
1200	Test the general alarms and whistle.
1230	Secure from noon meal.
1255	Officers' call.
	All hands fall in for quarters.
1300	Quarters for inspection and drills.
1350	Secure from drills.
1400	Turn to.

1545	Relieve the watch.
1600	Clean sweepdown fore and aft.
	Empty all trash receptacles.
1630	Liberty for all hands.
1645	Evening meal for all hands.

One hour before sunset—electrician test anchor, gangway, deck lights, and searchlights.

Five minutes before sunset—first call to colors, stand by ensign and jack, stand by anchor and deck lights.

Sunset—colors (moored or anchored). Energize lights.

1900	Set material condition Yoke about the ship.
1945	Muster eight o'clock reports.
	Muster restricted personnel.
1945	Relieve the watch.
1955	Report the ship secure to the commanding officer.
2000	Strike eight bells.
2200	Taps. Lights out and silence about the decks.
	Start hourly security patrols.
2330	Call the watch relief.
2345	Relieve the watch.

Time	Daily routine under way
0150	Relieve the wheel and lookout.
0330	Call the relief watch.
0345	Relieve the watch; relieve the wheel and lookout.
0500	Call the galley force.
0545	Call the MAA, messcooks, and other early risers.

Five minutes before sunrise—stand by the running lights.

Sunrise—secure running lights.

0550	Relieve the wheel and lookout.
0600	Reveille. Up all hands.
	Trice up racks.
	Announce weather.
	Boatswain's mate of the watch secures hourly rounds.
0630	Turn to field day.

Scrub down weather decks.

Wipe down deckhouses with fresh water.

Sweep down compartments.

MAA report "crew turned out" to OOD.

0700   Knock off field day.

Morning meal for all hands.

Publish uniform of the day.

0745   Relieve the watch.

Relieve the wheel and lookout.

Boat crew of the watch to muster.

0755   Report the approach of eight o'clock to the captain.

Request permission to strike eight bells.

0800   Turn to ship's work.

0950   Relieve the wheel and lookout.

1000   Up all late sleepers.

CO's material inspection (Saturday only).

1045   Division officers inspect compartments.

1100   Inspection of messcooks and food handlers.

Report mast.

Request and complaint mast.

1115   Knock off the relief watch from work.

Noon meal for the watch.

Knock off all hands from work.

Clean sweepdown fore and aft.

1130   Noon meal for crew.

1145   Relieve the watch.

Relieve the wheel and lookout.

Boat crew of the watch to muster.

1155   Wind ship's chronometers.

Report chronometers wound to captain.

Request permission to strike eight bells.

1200   Test the general alarms and whistle.

1255   Officers' call.

All hands fall in for quarters.

1300   Quarters for inspections and drills.

1330   Secure relief wheel and lookout from drills.

1350   Relieve the wheel and lookout.

Secure from drills.

1400	Turn to ship's work.
1545	Relieve the watch.
	Relieve the wheel and lookout.
	Boat crew of the watch to muster.
1600	Knock off ship's work.
	Clean sweepdown fore and aft.
1630	Evening meal for all hands.
1750	Relieve the wheel and lookout.

One hour before sunset—electrician test the running lights and searchlight.

Five minutes before sunset—stand by running lights.

Thirty minutes after sunset—division officers verify that designated material condition is set.

1945	Eight o'clock reports. Department petty officers report the department areas "secure for the night" to their department heads, who in turn report the department "secure" to the executive officer.
1945	Relieve the watch.
	Relieve the wheel and lookout.
	Boat crew of the watch to muster.
1955	Executive officer reports the vessel secure for the night to the captain.
	Report approach of eight o'clock.
	Obtain permission to strike eight bells.
2000	Movie call.
2150	Relieve the wheel and lookout.
2200	Taps. Lights out and silence about the decks.
2315	Midwatch rations served on the mess deck.
2330	Call the relief watch.
2345	Relieve the watch.
	Relieve the wheel and lookout.
	Boat crew of the watch to muster.

Holiday Routine

Some variation from the daily ship's routine is employed, when possible, on weekends and designated holidays, and when otherwise authorized by the commanding or executive officer. Holiday routine includes a later rev-eille, extended galley hours, and the cessation of nonessential ship's work for crew members not on watch at sea or in the duty section in port.

Plan of the Day

The executive officer prepares and issues daily a plan of the day, which includes such things as

- The schedule of the routine and any variations or additions;
- The orders of the day, detailing drills, training schedule, duty section, liberty sections, working parties, time and location of movies, and so forth;
- Notices of matters to be brought to the attention of officers and crew; and
- Reprints of regulations and orders to be brought to the attention of officers and crew.

Copies of the plan of the day are distributed to the OOD, all offices, officers, and division bulletin boards of the ship. Everyone is responsible for knowing what is in the plan of the day, as any instruction or direction it provides constitutes an official order.

The plan of the day is carried out by the OOD and the division officers unless modified by the authority of the captain or the executive officer. In case of necessity, the OOD may make immediate changes and report that action later to the executive officer.

Passing the Word

In the case of certain emergency situations, special signals are employed, such as the general alarm for battle, the collision alarm, and the chemical attack alarm. The main reliance for passing the word is placed on the loudspeaker system, when installed. This system consists of a transmitter on the bridge and others at selected spots. Speakers are located at useful and conventional places throughout the ship. Most ships also have a 21MC or "Captain's command" circuit, which consists of several transmitter-receiver units similar to an office intercommunication system. When this system is used, there is always an instrument on the bridge, one in the cabin, and one in the wardroom. There are usually at least three other command circuit transmitter-receiver stations on the ship.

The Boatswain's Pipe

Even though there are loudspeakers and intercommunication systems on modern ships, the boatswain's (pronounced "bosun's") pipe is sometimes used for calling and passing the word. The purpose of these calls is

- To command silence before passing an order or information;
- To call all hands' attention;
- To call away a boat;
- To call a division or divisions to quarters;
- To assemble the boatswain's mates;
- To rouse quick notice or attention from a working group;
- To announce meals; and
- To signal that official personages are coming on board or departing.

Tattoo and Taps

Tattoo is the signal for all hands to turn in and keep silence about the decks. *Taps* is sounded five minutes after tattoo except when circumstances warrant a longer delay.

Special Sea Detail

In getting under way or entering port, a ship has a *special sea detail*. This detail consists of stationing various personnel at locations necessary throughout the ship to moor and unmoor the ship. The special sea detail is set about one-half hour before getting under way, and remains there until the ship clears the harbor; or it may be set before entering port and remain so until the ship is properly docked or anchored. Special sea detail duty takes precedence over everything else for each member of the detail assigned to it.

Sea Watch

Once the ship has cleared the harbor, the special sea detail is relieved and the *sea watch* is stationed or *set*; that is, the cruising watches begin, and the regular sections of the watch take over their prescribed duties.

Smoking On Board Ship

Smoking is permitted only on the weather decks as authorized by the commanding officer.

Lights

All lights, except those in the cabin, offices, officers' quarters, and those designated as "standing lights," are extinguished at tattoo. The lights on the lower decks are reduced in number before tattoo unless they are required for the comfort of the crew. All lights in the holds, storerooms, and orlops (lowest decks) and all open lights in the ship, except those in officers' quarters, are, in some ships, extinguished before 1930 or at the time of evening inspection by the executive officer.

During the night, a sufficient number of lights throughout the open part of the ship, generally red in color to preserve night-vision capability, are kept lighted to enable the officers and crew to turn out, to get to the upper decks, or to attend to any duty arising from an emergency. These are known as "standing lights."

Such lights and fires as the captain may deem dangerous are extinguished when the magazines are opened or when handling or passing powder, explosives, or other combustibles. In time of war or when necessary to conceal a ship from an enemy, only such lights are used as are deemed advisable by the senior officer present.

Eight O'Clock Reports

Each evening, the department heads review the security and readiness of their departments. At 1945 the department heads or their representatives muster and report their finding as to the condition of the ship to the executive officer (under way) or the officer of the day (in port). Under way, the executive officer in turn passes these reports to the captain. Thus, at the end of each day, the captain, the executive officer, and all department heads are aware of the exact state of the entire ship.

Visitors

There are two types of visitation: *routine visiting* and *general visiting*. Routine visiting means the everyday, informal visits of friends. General visiting occurs on days such as Armed Forces Day, when the ship is open to the public.

Routine visiting. Members of the crew may have civilian visitors on board ship after working hours on weekdays, and on Saturdays, Sundays, and holidays during the hours specified within ship's regulations. A member of the crew who has guests must meet them at the gangway and escort them at all times while they are on board or within the shipyard. The OOD must be advised of the impending presence of all routine visitors. In the event of sickness or accident to any visitor, a report must be made immediately to the OOD. Guests are permitted only in the spaces authorized by the commanding officer and may be subject to other restrictions as the commanding officer may direct.

General visiting. Each ship lays out a plan for handling visitors—the general public as well as U.S. and foreign dignitaries who may be expected on days of open ship. Prior to visiting hours the word is passed, such as, "Rig ship for general visiting. All hands shift into uniform of the day." While visitors are on board, all officers and crew must remain alert that no visitors be permitted in an unauthorized part of the ship, that no visitors endanger themselves, and that all guests be treated with special courtesy.

Pets

Pets are not generally allowed on board ship. A pet may be adopted as a ship's mascot only by direct authorization of the commanding officer.

Officers' Quarters and Bridge

Crew members are allowed on the quarterdeck only when on duty. This is also true of officers' country (wardroom and staterooms), and the bridge, wheelhouse, and charthouse. Personnel should uncover when entering the cabin or wardroom; conversely, all personnel should remain covered on the bridge.

Prisoners at Large (PAL)

For certain reasons, personnel may be denied liberty or leave. Such members may not leave the ship without specific authority of the executive officer. This group includes those who are

- Confined to the limits of the ship by sentence of court-martial;
- Awaiting trial by, or approval of, court-martial;
- Awaiting mast for serious cases (designated PAL by the commanding officer);
- Restricted to the ship by punishment assigned at mast;
- On report and awaiting mast for other than serious offenses but not yet made PAL by order of the commanding officer; or
- On the medical restricted list.

Restricted Personnel

Restricted personnel are under restraint according to the Uniform Code of Military Justice. They are either under *restriction* or under *arrest*.

Restriction is a restraint of the same nature as arrest, imposed under similar circumstances and by the same authorities, but it does not involve the suspension of military duties.

Arrest is the moral restraint of a member, by an order, oral or written, to certain specified limits pending disposition of charges against the individual. It is not imposed as punishment, and it relieves the member of all military duties other than normal cleaning and policing.

Liberty

Normally, liberty commences at 1600 or 1630 daily and expires on board at 0745 the following day. Liberty is normally granted to three-quarters or two-thirds of the crew, depending upon circumstances. Personnel on continuous running watches, such as operations specialists on board some ships, may be granted liberty commencing at 1400 if the head of the department requests it for them.

Special liberty will be granted only for good reasons. A request for such liberty must be made on a form provided, signed by the division officer and the head of the department, and delivered to the executive officer's office prior to 1000. The request, marked "approved" or "disapproved," will be returned to the requesting member.

Liberty may be exchanged in special cases if a member obtains a volunteer relief from someone in the member's division who rates liberty. Such a relief, called a *standby*, must be of similar rating or with similar qualifications. The member desiring liberty submits a written request to the member's division officer. If the division officer approves, the request is delivered to the executive officer for final action.

Liberty may be granted more frequently in commands that have enough personnel to allow it and still maintain security. *Meritorious liberty* may be granted to personnel for superior conduct, performance of duty, or achievement. Remember: liberty is an earned privilege, not a right.

QUARTERS FOR MUSTER AND INSPECTION

Division Parades

Each division is assigned a *parade*—a space on deck where division personnel assemble for quarters or inspection. The fair-weather parades are on the uncovered decks; the foul-weather parades are on the covered decks and in the living spaces as designated.

Where possible, the departments form with all divisions facing in board. Divisions that are aligned athwartships face forward if in the forward part of the ship and aft if in the after part.

Mustering the Divisions

At the command FALL IN, each section forms in line. Section leaders muster the sections while the division is being formed. When the division is formed, the division's leading petty officer commands REPORT.

The section leaders in succession from front to rear, without moving from their assigned positions, report, "All present or accounted for," or "All present or accounted for with exception of (identity of absentees)."

The division's leading petty officer commands: 1. Dress right, 2. DRESS. After dressing the division, the division petty officer commands: 3. Ready, 4. FRONT. The division's leading petty officer posts in front of the division, faces the division officer, salutes, relays the report of the division to the division officer, and, without command, faces about and moves by the most direct route to post on the right flank of the front rank of the division.

The division officer proceeds with the inspection of the division. After the inspection, the division officer leaves the division in the charge of the junior division officer and falls in with other division officers to report to the executive officer.

Division Inspections

Ranks are opened for inspection unless the division parade space is too limited. The division officer posts on the right flank of the front rank. Upon the approach of the inspecting officer, the division officer presents the division by executing the hand salute and reporting, "Sir/ma'am, the division is ready for inspection." The inspecting officer inspects the division rank by rank, in succession, passing from right to left along the front of each rank, and from the left to right along the rear of each rank.

Preparing for inspection. The executive officer or department heads issue the command PREPARE FOR INSPECTION. The division officer or the leading petty officer of each division, posting at center and facing the division, commands: 1. Open ranks, 2. MARCH. At the command MARCH, the front rank takes two steps forward, halts, and executes dress right; the second rank takes one step forward, halts, and executes dress right; the third rank stands fast and executes dress right, and the fourth rank, if any, takes two back steps, halts, and executes dress right. The officer in charge aligns each rank in succession, moves three paces beyond the right flank of the front rank, halts, faces to the left, and commands 3. Ready, 4. FRONT.

To close ranks, the commands are 1. Close ranks, 2. MARCH. At the command MARCH, the front rank stands fast; the second rank takes one step forward and halts; the third rank takes two steps forward and halts; and the fourth rank, if any, takes three steps forward and halts. Each element covers their file leader.

To form for inspection, where the restricted space of the parade does not permit normal opening of ranks, the commands are 1. Form for inspection, 2. CLOSE DISTANCE. The division stands fast and, upon the approach of the inspecting officer, is presented to the inspecting officer in the standard manner.

When the front of the first rank has been inspected from right to left by the inspecting officer, the division officer commands 1. First rank one pace forward, 2. MARCH. The rear of the first rank is inspected from left to right, followed by the inspection of the front of the second rank from right to left. The same procedure is followed for the second, third, and fourth ranks.

General Muster (All Hands Aft)

The general muster or dress parade (sometimes called "all hands aft") is a massed formation of the ship's company at a designated place. Dress parade is formed with standard close-order drill commands, though with some variation from ship to ship.

Enlisted Details

Many jobs involving work and responsibilities are not covered by any particular rating description. Such duties may involve responsibility for a division's damage-control fittings and equipment and the orderliness of ship's berthing and messing spaces. Personnel may be detailed to these jobs by a division officer, department head, or the executive officer.

Inspections

Inspections are made on a daily, weekly, monthly, quarterly, semiannual, annual, and as-required basis. They are made to verify the material, administrative, and operational readiness of the ship to perform its missions and related functions.

Daily Inspection

At quarters for muster, division officers check their personnel for general appearance and cleanliness. They also make frequent inspections of the spaces, equipment, and supplies that belong to their division. Heads of departments, either by personal or delegated inspections, ensure that all aspects of their departments are in proper condition or that steps are being taken to make them so.

The executive officer likewise is required to make frequent inspections; usually the XO inspects a different part of the ship each day. Division officers and department heads or their representatives are required to inspect all parts of their respective departments each day as far as may be feasible.

Captain's Material Inspection

Captain's material inspection is usually held on Friday in port, and on Saturday mornings when under way, and consists of a formal inspection of the ship by the commanding officer or the commanding officer's delegates.

Personnel Inspection

Normally conducted once a month, personnel inspection is conducted by the commanding officer or the commanding officer's delegates to ensure that all members of the ship's company are in compliance with established uniform and grooming standards.

General Mess

At least once daily, the OOD or assistant inspects the general mess and samples a ration to determine that it is wholesome and properly served.

Messcooks

Messcooks and food handlers are inspected daily by the chief master-at-arms or the OOD, and weekly by the medical officer or the medical officer's assistant.

Locker Inspection

Locker inspection usually is made by the division officer but may be combined with another inspection. Such an inspection, normally conducted for personnel E-5 and below, is to determine that each has a complete uniform outfit, properly marked, cleaned, and stowed.

Change of Command Inspections

When the commanding officer of a ship is to be relieved, the commanding officer makes a thorough inspection of the ship—personnel, matériel, and administration—in company with the incoming commanding officer.

Customs and Health Inspections

When a ship returns to the United States from foreign waters, the ship is liable to inspection by U.S. customs and health authorities to ensure that no unauthorized or contraband items are on board and that no contagious or communicable diseases are among the crew. These inspections can be routine, but inspectors can turn a ship "upside down" if they have cause to suspect that anything is being hidden.

Figure 24-1.
The basic business of a professional afloat is carrying out the watch routine.

CHAPTER 24

Shipboard Watches

For the proper performance of duty, every unit in the Coast Guard must be organized. The organization for similar units throughout the service is similar; to make this organization work, routines and duties in similar units are standardized. Even time has been standardized.

In the Coast Guard, all routines are based on a twenty-four-hour day. Time is counted in hours 0 to 24 and minutes in each hour 01 to 59. The day begins at midnight, when the time is 0000, and ends with 2359. One minute later, the time is 0000, or midnight, and the next day starts. On board ship, the day is also counted in half-hour periods, marked by the ship's bell, and by watches, which regulate the time on and off duty.

WATCHES

A ship in commission always has part of its crew on watch. Even when she is tied up in port and receiving electricity from the pier or another ship, it is necessary to have crew members on watch for communications, security, and safety. Personnel assigned to watches are called *watchstanders*. Traditionally, the twenty-four-hour day is divided into seven watches, as follows:

0000–0400	midwatch
0400–0800	morning watch
0800–1200	forenoon watch
1200–1600	afternoon watch
1600–1800	first dogwatch
1800–2000	second dogwatch
2000–2400	first watch

The two "dogwatches," from 1600 to 1800 and 1800 to 2000, serve to alternate the daily watch routine so that personnel who have the midwatch one night will not have it again on the following night.

Watch may refer to the *location* of the person on watch, such as the forecastle watch or bridge watch, or to the *section* of the ship's crew on duty: "Section two has the watch."

Watch Section

Each crew member is assigned to a section of the watch. These sections are numbered; small ships usually have three sections, and large ships may have four sections. Thus, when the word is passed that the first section (or the second or the third) has the watch, each person in that section immediately reports to his or her assigned watch station. At sea, watches may be either two or four hours in length.

On some ships the divisions may be divided into a *starboard watch* and a *port watch*. Each of these watches is divided into two sections: the odd-numbered sections, 1 and 3, are in the starboard watch; the even-numbered sections, 2 and 4, are in the port watch.

Crew members in a three-section watch are spoken of as having a one-in-three watch; those in a four-section watch have a one-in-four watch, and so on. Thus, a crew member standing a one-in-three is on watch for four hours, off for eight hours, and then back on watch for four more. In all, that person stands eight hours' watch in twenty-four hours.

Relieving the Watch

Watches must be relieved on time. This does not mean at the exact minute the watch changes, but usually fifteen minutes before so the relief can receive information and instructions from the person on watch. Some ships muster the oncoming watch to make sure that each crew member is ready ahead of time.

When reporting directly to the person relieved, an oncoming watchstander says, "I am ready to relieve you." The person on watch passes on any pertinent instruction or information relating to the proper standing of the watch. When relieving watchstanders are sure that they understand conditions and instructions given, they say, "I relieve you." Thereafter they are completely responsible for the watch.

Remember that you must report to the responsible officer (officer of the deck or engineering officer of the watch [EOOW], for example) that you have been relieved.

Watch Officers

A watch officer is placed in charge of a watch or a portion of a watch. The commanding officer assigns as a watch officer any commissioned or warrant officer who is considered qualified. The CO may, when conditions require, assign a petty officer to such duty. The stations of the officers or petty officers in charge of the watch are where they can best perform their duties and supervise and control the performance of those on watch under them.

Logs

The *deck log* is a complete daily record, by watches, in which is described every circumstance and occurrence of importance or interest that concerns the crew and the operation and safety of the ship, or that may be of historical value. The navigator has general responsibility for preparation and care of the deck log. Each officer of the deck (OOD), however, is responsible for the proper completion of the log during the OOD's underway or anchor watch or for the entire duty day if in port.

The *quartermaster's rough log* is a record of events occurring during the watch. Entries in this log are made at the time of occurrence or when knowledge of such occurrence is first obtained. The quartermaster's rough log is signed by the quartermaster of the watch on being relieved. It is used in preparing the deck log, also called the *smooth log*.

ENLISTED WATCHSTANDERS

The two most interesting aspects of Coast Guard life are what you do and what the unit does. There are literally hundreds of different jobs in a large command, each one important to the mission. Because organization and routine are standardized as much as possible, the jobs you perform are also generally standardized.

A day's work on board ship, or ashore, involves both professional and military duties. "Turn to ship's work" means that you carry out the professional duties of your rating. During general drills, and when on watch, you may perform both general military assignments and professional duties. The daily routine, at sea, in port, or ashore, also requires crew members on duty at all times as watchstanders. The following is a list of the most common enlisted watchstanders, with a brief description of their duties.

Boatswain's Mate of the Watch

The petty officer in charge of the watch is one of the most important enlisted assistants to the OOD. The status of the boatswain's mate of the watch (BMOW) in these respects is the same regardless of the readiness condition in effect.

At sea, the normal peacetime deck watch for which the BMOW is responsible consists of the *helmsman, lee helmsman, OOD messenger, lookouts,* and *lifeboat crew of the watch* (also called *ready boat crew*). Besides being a principal enlisted assistant to the OOD, the BMOW is the watch petty officer. It is the BMOW's responsibility that all deck watch stations are manned and that all hands in the previous watch are relieved. The BMOW reports to the OOD when the deck watch has been relieved.

Main Engine and Auxiliaries Watches

The main engine and auxiliaries watches are stood by qualified personnel in the engineering division and under the supervision of the EOOW. These watchstanders ensure that all machinery is properly maintained and operating properly.

Bridge and Signal Watch

When stationed, in addition to their regular duties, bridge and signal watches keep the OOD informed of notable changes in the weather, boats approaching the ship, unusual disturbances or signs of distress in the harbor, and movements of other ships.

Cold Iron Watch

The cold iron watch inspects main machinery spaces that are secured and have no regular watch posted; this watch reports hourly to the OOD.

Combat Information Center (CIC)

Watchstanders perform duties under the supervision of the CIC officer and as required by the OOD.

Helmsman

The helmsman steers the ship as directed by the OOD or the conning officer.

Lookouts

Every ship will have at least one lookout. On larger ships, additional lookouts may be stationed, when circumstances demand, on the flying bridge, on the fantail and bow, or elsewhere. Bridge lookouts may report in person to the OOD; others will report via telephone. This is a very important job; it may look to an outsider as if all a lookout has to do is stand around and enjoy the view, but actually it takes considerable skill to stand a good lookout watch, and the safety of the ship may depend on just how well the lookout attends to duties.

The first duty of the lookout is to sight an object and report it. The second duty is to identify it. The report of an object sighted should not be held up until the object is identified. After reporting, the lookout must be certain that the OOD acknowledges the report.

Bridge lookouts. These lookouts, sometimes port and starboard, stand watch in the wings of the bridge or on the flying bridge. They report ships, aircraft, land, or any object sighted, and at night they make half-hourly reports on the ship's running lights.

Forecastle lookout. These lookouts may be stationed during periods of low visibility—fog or snow—and while entering harbor.

Fog lookout. This lookout may be the forecastle lookout. Other fog lookouts may be stationed on the wings of the bridge to keep watch astern. Their primary job is to listen carefully.

Sky and surface lookouts. On larger ships, when enemy operations are expected, additional lookouts are assigned certain sectors of the sky and surface all around the ship.

Anchor and Security Watches

Crew members are assigned to the anchor watch to assist the OOD in such tasks as reporting signs of dragging the anchor, veering chain, or other emergency work. The number of people assigned and their specific duties vary from ship to ship.

A security watch patrols topside and below decks, whether at sea or in port. At sea, the BMOW is normally assigned this responsibility. In port, it usually falls upon the *messenger*, who will make these rounds between taps and reveille.

Messenger

The crew member assigned to the OOD, either on the bridge when under way or on the quarterdeck in port, is commonly called the OOD messenger. The messenger should know the names and locations of the various parts of the ship, department offices, and department heads; names and duties of the various officers and senior enlisted; and where to find various crew members. The messenger must be familiar with standard naval phraseology to understand and repeat messages without error or confusion.

Phone Talker

There may be one or more phone talkers, depending on the ship and the operation. Phone talkers must use correct phone procedure, know all the stations on their circuit, transmit orders to the stations as directed, and inform the OOD of all information received.

Petty Officer of the Watch (In Port)

This is one of the most responsible jobs an enlisted person can have. There should be no confusion as to the official status of these petty officers; they are assistants to the *officers of the deck*, subject only in the performance of their duty to the orders of the commanding officer, executive officer, OOD, and junior officer of the deck. The assignment of petty officers as petty officer of the watch is made in writing, either in the ship's organization book, in the senior watch officer's watch list, or in the plan of the day.

Quartermaster of the Watch (Under Way)

The quartermaster of the watch assists the OOD or conning officer in navigation; supervises the performance of the helmsman; instructs the OOD messenger in calling officers and crew for the watch; reports to the OOD all changes in weather, temperature, and barometer readings; and makes appropriate entries in the ship's log. The quartermaster of the watch often acts as the junior officer of the deck when there are insufficient officers on board.

Operations Specialists of the Watch

The operations specialist of the watch maintains required communications in the main radio room, receiving, transmitting, and routing traffic as required.

Sounding and Security Patrol

The sounding and security patrol is a crew member in the engineering department who maintains a continuous patrol of unmanned spaces below decks. The patrol takes periodic soundings of designated tanks and spaces, checks damage-control closures, and is alert for evidence of sabotage, theft, fire, and fire hazards. In addition to making an hourly report to the EOOW or OOD, the patrol frequently relays the report of the cold iron watch. Both of these watches are usually manned during nonworking hours at sea and in port.

Late Sleepers

Crew members on watch from 0000 to 0400—the midwatch—have a difficult time getting enough sleep. For this reason, they are permitted to sleep in after "all hands" are called. They are the "late sleepers" referred to in the daily routines.

STANDARD COMMANDS AND TERMINOLOGY

There is a correct and proper name for everything and a standard form of order or command for everything that is done. Disregard for proper terms or orders may lead to mistakes, confusion, and perhaps disaster.

Commands to the Helmsman

All commands to the helmsman that specify a change in course or heading will refer to the compass by which the helmsman is steering and will give the new course or heading in degrees: "Come left to course two seven zero," or "Steer zero zero five." Helmsmen repeat most commands that are given to be sure they understand them. Only the following commands are used:

- *Right (left) standard rudder.* Put the rudder over to the right (left) the specified number of degrees necessary for the ship to make its standard tactical diameter.

- *Right (left) full rudder.* Put the rudder over to the right (left) the specified number of degrees necessary for the ship to make its reduced tactical diameter.

- *Right (left) five (ten, etc.) degrees rudder.* Turn the wheel to right (left) until the rudder is placed at the number of degrees ordered. (This command is used in making changes of course. The helmsman would then be ordered to steer the new course by such command as, "Steady on course ————," in time to permit the helmsman to meet it on the new course. The complete command would be, for example, "Right five degrees rudder. Steady on course two seven five.")

- *Increase your rudder to* ——— *degrees.* Increase the rudder angle. (Given with the rudder already over, when it is desired to make the ship turn more rapidly. The command must be followed by the exact number of degrees of rudder desired.)
- *Ease your rudder to* ——— *degrees.* Decrease the rudder angle. (Given when the ship, turning with right (left) rudder, is turning toward or nearing the heading desired. The command can be given, for example, "Ease to fifteen.")
- *Meet her.* Use rudder as may be necessary to check the ship's turn. (Given when the ship is nearing the desired course.)
- *Steady,* or *steady as you go.* Steer the course on which the ship is heading when the command is received. (Given when, in changing course, the new course is reached or the ship is heading as desired. The helmsman responds, "Steady as you go. Course ——— sir/ma'am.")
- *Rudder amidships.* Rudder angle zero. (Given when the ship is turning, and it is desired to make her swing less rapidly.)
- *Shift your rudder.* Change from right to left rudder (or vice versa) an equal amount.
- *Mind your helm.* A warning that the ship is swinging off the course because of bad steering. It is also a command to steer exactly, using less rudder.
- *Nothing to the left (right) of* ———. The helmsman is not to steer to the left (right) of the course ordered.
- *Mark your head.* Course currently being steered by the helmsman. The helmsman should reply, "Mark ——— degrees."
- *Keep her so.* A command given when the helmsman reports the ship's heading and it is desired to steady her on that heading.
- *Very well.* Given after a report by the helmsman to let the helmsman know that the OOD understands the situation.
- *Come right (left) to* ———. Put over the rudder right (left) and steady on new course.
- *Command.* Asked by the helmsman when the command from the OOD is unclear or misunderstood.

Lookout Reports

The proper way for a lookout to let others know exactly where an object is requires that the lookout give its direction with reference to the bow of the lookout's ship, that is, by means of *relative bearings*. To locate surface objects, only the direction and the range or distance from the ship need be given. To locate aircraft, the direction and the elevation, or height above water, are required. The elevation is known as the *position angle*.

In reporting objects, a strict form of wording is used. First the object is named: "ship," "buoy," "discolored water," "floating log," "periscope wake," "aircraft." If the object cannot be identified, the lookout says

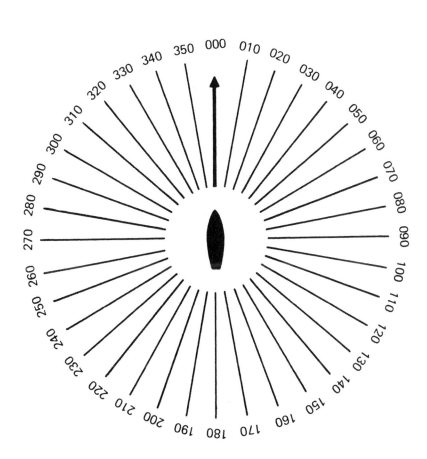

Figure 24-2. Relative bearings, measured clockwise from the ship's head, locate an object in relation to the ship. They have nothing to do with geographical directions.

"object." Next, the direction is given as explained below. Last, if the object is a surface craft, the range is given; if an aircraft, the elevation, or position angle, is given. Thus, a full report on a surface object would go something like this: "Buoy" (the object); "zero niner zero" (the bearing); five hundred" (the range).

When reporting relative bearings and ranges, first, say the word "bearing," then give the degrees of the bearing in three digits. Correct wording is "Bearing three five zero" and "Bearing zero one zero." There are no "ohs" in relative bearings; only "zeros" are used. The word "bearing" may be omitted. Do not add the word "degrees." Since the bearing comes first and is always in three digits, the receiver knows that degrees are intended and it is not necessary to say so.

The range is always given in yards. In making estimates, it might be useful to remember that two thousand yards equals one nautical mile. When the range is an even hundred or thousand, it is spoken so: "Range eight hundred (800)," "Range twenty thousand (20,000)." When the range is below one thousand, and not in even hundreds, it is spoken in hundreds and the number of digits. Ciphers are pronounced "zero": "Range nine seven zero (970)." When the range is above one thousand and in even hundred, it is spoken in thousands and hundreds: "Range four seven hundred (4,700)," "Range five zero two hundred (50,200)," "Range one one zero six hundred (110,600)." Where the range is above one thousand and not in even hundreds, it is spoken in number of thousands, the balance given as in ranges below a thousand: "Range three three five zero (3,350)," "Range three zero two five zero (30,250)."

Distance of Visibility at Sea

Estimating distances at sea takes practice. The table below will be of assistance. As an example, consider that the eyes of a 6-foot person standing on a bridge deck 30 feet above sea level are a little more than 35.5 feet above sea level, so the horizon is 6.8 miles away. But objects high enough to be seen above the horizon can be seen more than 6.8 miles away. A lighthouse 65 feet high has a horizon of 9.2 miles; this means that the person could see the light just at the horizon at a distance of 6.8 + 9.2 miles, or 16 miles.

Height (feet)	Nautical miles		Height (feet)	Nautical miles
1	1.1		31	6.4
5	2.5		35	6.8
10	3.6		40	7.2
15	4.4		45	7.7
20	5.1		50	8.1
25	5.7		75	9.9
30	6.3		100	11.5

Position Angle

The position angle is the height in degrees of an aircraft above the horizon (see figure 24-3). The horizon is zero degrees (0°) and directly overhead is ninety degrees (90°). For position angles, the ordinary way of expressing numbers is used; that is, say "one," "eight," "twenty," "sixty." As with bearings, the word "degrees" is omitted since the receiver automatically understands that, with aircraft, the numerals stated mean degrees of elevation. Examples of reporting position angles are

08° "position angle, eight"
10° "position angle, ten"
30° "position angle, thirty"

Note that the term "position angle" precedes the numerals of the report.

An example of a sighted aircraft report might be: "Bridge, sky aft, plane, bearing one four five, position angle twenty-five, moving right."

SHIP CONTROL

One of the most important people in a bridge crew is the *helmsman*, who steers the ship. The helmsman must know the course ordered and the rudder angle required to keep the ship on that course. The helmsman must understand the use of the gyro and magnetic compasses, wheel, and rudder, the principles of steering a ship, and the proper helm commands and responses.

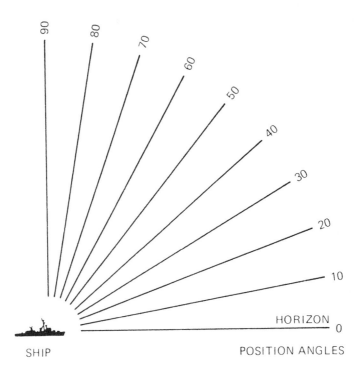

Figure 24-3. Position angles locate an object in the sky. They measure up, not down.

Compass

The *magnetic compass* consists of a magnetized needle attached to a circular compass card; both are supported on a pivot set in a cast bronze bowl filled with a mixture of alcohol and water. This liquid supports the card and magnet, thus reducing the friction and letting the card turn more easily on the pivot. It also slows the swing of the card and brings it to rest more quickly. The lubber's line marked on the bowl agrees with the fore-and-aft line of the ship. The heading (direction the ship is going) is read from the compass card at the point nearest the lubber's line. A magnetic compass can be thrown off the correct heading by the attraction of any metal near it. Never go on watch as helmsman carrying a large knife or other metal articles.

A *gyrocompass* is basically a heavy spinning flywheel that, because of the rotation of the earth, will seek a position in which its axis always points true north and south and is unaffected by variation or deviation.

Gyro repeaters are set to point the same way as the master gyro and are located in the wheelhouse, steering stations, and on the bridge wings. If the ship has two master gyros, there will be two repeater stands in the steering station, but only one of these will be used. Be certain that you are steering from the repeater in use, and be especially careful not to confuse them.

The course by gyro, when corrected for any slight mechanical error, is always the true course, but the gyro is subject to mechanical and electrical failure. Report to the OOD at once if it stops swinging as the ship's head changes, or if it starts to jump.

Rudder

All ships or boats are steered by one or more rudders at the stern. When the rudder is moved to one side or the other, the headway of the ship causes water to push against the side of the rudder, creating a force that swings the stern of the ship in the opposite direction. The faster a ship is moving, the greater the pressure against the rudder and the quicker the turning effect. A ship always "answers" its rudder more quickly at high speed than at low speed, and it takes more rudder to turn a slow ship than a fast one. Large rudders are balanced, with part of the rudder forward of the pintle on which it hangs. This is to balance some of the force of the water against the after part and thus make the rudder easier to turn.

The rudder of a small boat is moved mechanically by rudder ropes (usually of wire) as the wheel turns. Ships have an electric motor called a *steering motor* that pumps hydraulic fluid through a system that moves the rudder.

Steering Sense

Many an inexperienced helmsman gets into trouble when bringing a ship to a new heading, trying to make the compass card turn by turning the ship's wheel. *It won't work!* The compass card floats and is not fast to the compass. The compass stand is fast to the ship; the lubber's line moves as the ship moves, and always in the direction the wheel is turned. Remember: turn the wheel the way the bow is to swing; if the wheel turns left, the lubber's line moves left, and the ship turns left, everything moves in the same direction, left, or right, from bow to rudder.

Steering a ship is not as simple as driving an automobile, where control is quick and positive. A ship is heavy, slow to start turning, and sometimes reluctant to stop turning. When it reaches the proper heading, you may have to give it a touch of rudder to finally steady it, and you must get the rudder off it before it starts to swing again.

The most common error is using too much rudder. The new helmsman turns the wheel a few degrees, but nothing happens; it takes a little time before the ship begins to answer the rudder. After a second or two, the helmsman turns the wheel a little more. When the lubber's line reaches the course desired, the helmsman puts the rudder amidships, but the ship keeps on going. The helmsman puts on opposite rudder, but has to increase it considerably before the swing stops, and starts again immediately in the other direction with increased speed. Remember, the less rudder you use, the better; much less rudder is required to head off a swing than to stop one that has started.

Have the ship steady on its course before you surrender the wheel to your relief. Do not turn it over in the middle of a swing. Tell your relief the course and which compass or repeater you are steering by. Tell your relief about any steering peculiarity you know of—for example, "Taking a little right rudder," "Taking mostly left," and so on. Relay any orders you received that still are standing, as, "Nothing to the left," "Steady as you go," and so forth. Before leaving the steering station, report to the OOD that you have been relieved.

Engine Order Telegraph

Speed orders to the engine room are handled by an engine order telegraph, which generally has sectors marked AHEAD: FLANK, FULL, STAND (standard) ⅔, ⅓—STOP—BACK: ⅓, ⅔, FULL. A lever, fitted with an indicator, travels over the circular face of the instrument. When moved to the required speed sector, an answering pointer follows to the same sector as soon as the engine room has complied with the order.

A ship with one engine has a telegraph with a single handle. Two-engine ships generally have a handle on the port side and another on the starboard side of the telegraph. Be sure you grasp the handle for the correct engine. If the answering pointer moves to a line between two sectors so that you are in doubt about the speed set on the engine, repeat your operation on the lever. Report immediately to the OOD if the pointer does not move as ordered.

Orders to the person operating the engine order telegraph are given in this form: First, the engine to be used, *port* or *starboard*; next, the direction, either *ahead* or *astern*; then the speed, *standard*, *one-third*, and so forth. Make sure you have heard an order correctly, and repeat it aloud distinctly before you act, thus: "Starboard engine ahead two-thirds, sir/ma'am." When the answer appears on the pointer from below, sing out, "Starboard engine answers ahead two-thirds, sir/ma'am."

Ship Control Console

Ship control consoles have controls for the main engines, the bow thruster (an athwartships propeller forward to help steer the ship), and the rudder. Indicators are located on the panel to show engine revolutions per minute, direction, and degrees of thrust (on variable-pitch propellers), and rudder angle and direction of the bow thruster. Commands relating to propeller pitch are given in so many inches or feet of thrust as "zero pitch," twelve inches ahead, twelve astern or any desired pitch. In ships with control consoles, the conning officer usually handles all the controls.

Figure 25-1.
Crew member dons a chemical, biological, radiation, or nuclear (CBRN) mask during a general quarters drill on board USCGC *Bertholf* (WMSL-750).

CHAPTER 25

Shipboard Drills and Emergencies

General drills on board ship are designed to prepare a crew to handle emergencies. At shore stations, drills are based on shipboard practices. By using a standard system of drills throughout the Coast Guard, it is possible for a crew member to move from one ship or unit to another and still know generally what to do in an emergency.

On board ship, a general drill means all hands—everyone. You must know exactly where your station is and what your job is for each drill or emergency. This important information is clearly stated on the Watch, Quarter, and Station Bill for each division and is usually posted in the division's living compartment. Your job may be important, or perhaps you may be required only to stand at quarters waiting to be assigned as a relief. No matter what it is, no one is excused from any drill unless permission has been granted by the executive officer through the department head.

General drills develop teamwork so that every crew member in every department, division, and section knows and does exactly what must be done in every emergency. Every job has been assigned for a purpose. By drilling together, the ship's crew develops unity of action and teamwork. For this reason, all hands must participate wholeheartedly in all emergency drills.

ALL-HANDS DRILLS

All-hands drills are the emergency drills: General Quarters, Fire, Man Overboard, Collision, Rescue and Assistance. When the alarm for one of these sounds, go to your station with a sense of urgency—FORWARD and UP the STARBOARD side, DOWN and AFT the PORT side. Maintain silence; only personnel in charge will speak.

ALARMS

General Quarters (GQ) or General Emergency (GE)

Sounding of the general alarm plus the words "general quarters" or "general emergency" passed twice over the general announcing system, for example, "General quarters! General quarters! All hands man your battle stations."

Fire

Sounding of the general alarm plus the word "Fire" passed twice over the general announcing system, giving class of fire and compartment location, for example, "Fire! Fire! Class Bravo fire in compartment 1-10-3, Paint Locker."

Man Overboard

Word passed twice "Man overboard, man overboard, port (starboard) side," followed by five or more short blasts of the whistle. Do not sound the general alarm as stations for man overboard are normally different from those for GQ.

Collision

Sounding of collision alarm plus words "Collision, port (starboard) side, forward (aft, amidships), all hands brace for collision." After a collision, all hands go to GQ.

Rescue and Assistance

Word passed over the general announcing system, as appropriate: "Plane crash, port (starboard) side," "Away, the rescue and assistance detail, rescue survivors." Do not sound the GQ alarm.

EMERGENCY BILLS

Emergency bills detail procedures and the specific assignment of all hands to perform an evolution on short notice when there is danger of loss of life or of the ship.

- General emergency
- Man overboard

- Emergency destruction
- Abandon ship

General Emergency

The general emergency bill organizes the crew to handle the effects of a major emergency, such as collision, grounding, explosion, earthquake, storm, or battle damage. It also provides for an orderly process of abandoning ship, if necessary, and for salvage of the ship, if feasible. It is set up to assign crew members to necessary duties, whether a full crew or partial crew is on board.

The general emergency bill will not give detailed duties for every emergency because it is impossible to plan for all emergencies. The one thing all hands must remember is that they must carry out all orders from persons in authority with speed and precision. As it is possible that any emergency may result in casualties, all hands must learn the general duties and responsibilities of those senior to them because they may have to take them over.

Man Overboard

The first necessity when a person goes overboard is prompt action. Anyone who sees someone go overboard should immediately give the alarm, "Man overboard, port (starboard) side," drop life rings or life jackets if possible, try to keep the person in sight, and point. If a smoke float and dye marker are available, they should be dropped. Make sure the officer of the deck (OOD) is informed immediately. Every underway watch is organized to handle this emergency. The OOD will maneuver the ship to reach a recovery position, at the same time having the word passed twice, sounding five or more short blasts on the whistle, hoisting OSCAR by day or showing two pulsating red lights by night. These signals apply to naval task force operations and are practiced by all cutters. The ready-boat crew will stand by to lower away when directed. A helicopter may be launched, if available, as the helicopter can quickly spot people in the water and can pick them up even if they are unable to help themselves. A muster of the crew is held in order to find out who is missing.

There is always the possibility that the person overboard may be *you*. If this should happen, keep your head. Do not panic or despair. Hold your breath when you hit the water and the buoyancy of your lungs will bring you to the surface. Do not swim frantically away from the ship; the screws will not suck you under because they are too deep in the water. Just keep afloat and try to stay right where you went in. The ship will maneuver to come right back down its track to you.

Even if no one saw you go over, keep afloat. When a person is missed, ships and aircraft commence a search. People have been found and picked up after several hours in the water. In at least one case, a person was picked up after an entire day.

If a person goes overboard in port, the alarm is sounded as usual, and the OOD uses the best available method of rescue. Boats in the water will assist in any emergency without orders.

Emergency Destruction

Emergency destruction is action ordered by the commanding officer to burn or otherwise destroy registered publications and cryptographic material to avoid their falling into enemy hands. Personnel assigned to duties under this bill are given specific and detailed instructions. In shallow water (less than one hundred fathoms) papers must be burned; in deep water (more than one hundred fathoms) papers may be permitted to go down with the ship.

Abandon Ship

Abandon ship is one emergency in which many senior officers and petty officers may be lost as battle casualties, and full responsibility for leadership may fall on the shoulders of very junior personnel. Abandon ship stations and duties are noted on the Watch, Quarter, and Station Bill. Additional details are listed in the *Coast Guard Organization Manual*. Careful planning takes care of who goes into which boat or raft and what emergency equipment is to be supplied and who supplies it. Know your abandon ship station and duties. Know *all* escape routes from berthing spaces or working spaces below decks to topside, how to inflate a life-jacket, how to lower a boat or let go a life raft, how to handle survival gear, and how to do it all in the dark.

Only the commanding officer can order the ship abandoned, and the commanding officer will do so only after all efforts to save the ship prove futile. When the abandon ship alarm sounds, act fast. It is your last chance. Survival at sea depends on knowledge, equipment, training, and self-control; it is your only aim after you do abandon ship, but you have to prepare and train for it before the emergency comes. Disaster can strike suddenly at sea; a ship can go down within three minutes after a collision or explosion. On the other hand, people with very little equipment but plenty of self-control have survived for more than forty days in an open boat.

Remember that it is a seaman's duty to "maintain hope, perseverance, and obedience." If personnel cooperate in keeping high morale, they will be happier and more comfortable, and most important, they will survive. When you have to abandon ship, that is the name of the game—survive!

THE BATTLE BILL AND THE WATCH, QUARTER, AND STATION BILL

Two documents devoted to organizing and assigning the crew on a ship that are common to all ships but not included in the *Coast Guard Organization Manual* are the battle bill (or the Ship Manning Document) and the Watch, Quarter, and Station Bill.

The Battle Bill

The battle bill assigns personnel with certain qualifications to specific jobs on a ship while at GQ and complements the organization of watches for conditions of lesser readiness when all hands are not at a battle station.

In the battle bill, each station and duty is assigned to an enlisted person by a *billet number*—a combination of numbers and letters indicating a person's division, section within the division, and seniority within the section. All hands reporting on board ship, when assigned to a division, will be given a *billet slip* by the division officer, listing the billet number and duties for the various bills. It is each person's responsibility to know the stations and duties for each bill.

The Watch, Quarter, and Station Bill

The Watch, Quarter, and Station Bill displays in one place the duties of each person in each emergency and watch condition. It also shows the duties in the administrative and the operational bills. The readiness conditions are

Condition I. General quarters: All hands at battle stations.

Condition II. Modified general quarters: Used only in large ships, it permits some relaxation among personnel.

Condition III. Wartime cruising: Usually only one-third of the crew on watch and only certain stations staffed or partially staffed.

Condition IV. Peacetime cruising: Only necessary personnel on watch, the remainder available for work and training.

Condition V. Peacetime watch in port: Enough personnel on board to get ship under way if necessary or to handle fires and similar emergencies.

Condition VI. Peacetime in port: No armament staffed.

Variations in these conditions include

Condition 1A. Amphibious battle stations: All hands on station to conduct amphibious operations and limited defense of the ship.

Condition 1AA. All hands at battle stations to counter an air or surface threat.

Condition 1AS. All hands at battle stations to counter a submarine threat.

Condition 1E. Relaxation from general quarters for brief periods of rest and the distribution of food at battle stations.

Condition 1M. All hands at battle stations to take mine countermeasures.

General Quarters

The ship is in Condition of Readiness I, with all hands at battle stations and all equipment ready for instant action; GQ is sounded whenever battle is imminent or when the highest state of readiness to meet an emergency is desired. For example, the OOD should immediately sound the GQ alarm when a lookout sings out that a periscope has been sighted; when CIC reports that an unidentified plane has been picked up by radar; or for precaution at dawn and dusk, when there is an increased chance of enemy attack.

When personnel reach their stations, they throw off gun covers, break out ammunition, or prepare for action the equipment to which they are assigned. GQ must be set in seconds, not minutes. When each station is manned and ready, the person in charge notifies the control station. When all stations in the control group are ready, the bridge is notified.

When the danger has passed or the drill is over, "Secure from GQ" is announced over the general announcing system and sound-powered telephone circuits. No person leaves a station until "Secure from GQ" is sounded or until permission from a control station has been granted. All gear on hand must be stowed or secured as necessary.

Fire

Alarm for a real fire may be given at any time. For drill purposes, a fire may be assumed to be in a specified place—for example, in an ammunition space.

The crew member who discovers an actual fire must give the alarm. The most important thing is to notify at least one other person who can go for help. Too often a fire has gotten out of control because someone tried to put it out alone without calling for help.

Use any means at hand to spread an alarm. Use the telephone, messenger, or word of mouth to notify the OOD in port or the OOD or damage-control central when under way. Once the alarm has sounded, personnel nearby should act promptly to check or extinguish the fire using the means nearest at hand. All other crew members respond to the alarm in accordance with the Watch, Quarter, and Station Bill. If you and several others have begun to fight the fire, do not leave the scene until the fire or repair party arrives.

Fire during Condition I

When the ship is in Condition I, the damage-control organization is ready to fight several fires at once, and little assistance will be required from other stations. Damage-control central will direct a repair party to fight the fire and will keep the captain and the OOD informed as to progress.

Fire during Condition II, III, IV, or V (Under Way or In Port)

When the ship is in any of these conditions of readiness with all hands on board, Condition I is set immediately so that the highly trained damage-control organization may be used most effectively to fight the fire.

In port when Condition IV or V is set, the duty section furnishes the fire party. This fire organization is made up of key people from the regular damage-control parties.

Special steps must be taken when the fire alarm is sounded:

- Main engine control must maintain fire-main pressure. All electrical circuits in the vicinity of the fire are shut off.

- Supply department sends a storekeeper to the scene of the fire, with keys to the supply storerooms in the area.

- Medical department sends a health services technician with first-aid kit to the scene of the fire.
- Air crews stand by their helicopters if any are on board.
- Weapons department stands by all magazine spaces and inspects bulkheads for rise in temperature. The magazine sprinkling-system valves are constantly staffed.

CBRNE DEFENSE

CBRNE warfare refers to the use of *chemical, biological, radiological, nuclear,* and *high-yield explosive weapons*, which are termed mass weapons because they can affect great numbers of people. CBRNE defense is planned for in the CBRNE bill, which provides an organization and procedures to follow in case an enemy attacks with any of these weapons or in the event of an accident involving any CBRNE weapon in use by friendly forces.

The effects of a CBRNE attack may extend over large areas of land or water. Some effects—air blast, thermal radiation, and ionizing radiation of nuclear weapons—end almost instantly. Other effects—base surge and fallout of nuclear weapons—can contaminate an area for a long time and spread to other areas if carried by wind or water currents. The hazards of biological and chemical agents can also move with the winds. All CBRNE contamination remaining in an area can affect personnel in the area where it occurs or those entering the area later.

Chemical Warfare

Chemical warfare (CW) is the military use of any chemical that can produce powerful casualty or harassing effects, including screening smokes and incendiaries. With proper protection, such as clothing and gas masks, you have an excellent chance of surviving any chemical attack. Chemical smoke screens are used to hide enemy targets or troop movements, but they might also disguise other types of CW agents. Incendiaries—chemical compounds that burn with terrific heat—may be dropped by aircraft, fired in shells, or used with flamethrowers, and include napalm, jellied gasoline, and thermite and magnesium bombs. The most dangerous aspect of chemical warfare is the use of casualty or harassing agents against troops or civilian populaces.

Casualty Agents

These agents can kill or seriously injure large numbers of people. They include nerve agents, blister agents, choking agents, and blood agents.

Harassing Agents

These agents are not as dangerous as casualty agents but produce effects that harass an enemy by reducing efficiency. Principal types are tear gases and vomiting agents.

Detection of CW Agents

The most important CW agents to detect are the nerve gases, which can kill very quickly. Several chemical-agent detector kits are in use, but some of them identify only certain gases, and some do not give warning soon enough for personnel to don protective clothing.

Your best defense is to learn to note clues that might indicate CW agents: the presence of oily liquids after an explosion; drying and browning vegetation; a spray from aircraft; a vaporous cloud that remains close to the ground or water. Any unusual enemy operation that does not involve conventional projectiles and explosives should be treated as possible CW or biological warfare (BW) attacks.

Defense against CW attack is much the same as for radiological warfare (RW) or BW attacks. The main difference will lie in the speed with which nerve and blood agents act. At the first sign of a nerve gas, *hold your breath* until your mask is on and properly adjusted. Any liquid on your skin should be blotted with cloths from a protective ointment kit or some other cloth. Do not rub; this will only increase skin absorption. Flush exposed skin with water for half a minute and apply the ointment from the kit, but *not near or in the eyes*. If drops of nerve agent get into your eyes, flush with water immediately, even if you have to remove your mask. While unmasked, do not breathe contaminated atmosphere. Symptoms of nerve poisoning are local sweating and muscular twitching of contaminated skin area, or contraction of the pupils of the eyes. If these are observed, use the atropine syrette in the protective ointment kit, but *not* until you are *sure* that the symptoms are caused by nerve agents.

Biological Warfare

Biological warfare is the military use of living organisms or their toxic products to diminish the ability of an enemy to wage war through destruction or contamination of food sources, such as crops and domestic animals, and through infliction of disability or death on the populace by spreading an epidemic. A BW attack is most probably conducted by producing a biological agent that can be spread by an aerosol released into air currents or water supplies; this can be done by aircraft, bombs, long-range missiles, or even by people trained to infiltrate enemy territory, as the quantity of material needed could be very small.

It is nearly impossible to see, smell, or taste any BW agent, and because of their various incubation periods, there may be no immediate effects on people, animals, or crops to indicate that an attack has been made. The only way to detect and identify a BW agent is for trained experts to examine specimens of air, food, water, or human and animal blood and excretions. Most BW agents are not fatal; they will cause illness if taken into the body, but the victims will recover. Depending on their type, BW agents may be absorbed through the skin, taken in with food or water, or inhaled. The best defense against them is to keep yourself and your living areas clean, and report any sickness immediately. Action under known attack is much the same as for RW: wear protective clothing and a gas mask, or breathe through a folded cloth. Decontamination and wash-up procedures are the same as for RW.

Most BW agents will die or lose their effectiveness after a few days of exposure to ordinary weather conditions, but there are some that may remain dormant for a long period. All clothing should be boiled in soapy water or exposed to the sun for a few days before wearing. Food in sealed containers should be safe, but food or water that has been exposed should be avoided until tested. Play it safe; do not assume danger is past after a BW decontamination team has done its work.

Radiological Warfare

Radiological or nuclear warfare, the use of a weapon armed with a nuclear warhead, may produce these effects:

- *Shock wave*, a blast, as in any explosion, followed by rapid movement of air or water away from the explosion, the force of which can wreck ships, smash buildings, and kill or severely injure people at a considerable distance. Shock wave protection is similar to general damage-control precautions against any explosion: keep loose gear secured, be prepared for the effects of heavy shock and blasts, maintain watertight integrity, and make repairs as fast as possible.

- *Thermal radiation*, heat so intense that everything touched by the fireball melts, and buildings at a great distance burst into flame.
- *Radiation (both initial and residual)*, which cannot be seen or felt, is more dangerous than the shock or heat. The radiation is composed of invisible alpha and beta particles of energy, neutrons, and gamma rays. Alpha particles can be stopped by a thin sheet of paper; the other particles travel at the speed of light and therefore are more difficult to stop.

Initial radiation from the detonation of a nuclear weapon consists mostly of neutrons and gamma rays, which are both injurious to human tissue. It lasts only one-tenth of a second. Heavy shielding is required as a protection, depending on distance from the burst and the "yield" or size of the bomb. *Residual radiation*, the radioactive "fallout" from a nuclear explosion, can contaminate large areas. (Fallout from bomb tests in China can be detected in the United States.) This contamination can be detected and measured only by special instruments; if it is hazardous, the area must be decontaminated (washed clean) before unprotected personnel may enter.

Ships may be exposed to three types of bursts: *air burst*, in which the fireball does not touch the earth; *surface burst*, in which the fireball touches the surface; and *subsurface burst*, in which the explosion is underground or underwater. An air burst produces blast, heat, intense light, and initial radiation but no fallout requiring decontamination of ship or personnel. In an air burst, 85 percent of the damage is due to shock, 10 percent due to thermal radiation, and 5 percent due to initial or residual radiation. A surface burst will produce shock, heat, intense light, and initial and residual radiation. The fallout, especially over water, will spread radioactive contamination over a much wider area than the base surge of an underwater (or subsurface) explosion. A subsurface blast normally produces no heat or light and very little initial radiation. The danger in this type of explosion is in the intense underwater shock and heavy residual radiation from the highly contaminated base surge of earth or water created when the column formed by the explosion falls back to the surface.

Nuclear Attack at Sea

The alarm for a nuclear attack is the one-thousand-cycle chemical alarm. When it sounds, *take shelter* at once. The actions of all hands on duty, of course, must be determined by circumstances and their orders.

The air blast can knock people around like Ping-Pong balls, so *hang on* to something solid to reduce the chance of injury. The same warning goes for personnel below decks, as the shock wave transmitted to the hull by water can smash them against bulkheads and overheads. Never lie flat on the deck, but rest on the balls of your feet and flex your knees. Thermal radiation from an air or surface burst will produce skin burns and eye damage only among those directly exposed. Personnel can reduce the chance of burns by dropping out of direct line of sight of the fireball and covering exposed skin surface.

Attack without alarm is always possible, and the only warning will be a very bright flash or the sky lighting up. When you see the flash, close your eyes, cover your face with your hands, drop to the deck if you can, or crouch and bend down, and after two to five seconds, or after you feel a sudden wave of heat, grasp something solid and hang on. Otherwise, the air blast could blow you over the side. You may suffer from flash blindness but will be able to see perfectly well in thirty minutes.

Radiation Measurement

The effect of nuclear radiation on any living thing depends on the intensity of radiation and the time exposed to it, and can be determined by multiplying one by the other. The result, called *dosage*, is measured in RADs or roentgens and is used to determine the treatment required after exposure to radiation. It is important that no one is exposed to a greater radiation dose than the body can absorb without permanent damage. Some individuals have greater resistance to radiation injury than others. Personnel previously exposed may require less radiation to become ill. Radiation injury normally does not become apparent for some time. This delay is called the "latent period." The amount of radiation received affects the degree of injury and the length of the latent period—the larger the radiation dose, the quicker symptoms are noticeable. The latent period is important because people exposed to a large amount of radiation may still be able to perform their duties for hours and even days. Killing doses of radiation may reduce the period to less than an hour.

Radiation sickness is *not* a communicable disease; you get it only by exposure to radiation. Sometimes, for psychological or other reasons, an individual may develop symptoms that are similar to radiation sickness, such as nausea and vomiting, but that are actually not due to radiation exposure.

Maximum Permissible Exposure

There is a limit on radiation exposure, established by competent authority. Ships will use appropriate maneuvers and countermeasures (wash down, shelter, rotation of personnel between topside and deep shelter stations, etc.) to limit radiation exposure as much as possible.

Detection Equipment

Detection equipment is designed to be worn by individuals in certain situations, or to be used by monitors. Individuals will use a *pocket dosimeter*, which looks like a fountain pen and records radiation by deflection of a needle on a roentgen scale; a *film badge*, made of lead or cadmium, in which the film records radiation but must be removed and developed to determine the amount; or a *glass dosimeter*, which replaces the film badge. The glass dosimeter is a plastic disk worn on a dog-tag chain. The important fact about the dosimeter is that it measures *total dose* received.

Monitoring Equipment

Monitoring equipment is usually handled by trained teams and is used to locate and monitor "hot spots." The purpose in monitoring is to measure the *instantaneous* intensity of radiation; once this is known, it is possible to control the amount of time any person spends in any area so as to keep the total dose below the maximum permissible exposure. Monitoring equipment, generally known as *radiac sets*, has no popular names but carries "short titles" such as AN/PDR-45 or AN/PDR-56. The various types are designed to measure high-range gamma radiation, lower-range gamma and beta radiation, alpha radiation, and neutron radiation. A monitoring team consists of a person in charge, called the *monitor*, who measures intensity of radiation; a *recorder*, who records intensity of readings and time and location; a *marker*, who marks intensity and time of contamination and places warning signs; and a *talker* or *messenger*, who relays readings to damage-control central.

Shipboard Decontamination

Tactical decontamination takes place immediately at sea and reduces contamination so that the ship can carry out its mission without subjecting the crew to more than the maximum permissible exposure. *Phase one—primary gross decontamination*—consists of a saltwater washdown of the entire topside. This will remove 98 percent of contamination if the surface was wet prior to contamination and washdown commences while it is still wet; if the surface has dried, only 50 percent of contamination will be removed. Washdown should be completed within fifteen minutes after contamination occurs. Hose squads wear protective clothing and gas masks. Saltwater washdown commences even if the sea is contaminated, in which case it will continue until the ship reaches a clear area.

Phase two—conducted by individual ship departments—involves reducing contamination of equipment vital to the ship's mission by using a steam lance and detergents, scrubbing paint work, scraping with abrasives, or removing with flame.

Personnel Decontamination

Preliminary decontamination is done immediately; exposed personnel wash down under fire hoses or topside showers. Detailed decontamination is done in stations divided into three sections where all possible precautions are taken to prevent contamination spreading from one section to the other. In the undressing area, all contaminated clothing is removed; in a washing area, each person scrubs thoroughly with soap and water; in the clean dressing area, uncontaminated clothing is provided. Radiation detection devices at each section ensure that contamination is not spread throughout the center. Decontamination priority is as follows: repair party and damage-control teams, uninjured personnel, injured personnel. Injured personnel are usually handled last because of the extra time required for medical personnel to assist them.

Protective Mask

This important part of CBRNE defense is designed to protect your face, eyes, nose, throat, and lungs by filtering the air; removing particles of dust and smoke that may be radioactive or contaminated with BW or CW agents; and purifying the air of many poisonous gases. It will not protect against ammonia, carbon monoxide, carbon dioxide, cooking gas, sulfur dioxide, and certain other industrial gases or fire-produced fumes. It will not produce oxygen and is useless where there is not enough oxygen to breathe, even after the mask has purified what air there is. In such places the self-contained breathing apparatus, or air-line hose mask, must be used.

Because of the great importance of the protective mask, it should receive careful attention. Perspiration should be dried out completely as moisture will cause rotting, corrosion, deterioration of the canisters, and mildew.

Only authorized equipment, such as the protective ointment kit, should be stowed in the carrier. The mask should be inspected periodically, tested in a gas chamber, and stored in a cool, dry place away from solvents, such as cleaning fluid, and their vapors.

Protective Clothing

Protective clothing, provided for monitoring and decontamination teams, must have the best possible care. Tears, holes, broken or missing buttons, broken or jammed zippers, and other such defects might leave you unprotected during an attack.

Special protective clothing made of rubberized cloth is airtight and provides the most complete protection against BW and CW agents. Liquid CW agents will penetrate it after a few hours, so wearers should go through decontamination as soon as possible. Permeable clothing is chemically treated to neutralize blister agents but provides little protection against nerve gas and only limited protection against any other contamination. Regular stock issue wet-weather clothing and ordinary work clothing will keep some contamination away from the skin and is easily decontaminated by washing. All contaminated clothing should be removed as soon as possible.

Figure 26-1.
Members of a Coast Guard boarding team training for a law enforcement boarding.

CHAPTER 26

Boardings

O ne of the most important and common evolutions is the boarding of vessels at sea. Because law enforcement is carried out under a wide range of circumstances and conditions, actual equipment, boarding situations, and procedures may vary throughout the Coast Guard. Nonetheless, there are a number of fairly standard elements for every boarding. The following section describes a "standard" Coast Guard boarding.

BOARDING UNITS

Boardings may be conducted by any Coast Guard unit, but they are generally conducted by cutters and small boats. Larger cutters, such as medium and high endurance cutters (WMEC and WHEC) often conduct extensive boardings well offshore using their small boats. Patrol boats (WPB) conduct a number of boardings from inshore waters to areas hundreds of miles off the coast, also using attached small boats. Small boats from stations and groups usually conduct their boardings in inshore waters. All Coast Guard units board a wide range of vessel types, in all sorts of circumstances.

Types of Vessels Boarded

The Coast Guard boards virtually every type of vessel afloat, from small rowboats to large supertankers. Most vessels boarded in the course of general maritime law enforcement, though, are either pleasure craft such as sailing vessels or sports fishing boats, fishing vessels, or small freighters.

Boarding Teams

A boarding team consists of at least two Coast Guard people and may be as large as eight or more persons, depending upon the size of the vessel to be boarded, the size of the unit doing the boarding, and the situation involved. For example, while a two-person boarding team may be fine for a small powerboat in inland waters, an eight-person boarding team may be necessary to board a coastal freighter off the Bahamas. The person in charge of the boarding team is called the "boarding officer." Although all Coast Guard officers and petty officers have law enforcement authority, the boarding officer will be the person whose authority will be used for the actual boarding. All others on the boarding team assist the boarding officer.

Equipment

Every person in the boarding team must wear the proper Coast Guard uniform. In addition, boarding team members wear specialized equipment for their law enforcement activities

Soft body armor. Soft body armor is a vest designed to minimize injuries from shots fired at the body of the wearer.

Equipment belt. Also known as "gun belt," the belt holds the person's weapons and equipment.

Chemical irritants. Similar to mace or tear gas, chemical irritants are chemical agents contained in a spray can; they create a burning sensation on the skin of the target and lead to severe eye, nose, and throat irritation.

Service pistol. All Coast Guard boarding team members are armed with the P229R DAK personal defense weapon during law enforcement boardings.

Impact weapon. This is an expandable steel baton.

Handcuffs and key.

Flashlight.

Head gear. Boarding personnel may wear either a blue ball cap with the words "U.S. Coast Guard" and appropriate Coast Guard cap insignia according to their rank or a white plastic safety helmet with a Coast Guard emblem.

Personal flotation device. Because of the amount of equipment carried by boarding team members, all boarding team members must wear an approved flotation device to ensure their safety if they should go into the water.

Communication equipment. The boarding team will carry a radio to communicate with the unit. Those radios are usually "scrambled" so that others cannot listen in to the law enforcement conversations. In addition, many boarding teams have small personal radios so that team members may communicate with each other during the course of the boarding.

Anti-exposure coveralls. In cold weather, boarding team members wear specialized coveralls that combine protection against the cold with flotation.

Riot shotgun. In some instances, when the boarding officer has specific reasons for suspecting a greater than normal degree of risk in the boarding, one of the boarding team members may carry a riot shotgun.

Boarding bag. In addition to the personal equipment, the boarding officer will carry a bag with miscellaneous forms and equipment required for the boarding. The bag will include boarding forms, check-off lists, hand tools, tape measures, mirrors, narcotics identification kits, a camera, and other specialized gear.

Access kit. In cases in which the boarding officer suspects that the boarding vessel has hidden compartments concealing drugs or other contraband, the boarding team may bring a kit containing equipment to locate, open, and inspect compartments. The kit may include such things as a power drill, a "borescope" for looking through a drill hole into a dark compartment, saws, crowbars, and sledges.

Boarding Procedure

The boarding process begins with the Coast Guard unit establishing the identification and nationality of the vessel to be boarded. The unit will send the name and description of the vessel to a communications center ashore, where they are checked in a federal computer system to see if there is any information available on the vessel or the crew. Once the person in charge on the Coast Guard unit decides to conduct a boarding, the vessel is told to stand by for a boarding, and the boarding procedure is explained.

The boarding team forms up on board the cutter or small boat, checks the available intelligence, double-checks equipment, and then proceeds to the vessel, often by rigid-hull inflatable small boat.

When the boarding team arrives on board the other vessel, the team members take prearranged stations about the vessel to establish and maintain control for the boarding. The boarding officer will meet with the vessel's master, explain the purpose and the process of the boarding, and answer any questions. If the boarding officer has any specific, justifiable concerns about the safety of the boarding, due either to dangerous conditions on the boat or suspicions about the crew, the boarding team may conduct an initial safety inspection. In this inspection, team members conduct a brief survey of the vessel to ensure that the vessel is safe enough to continue with the boarding.

The boarding officer then conducts the boarding, checking for compliance with a wide range of federal laws and regulations. If, during the course of the boarding, the boarding officer develops probable cause to believe that the vessel is engaged in illegal activity, the scope of the boarding may be expanded to include a complete search of the vessel for evidence. If evidence of criminal activity is found, such as illegal drugs or other contraband, the crew will be arrested and the vessel seized, brought into port, and turned over to other agencies. If only minor safety violations are found, the master will be given a "citation" listing the violations, much like a traffic ticket. If there are no violations detected during the boarding, the master will be thanked for cooperating, and the boarding team will depart.

USE OF FORCE

As federal officers, Coast Guard personnel are armed when carrying out their law enforcement duties. They are authorized to use appropriate force to defend themselves and others and to carry out their duties. The use of force by Coast Guard personnel is closely regulated by federal law and by Coast Guard policy.

Policy

Only that force reasonably necessary under the circumstances may be used. Excessive force may never be used. Force shall not be used where assigned duties can be discharged without it. Nothing in the application of the Coast Guard use-of-force policy shall be construed as to necessarily require personnel to meet force with equal or lesser force. Although personnel are encouraged to consult with higher authority before using force, in most instances, they will have to make on-the-spot use-of-force decisions based on their training and experience. In general, Coast Guard personnel may use force in the following circumstances:

- Self-defense
- Prevention of a crime
- Effecting an arrest
- Protection of property

Weapons Qualifications

Coast Guard personnel must be highly trained in the use of the weapons they will carry as law enforcement officers. All personnel receive weapons training in the use of the P229R DAK personal defense weapon and the M16 during their initial training. In addition, personnel at operational units must requalify every six months to be considered qualified to carry a weapon. Coast Guard personnel also receive training in the use of riot shotguns, chemical agents, and impact weapons. In addition, everyone who may carry a weapon in the course of duty must qualify every six months on a judgmental "shoot/no shoot" course in which the person must make split-second use-of-force decisions. The commanding officer/officer-in-charge of the unit will make the final determination as to who may carry a weapon.

Figure 27-1.
USCGC *Waesche* (WMSL-751) as decorated for her christening on 26 July 2008.

CHAPTER 27

Ship Construction

J oining the Coast Guard is somewhat like entering any other business or profession—you will have to learn the language that is used there. Most of the words will be familiar, but they will have new meanings. And along with learning a new language, you will have to learn a good deal about ships. A modern cutter is an extremely complicated craft.

"Ship" is a general term for any large floating vessel that moves through the water under its own power. A *boat* is essentially the same as a ship, only smaller, and can be hoisted and carried on board a ship.

PARTS OF A SHIP

No matter how specialized your professional training may be, you must be thoroughly familiar with basic nautical terminology and ship construction. This is necessary to carry out routine orders and commands and to act quickly during combat or in emergency conditions.

In some respects a ship is like a building. It has outer walls (called the *hull*), floors (called *decks*), inner walls (called *partitions* and *bulkheads*), corridors (called *passageways*), ceilings (called *overheads*), and stairs (called *ladders*). But a ship moves and is never in the same place twice, so you have to learn new terms for directions and getting about. When you go up the stairs from the pier to a ship, you use the *accommodation ladder* to go on board, and what might be an entrance hall or foyer in a building is the *quarterdeck* on a ship.

The forward part of a ship is the *bow*; to go in that direction is to *go forward*. The after part is the *stern*; to go in that direction is to *go aft*. The topmost open deck of a ship that runs from bow to stern is the *main deck*; anything below that is *below decks* and anything above it is *superstructure*.

The forward part of the main deck is usually the *forecastle* (pronounced "FOKE-sul"); the after part is the *fantail*. The forecastle and fantail are on the same (main) deck on a *flush-deck ship*; the fantail is one deck lower on a *broken-deck ship*. To proceed from the main deck to a lower deck, you *go below*; going back up again, you *go topside*. As you face forward on a ship, the right side is the *starboard* side and the left side is the *port* side. An imaginary line running full length down the middle of the ship is the *centerline*; the direction from the centerline toward either side is *outboard*, and from either side toward the centerline is *inboard*. A line from one side of the ship to the other runs *athwartships*.

If the interior hull of a ship were all one space, a single large hole made below the waterline would quickly cause the ship to flood and sink. To prevent this, the interior of the ship is divided by bulkheads and decks into *watertight compartments*. In theory, any large ship could be made virtually unsinkable if it were divided into enough small watertight compartments. There is a limit to this, however, since compartmentation interferes with the arrangement of mechanical equipment and with the operation of the ship. Engineering spaces must be large enough to accommodate bulky machinery and cannot be subdivided.

Hull

The hull is the main body of the ship below the main outside deck. It consists of an outside covering, or *skin*, and an inside structural framework to which the skin is fastened. In almost all types of modern ships, both the framework and the skin are made of steel. The steel skin is called the *shell plating*.

Keel

The keel is the principal structural part of the hull. It runs from the *stem* at the bow to the *sternpost* at the stern. *Frames* are fastened to the keel; they run athwartships and support the shell plating. *Bulkheads, deck beams*, and *stanchions* are joined together and fastened to the frames; they support the decks and resist the pressure of the water on the sides of the hull. The system of interlocking steel bulkheads and decks of a warship furnishes a large part of the hull strength.

Large ships have an outer and inner bottom, often called *double bottoms*. These are divided into many compartments. Many of the compartments on board a ship are used as tanks for fuel-oil storage, or fresh water, or for trimming the ship. Tanks at the extreme bow and stern, used for trimming ship fore and aft, are called *peak tanks*. A heavy watertight bulkhead just abaft the forward peak tank is called the *collision bulkhead*. All compartments and tanks have pump and drain connections for pumping out sea water and for transferring fuel or water from one part of the ship to another.

Hull Reference Terms

Waterline, freeboard, and draft. The line to which a hull sinks in the water is the *waterline*. The vertical distance from the waterline to the edge of the lowest outside deck is the *freeboard*. The vertical distance from the waterline to the lowest part of the ship's bottom is the *draft*. The draft may also be thought of as the least depth of water in which the ship will float. The waterline, freeboard, and draft will, of course, vary with the weight of the load carried by the ship; as freeboard increases, draft decreases.

Draft is measured in feet, and numbered scales are painted on the sides of the hull at bow and stern. The relation between the drafts at the bow and the stern is the *trim*. When the ship is properly balanced fore and aft, she is "in trim." When the ship is "out of trim," because of damage or unequal loading, she is said to be "down by the head" or "down by the stern." When the ship is out of balance laterally or athwartships, she has a *list*; she is said to be "listing to starboard" or "listing to port." Both trim and list are adjusted by emptying or filling tanks and compartments in various parts of the hull.

The bow and the stern. The part of the bow structure above the waterline is the *prow*, although this word is not used as generally as the terms *bow* and *stem*. The part of the weather deck nearest the stem is called the *eyes* of the ship. The general area of the weather deck in the forward part of the ship is the *forecastle*, even though the ship may not have a forecastle deck. The edges of the weather deck from bow to stern are usually guarded by removable light-wire ropes and stanchions called *lifelines*, or by extensions of the shell plating above the deck edge, called *bulwarks*.

The main deck area at the stern of the ship is the *fantail*. The part of the stern that literally hangs over the water is the *overhang*. The lower part of the bottom of a ship is called the *bilge*. The curved section where the bottom meets the side is called the *turn of the bilge*.

Propellers. The propellers, or *screws*, that drive the ship through the water are attached to *propeller shafts* and are turned by them. Ships with only one propeller are called *single screw*; ships with two propellers are *twin screw*. Ships with more than two propellers, usually four, are called *multiple-screw* ships.

In twin- or multiple-screw ships, the exposed length of propeller shafts is so great that they must be supported by braces extending from the hull called *propeller struts*. Because of the shape of the hull at the stern, the screws may be damaged when the ship is close by a pier. To prevent this, metal frames called *propeller guards* are built out from the hull above the water.

Decks

The decks of a ship correspond to the floors of a building. The *main deck* is the highest deck that extends over the entire ship from stem to stern. The *second deck*, *third deck*, *fourth deck*, and so forth are other complete decks below the main deck, numbered in sequence from topside down.

A partial deck above the main deck is named according to its position on the ship; at the bow it is called a *forecastle deck*; amidships it is called an *upper deck*; at the stern it is called a *poop deck*. A partial deck between two complete decks is called a *half deck*. A partial deck below the lowest complete deck is called a *platform deck*.

The term *weather deck* includes all parts of the main, forecastle, upper, and poop decks that are exposed to the weather. The *quarterdeck* is not a structural part of the ship but is a location on or below the main deck designated by the commanding officer as the place for masts and ceremonies. Any deck above the main deck, upper forecastle deck, or poop deck is called a *superstructure deck*.

Compartmentation Numbering

Almost every space on a ship is assigned a compartment number. This number is marked on a label that is secured to the door or hatch and painted on the bulkhead of the compartment. Most Coast Guard vessels follow these basic rules.

All compartments on the port side end in an even number. All compartments on the starboard side end in an odd number. A zero precedes the deck number for all levels above the main deck; for example, 01, 02, and so forth. Ships built after March 1949 have compartment numbers that consist of a deck number, frame number, the relation of the compartment to the centerline of the ship number, and a letter showing the use of the compartment. Using 4-75-3-M as an example, the compartment is located

4: Fourth deck
75: Forward frame number

3: Second compartment to starboard outboard of the centerline

M: Compartment usage (ammunition)

Deck Numbers

The deck numbers of a ship (with corresponding names) are shown in figure 27-2. Where a compartment extends down to the bottom of the ship, the number assigned to the bottom compartment is used. The deck number becomes the first part of the compartment number.

Frame Number

The frame number at the foremost bulkhead of the enclosing boundary of a compartment is its frame location number. Where these forward boundaries are between frames, the frame number forward is used. Fractional numbers are not used. The frame number becomes the second part of the compartment number.

Relation to the Centerline

Compartments on the centerline carry the number zero (0). Compartments completely to starboard are given odd numbers, and compartments completely to port are given even numbers. Where two or more compartments have the same deck and frame number and are entirely starboard or entirely port of the centerline, they have consecutively higher odd or even numbers (as the case may be) numbering from the centerline outboard. In this case, the first compartment outboard of the centerline to starboard is 1, the second, 3, and so on. Similarly, the first compartment outboard of the centerline to port is 2, the second, 4, and so on. When the centerline of the ship passes through more than one compartment, the compartment having

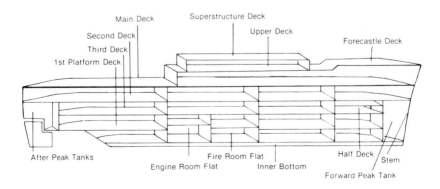

Figure 27-2. Deck arrangement of a typical ship.

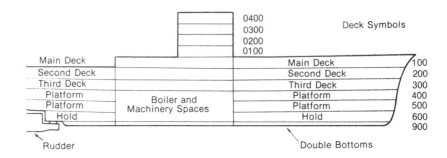

Figure 27-3. The system of naming and numbering decks of a ship.

that portion of the forward bulkhead through which the centerline of the ship passes carries the number 0, and the others carry the numbers 01, 02, 03, and so forth. These numbers indicate the relation to the centerline and are the third part of the compartment number.

Compartment Usage

The remaining element of the compartment number is the letter that identifies the primary usage of the space. On dry- and liquid-cargo ships a double-letter identification is used to designate compartments used for carrying cargo, as follows:

A   Storage area, such as
  - Store rooms
  - Issue rooms
  - Refrigerated stores
  - Ordnance-related stowage (non-hazardous)
  - Clothing and cleaning gear lockers

AA Cargo holds

C   Cutter and fire-control operating areas, such as
  - CIC, electronic operating spaces (manned)
  - IC rooms, radio rooms
  - Communication office
  - Pilothouse, plotting rooms

E   Machinery compartments, including
  - Main machinery spaces
  - Auxiliary machinery spaces
  - Evaporator rooms

- Pump rooms
- Refrigerating machinery rooms
- Steering-gear rooms
- Windlass rooms

F   Fuel compartment, including
- Fuel oil
- Lubricating oil

G   Gasoline compartments (including gasoline cofferdams)

J   JP-5 compartments

L   Living compartments, medical and dental spaces, and horizontal passageways, including
- Officers' quarters
- Crew's quarters
- Water closets and washrooms
- Medical and dental spaces
- Passages

M   Ammunition spaces, including
- Magazines
- Small-arms magazine

Q   All spaces not otherwise covered, such as
- Engineering, electrical, and electronics spaces that are not covered in "E"
- Galley
- Pantries
- Scullery
- Laundry
- Offices
- Shops
- Wiring trunks
- Fan rooms

T   Vertical access trunks

V   Voids and cofferdams, other than gasoline tank cofferdams

W   Water tanks, including
- Freshwater tanks, reserve feed-water tank
- Peak tanks, ballast tanks
- Bilge tanks, sump tanks
- Built-in sewage holding tanks

Superstructure

The superstructure of a ship is everything above the weather deck. There is a great deal of variation in the superstructure layout on different types of Coast Guard cutters, but the same elements will be found in all of them in one form or another. A large part of the superstructure in passenger ships consists of living quarters and recreation facilities. The superstructure of a fighting ship is made up of the actual armament and controls necessary for operating the ship.

Bridge

The bridge is the primary control position for the ship when under way, where all orders and commands affecting the ship, its movements, and routine originate.

The OOD is always on the bridge when the ship is under way; the captain will be on the bridge during GQ and during most operations. Some ships can also be handled from a secondary control station, called *secondary conn*, the GQ station for the executive officer; thus, if the bridge is knocked out or the captain disabled in battle, the executive officer can take over.

Pilothouse

The pilothouse, sometimes called the *wheelhouse*, contains equipment and instruments used to control the movements of the ship. Usually the bridge extends out on both sides of the pilothouse. Some pilothouse equipment is duplicated on the bridge.

Bridge and pilothouse equipment. Some ship control consoles consist of the engine order section and propeller order section that control speed and direction (ahead or astern) of the ship. The engine order section has a dial for each engine, divided into sectors marked flank, full, standard, ⅔ and ⅓ speed ahead, stop, and ⅓, ⅔, and full speed astern. When a hand lever is moved to the speed sector ordered by the OOD, the engine room watch sets the engine throttle for the same speed and notifies the bridge by moving an answering pointer to the same sector.

The ship control console is manned by a helmsman. The propeller order section enables the OOD to make minor changes in speed by ordering the engine rooms to increase or decrease the revolutions per minute of the propellers.

The steering control console contains the controls and indicators required to control the course of the ship. The steering wheel (helm) is operated by the helmsman. On the panel in front of the helmsman are various indicators and switches. The ship's course indicator is a gyrocompass repeater, which indicates the ship's true course. Another indicator shows the course to be steered. Two more important indicators show the rudder angle (number of degrees left or right of amidships) and the helm angle (number of degrees left or right of amidships).

Tachometer. A tachometer is the same type of instrument that is used on a sports car; it shows shaft revolutions per minute. On a ship, there is a tach for each propeller.

Lighting panels. The two primary lighting panels in the pilothouse are the signal and anchor light supply-and-control panel and the running lights supply-and-control panel. Lights installed on cutters usually include aircraft warning lights, blinker lights, breakdown, man overboard, and underway replenishment lights, and steering lights. Switches are located on the signal and anchor light supply-and-control panel.

The location of bridge equipment varies among ship types and may even be different on ships of the same type. A bridge watchstander has to learn where everything is in order to find it in the dark.

Charthouse

The charthouse is normally just abaft the pilothouse and on the same deck, but it may be on another deck some distance away. It contains charts, a chart desk, and the chronometers. On some ships, the dead-reckoning tracer may be in the charthouse. The charthouse also contains navigational instruments such as sextants, stadimeters, bearing circles and stopwatches, parallel rulers, protractors, position plotters, and navigational books and tables.

Secondary Conn

The secondary conn area contains steering equipment, engine order telegraph, phone circuits, and other equipment necessary for ship control in the event the primary control station is unable to perform because of battle damage. A magnetic compass may be located here.

The Signal Bridge

The signal bridge is an open platform near the navigational bridge and is equipped with yardarm blinker controls, signal searchlights, and flag bags, where trained personnel can maintain visual communication with other ships.

Radio

Radio is the station of the *communication watch officer*, where outgoing traffic is prepared for transmission and incoming messages are readied for local delivery. All messages, except tactical signals received and sent direct from shipboard control stations, go through the message center.

Combat Information Center (CIC)

The CIC is the nerve center of the ship. It has a fivefold function: to collect, process, display, evaluate, and disseminate information from sources both inside and outside the ship. A wide range of electronic equipment is installed in the CIC: radar, sonar, electronic warfare intercept receivers, IFF (identification friend or foe), radio and visual communications, PPI (plan position indicator) repeaters, display screens, and computers. Radar installations include both air and surface search and fire control.

Damage-Control Central

Damage-control central maintains damage-control charts, machinery charts, and liquid-loading diagrams and is responsible for seeing that the proper conditions of readiness are set and maintained. The conditions of stability and damage throughout the ship are known in damage-control central at all times and reported by direct communication to the bridge. All repair parties report to damage-control central.

Masts and Stacks

In some modern ships the masts are included as part of the main superstructure assembly. On older ships and on the smaller types of escort and patrol craft, masts still are a distinct feature of the superstructure. If a ship has two masts, the forward one is the *foremast* and the other is the *mainmast*. On single-masted ships, the mast is amidships or forward, is usually part of the main superstructure assembly, and is called the foremast or simply the mast.

Masts vary greatly in size and shape, some being built of several structural steel members and others being a single steel, aluminum, or wooden pole. On all ships, at least one mast, together with a spar running athwartships called a *yard*, is used for flags and other signal devices. These must be rigged in such a way as to be visible from other ships. The lighter types of masts are supported by *standing rigging*, consisting of wire rope *stays* running fore and aft and *shrouds* running athwartships down to the deck or bulwarks.

The small cap at the top of a pole mast or flagstaff is called a *truck*. The top of any mast may be called the truck. Above the truck there is frequently a slender vertical extension of the mast, called a *pigstick*. Almost every naval vessel has a pigstick on the mainmast, from which the *commission pennant* or an admiral's *personal flag* is flown. Extending abaft the mainmast is a small spar that is known as the *gaff*, from which the national ensign is flown when the ship is under way.

The small vertical spars at the bow and stern of a ship are the *jackstaff* and *flagstaff*, respectively. When a ship is at anchor or moored, it flies the Union Jack on the jackstaff and the national ensign on the flagstaff from 0800 to sunset.

Pipes for the expulsion of smoke and gases from the boilers are called *stacks*. There are many different varieties—short, high, vertical, raked, single, double, split trunk, and so forth. On most vessels, stacks are located on the centerline approximately amidships. Many new ships, especially those with diesel or gas-turbine propulsion, have twin stacks side by side. Stacks, their number, and their arrangement are important aids in correctly identifying ships at a distance.

Macks

In some new ships, the mast and stack are combined to form a large tower called a *mack*.

Accommodation Ladder

The accommodation ladder is a "stairway" suspended over the side of the ship with a platform at the bottom that serves as a landing for boats and is so designed that, as it is adjusted up and down, the steps remain level. A *boat rope* or *sea painter* is provided to secure boats alongside while they load and unload.

Brow

The "brow" is basically a gangplank used when the ship is moored alongside a pier or "nested" alongside other ships. Its size and construction will depend on the size of the ships and the distance from the ship to the pier.

Living Quarters

A considerable portion of the interior of a ship must be devoted to living spaces and facilities for the ship's complement of officers and crew. Officers' quarters and mess facilities are generally near the bridge. Quarters for the crew may be distributed throughout a large ship. Other facilities include the galleys, washrooms and heads, various storerooms, and sick bay. The location, size, and arrangement of living quarters are secondary considerations in the design of a ship. Its primary missions are considered first.

Ship's Equipment and Other Spaces

Much of the hull space of most ships is taken up by the engines, engineering equipment, and related piping and electrical systems. Storage and living space may be limited.

Engineering Plant

The engine rooms contain the main engines that drive the ship; these may be diesel, diesel-electric, or gas turbines. Auxiliary engine rooms contain generators that produce electricity and evaporators and condensers that convert salt water to fresh water. The steering room contains the machinery that powers the rudder. Fuel-oil tanks carry fuel for the engines.

Electrical System

A large ship has hundreds of electric motors driving everything from fans and tape decks to gun mounts. Every other system in the ship depends on electric motors. The main power supply is produced in the engineering spaces by diesel-driven generators. Emergency diesel-electric generators in other parts of the ship automatically cut in to supply power if the main generators are disabled for any reason.

Drainage System

The drainage system includes the piping, valves, and pumps that discharge water from the ship. This includes water in flushing systems, water used in firefighting, or seawater that enters the hull as the result of damage, collision, or heavy weather.

The main eductor is a large pipe in the bottom of the ship to which other drain pipes are connected; the secondary eductor is a smaller pipe running lengthwise in the ship. This system includes connections to all watertight compartments.

Ventilation System

The ventilation system includes air supply, exhaust, and air-conditioning equipment. There are many separate systems so that ducts do not have to run through watertight bulkheads.

Freshwater System

The freshwater system provides water for the crew—for drinking, showers, and cooking—and for the boilers. Freshwater tanks may be filled in port from shore supplies; at sea, fresh water is made from salt water by condensers and evaporators.

Saltwater System

The saltwater system provides water for fire protection, including turret sprinkling, magazine flooding, CBRNE washdown, and flushing. Flushing water may come directly from the fire main or from separate lines. The fire main is a large pipe running the length of the ship, with risers and branch mains connected to it.

Fuel-Oil System

The fuel-oil system includes fuel storage tanks, filling lines, feed lines to the boilers or diesel engines, and lines and connections for pumping oil from one tank to another to control trim or list when the ship is damaged.

Compressed-Air System

The compressed-air system includes compressors, storage tanks, and high-pressure lines used for testing and blowing out compartments, and for operating pneumatic tools and other equipment.

Magazines

Ammunition for all guns is stored in magazines, which are placed well below the waterline when possible. Projectiles and powder may be stored in separate compartments. All magazines can be flooded by remote control in case of fire. Ammunition is passed to handling rooms, where hoists take it up to the gun mounts or turrets.

Storerooms

Storerooms are spaces in which a ship carries her own supplies; these may be clothing, dry or refrigerated provisions, and various types of spare parts and supplies.

Crew Accommodations

There are many compartments throughout the ship designated as wardroom, officers' cabins, berthing compartments, pantries, messes, heads, washrooms, and sick bay. Other spaces provided for the health and comfort of the crew may include a barber shop, laundry, galley, library, and ship's store, depending on the size of the ship.

Figure 28-1.
A Coast Guardsman waves flags to represent fire and smoke during
a damage-control drill simulating a helicopter crash on the
flight deck of a Coast Guard cutter.

CHAPTER 28

Damage Control and Firefighting

Damage control includes all efforts to prevent damage to a ship as well as all action taken to reduce the harmful effects of damage after it happens. The primary purpose of damage control is to keep a ship in condition to perform its assigned mission; a secondary purpose is to return the ship to port after damage so it can be repaired. The need for damage control is not limited to combat operations; any routine operation can result in an accident calling for prompt damage control.

DAMAGE CONTROL

The three main objectives of the damage-control organization of a ship are as follows:

- Taking all necessary action before damage occurs. This means making the ship watertight and airtight in those parts that were so designed, removing all fire hazards, and maintaining and distributing necessary emergency equipment.

- Reducing damage by controlling flooding, fighting fires, and providing first-aid treatment to injured personnel.

- Providing emergency repairs or restoring services after damage occurs. This means manning essential equipment, supplying emergency power, and repairing important parts of the ship that were damaged.

The ship's ability to perform any assigned task depends upon the actions of its damage-control organization. It is important that you realize your responsibility in damage-control procedures. Damage control must be considered an offensive as well as a defensive weapon.

Damage control is concerned not only with battle damage but also with damage from fire, collision, grounding, or explosion. Damage control may be necessary in port as well as at sea. Learn all you can about your ship—the whole ship, not just the part to which you are assigned. In an emergency, you may have to work anywhere on the ship, depending upon the location and the extent of the damage.

The important damage-control systems are *drainage, ballasting, fire main, sprinkling and washdown, communications*, and *casualty power*. The most effective people in any repair party are those who can find their way to any compartment in the dark and close any valves, doors, or hatches by touch alone. Much of this you will learn from drills and practical experience.

Damage-Control Repair Lockers

When a ship is placed in commission, damage-control repair lockers are set up in the ship's organization. The number of such lockers depends upon the size and type of ship and the number of personnel available.

Each damage-control repair locker contains a number of different ratings in order to provide skilled personnel for any type of work. For example, an engineering repair locker may include machinery technicians, electrician's mates, and damage controlmen.

The engineer officer, as the *damage-control officer*, is responsible for damage control. The *damage-control assistant* (DCA), who works for the engineer officer, is responsible for establishing and maintaining an effective damage-control organization. Specifically, the DCA is responsible for the following:

- The prevention and control of damage, including control of stability, list, and trim. The DCA supervises placing the ship in the material condition of readiness (to be explained later) ordered by the commanding officer.

- The training of the ship's personnel in damage control, including firefighting, emergency repairs, and nonmedical defensive measures against chemical, biological, and radiological weapons.

- The operation, care, and maintenance of auxiliary machinery, piping, and drainage systems not assigned to other departments.

Duties of Repair Parties

All repair parties must be able to

- Control and extinguish all types of fires;
- Repair hull damage and remove flooding and firefighting water;
- Correctly and promptly evaluate the extent of damage in their areas and make accurate "on scene" reports;
- Repair electrical and sound-powered telephone circuits;
- Detect, identify, and measure dosage and intensity of radiation, and decontaminate after CBRNE attack; and
- Give first aid and transport injured to battle dressing stations without seriously disrupting other functions.

All available personnel are assigned to repair lockers. A main propulsion repair party will consist largely of engineering rates and firemen. Repair parties assigned to areas in which magazines are located will have gunner's mates as part of their personnel. Storekeepers are assigned to repair parties that have storerooms located in their areas.

The number and ratings of personnel assigned to a repair party or station, as specified in the battle bill, are determined by the location of the station, the portion of the ship assigned to that party, and the total number of personnel available for all stations.

Each repair locker has a *locker leader*, either an officer or chief petty officer. The second person in charge of a repair party is, in most cases, a senior petty officer who is in charge of the repair party at the scene and is designated *scene leader*.

Investigating and Reporting Damage

In order to make a complete, thorough investigation of any damage, you must know your ship and be familiar with the basic principles of investigating and reporting damage—be cautious, be thorough and persistent, and report all the damage you see.

Damage Repair

Repairs made to a vessel damaged in action, or otherwise in operation, are termed *damage repairs*. These are emergency, usually temporary, repairs necessary to allow the vessel to continue its mission and permit its return to port for permanent repairs. Such repairs are made with whatever material is at hand. The important thing is to keep the ship afloat and maintain as much of its operational capability as possible.

Damage-Control Kits

At each repair locker a number of repair kits are made up and stowed in canvas bags. These kits are kept ready to be taken to the scene of damage. The kits should be constructed and packaged so that they will fit through the smallest watertight scuttle on your ship. These kits are commonly called plugging kits, pipe-patching kits, shoring kits, investigator kits, electrical repair kits, and CBRNE monitoring kits.

Plugging and Patching Holes

There are two general methods of making temporary repairs to a hole in the piping or hull: put something in it or put something over it. In either case, the patches will reduce the area through which water can enter the ship or through which water can pass from one compartment to another.

Temporary repairs do not call for elaborate tools or equipment. They involve principles that can be applied when using wooden plugs, prefabricated patches, or other readily available materials.

Shoring

Shoring is often used on board ship to support ruptured decks, to strengthen weakened bulkheads and decks, to build up temporary decks

Figure 28-2. Deck crew members on board USCGC *Walnut* (WLB-205), a 225-foot buoy tender homeported in Honolulu, practice shipboard dewatering drills during annual training off the south shore of Oahu.

and bulkheads against the sea, to support hatches and doors, and to provide support for equipment that has broken loose. Shores are made of either softwood or metal.

Battle Dressing Stations

Most ships have at least two battle dressing stations equipped to handle personnel casualties. These stations are manned by personnel of the medical department and are so located that stretcher cases may be brought directly to the station by the repair party stretcher bearers. Besides the battle dressing station, emergency supplies of medical equipment are placed in first-aid boxes at various stations throughout the ship.

Decontamination Stations

Most ships have two decontamination stations, primary and secondary, which are provided in widely separated parts of the ship, preferably in the vicinity of battle dressing. These stations provide a place where, after CBRNE attack, personnel go through decontamination. Stations are manned by medical department and repair party personnel who make certain that proper decontamination procedures have been followed.

Damage-Control Communications

Effective communications are of vital importance to the damage-control organization. Without adequate means or proper procedures of communication among the units of the damage-control organization, the whole organization would break down and fail in its primary mission.

The normal means of damage-control communication on board large ships are

- Battle-telephone circuits (sound-powered),
- Interstation two-way systems (4MC intercoms),
- Ship's-service telephones,
- Ship's loud-speaker system (1MC general announcing),
- Voice tubes (where installed), and
- Messengers.

Watertight (WT) Integrity

One of the most important aspects of any operation at sea is to keep the ship watertight. It may sustain any degree of damage, but if the proper degree of watertight integrity is maintained, the ship will stay afloat.

Entering Closed Compartments after Damage

WT doors, hatches, manholes, and scuttles should be opened only after making sure that the compartment is dry or so little flooded that no further flooding will be produced by opening the closure.

Compartments should never be opened until permission is obtained from the damage-control officer. Extreme caution is always necessary in opening compartments below the waterline in the vicinity of any damage.

Closing and Opening Watertight Doors

Doors, hatches, and scuttles giving access to all compartments must be securely dogged. Double-bottom scuttle covers should be bolted at all times except when they must be open for inspection, cleaning, or painting. They must never be left open overnight or when crew members are not actually engaged in work. You must obtain permission from the OOD or damage-control central before opening any secured watertight fitting, and you must report back after you have secured the space. General-drill orders call for closing all watertight doors, hatches, and ventilator openings designed to be closed during maneuvers, in fog, or as a matter of routine at night.

Watertight doors and hatches will retain their efficiency longer and require less maintenance if they are properly closed and opened. To close a door, first set up a dog opposite the hinges, with just enough pressure to keep the door shut. Then set up the other dogs evenly to obtain uniform bearing all around. To open a door, start with the dogs nearest the hinges. This procedure will keep the door from springing and make it easier to operate the remaining dogs.

The strongest doors are those classified as *watertight* (WT) *doors*. Many of these are used in watertight bulkheads of the compartments in the second deck. They are designed to resist as much pressure as the bulkheads through which they give access and usually have eight dogs. Some doors have dogs that must be individually closed and opened; others, known as quick-acting watertight doors, have handles that operate all the dogs at once.

> *Nonwatertight doors* are used in nonwatertight bulkheads. Usually they have fewer dogs than WT doors and are made with dogs that require individual operation.

> *Airtight doors* are also fume-tight and gas-tight. When such doors are used in air locks, they usually have lever-type quick-acting closures, but most others have individually operated dogs.

Panel doors are ordinary shore-type doors that are made of metal and used to provide privacy closures for staterooms, wardrooms, and so forth.

Hatchways are access openings in decks; *hatches* are the coverings for the hatchways. Raised lips (coamings encircling the hatchways) keep water and dirt from entering the compartment when the hatch is not secured. Hatches operate with quick-acting devices or may be secured with individually operated drop bolts or individual dogs. Quick-acting escape scuttles are often provided for rapid access through a hatch.

Manholes are small openings to the water and fuel tanks and voids. They are usually secured by bolting steel plates (provided with gaskets) over them. Occasionally, however, manholes are provided with hinged covers and dogs or drop bolts. Manholes may also be placed in bulkheads.

Gaskets are rubber strips mounted in the covering part of doors or hatches to close against a fixed-position *knife-edge*. Gaskets of this type are either pressed into a groove or secured with retaining strips held in place by screws or bolts.

Watertight closures must have clean, bright (unpainted), and smooth knife-edges for gaskets to press against. A well-fitted WT door with new gaskets will still leak if knife-edges are not properly maintained.

MATERIAL CONDITIONS OF READINESS

In order to use compartmentation to its maximum advantage and to provide for maximum preparedness, all watertight doors, hatches, scuttles, access, valves, and fittings having damage-control value are classified and marked. Ships maintain different material conditions of readiness according to whether contact with an enemy is improbable, probable, or imminent. Each condition represents a different degree of tightness and guarantees the maximum protection against possible damage with proper regard for the health and comfort of personnel and the essential operation of the ship.

Three-Condition Ships

All ships are classified as three-condition ships. These conditions are the number of progressive steps required to achieve maximum protection for the ship and its personnel against the spread of fire, flooding, smoke, dangerous fumes and gases, and the effects of nuclear, biological, and chemical agents. These material conditions of readiness through which all ships progress are *X-ray*, *Yoke*, and *Zebra*. They are described and defined in the following paragraphs.

> *X (X-ray)*. This condition provides the least protection and is set when the ship is in no danger of attack. A black *X* is placed on all fittings that shall be closed at all times when not in actual operations.

> *Y (Yoke)*. This condition is set and maintained at sea and in port during wartime. For peacetime cruising, Yoke may be modified during regular working hours. A black *Y* is placed on all fittings that shall be closed for wartime cruising conditions.

> *Z (Zebra)*. This condition provides the maximum protection for ship and personnel for battle and emergency situations, such as fire, collision, or General Quarters, and is set when entering or leaving port during wartime. A red *Z* is placed on all fittings that shall be closed for battle or emergency conditions.

Yoke and Zebra fittings in condition X-ray, or Zebra fittings in condition Yoke, which are secured by lock or other obstruction that makes them inoperable, will be logged closed in damage-control central. Yoke and Zebra fittings in condition X-ray, or Zebra fittings in condition Yoke, which are secured but not locked or obstructed, need not be logged.

The setting of material conditions X-ray and Yoke is normally a ship division responsibility, and is accomplished by referring to the compartment check-off lists provided. At General Quarters, the setting of and maintaining of condition Zebra is the responsibility of personnel stationed within a compartment or area, and of the damage-control organization.

> *Designation W (William)*. Certain fittings that serve vital systems such as cooling water and fire-main systems shall be open and operating at all times. They are closed only to prevent the spread of damage. These fittings receive the letter classification W (William) which, in itself, is not a material condition of readiness. When contact with significant amounts of radioactive material in the water is expected, sea suction fittings (classified W), except those essential

to main engine operation, shall be closed. Particular attention is to be given to the timely securing of evaporators.

Special Classifications

Special classifications are placed on various fittings to meet special requirements. These are described and defined in the following paragraphs.

Circle X and Circle Y. These classifications are indicated by a *black letter within a black circle.* These classifications indicate that the fittings may be opened without special permission by personnel proceeding to battle stations or as required in routine inspection checks. The fittings shall be secured immediately after use.

Circle Z. This classification is indicated by a *red letter within a red circle.* This classification indicates that the fittings may be opened by permission of the commanding officer for health and comfort of the crew during prolonged periods of General Quarters. However, they will be guarded for immediate closure if necessary.

Circle W. This classification is indicated by a *black letter within a black circle* and shows fittings that are normally open and operating. The fittings are closed only to prevent the spread of CBR contamination or for defense.

Dog Zebra. This classification is indicated by a *red Z enclosed by a large black D.* It is placed on fittings that must be secured to prevent light escaping outside the ship.

FIREFIGHTING

The danger of fire is always present. Fire can start in the aviation-fuel stowage area of a cutter, or it can start in an office wastebasket. If not handled promptly and properly, one can be as dangerous as the other. There are three basic elements in firefighting:

- *Prevention.* Eliminate the possibility of fire before it starts.
- *Equipment.* Ships and stations are supplied with the items needed to fight fire. Know where they are.
- *Technique.* The best equipment is useless if not handled properly. Know how to use it.

No one wins against a fire. It may be fought efficiently and well, the damage may be minimized, and the fighters may receive commendations for their work. The fact remains that in any fire some property is lost, personnel may be injured, lives may be lost, and valuable time is lost for all concerned.

Fire Prevention

The best way to combat fires is to prevent them. Fire prevention must become a daily habit. Keep equipment squared away; keep working areas shipshape. When things are properly stowed and handled, they do not start fires. Remember:

- Keep containers of volatile liquids tightly closed.
- Prevent the accumulation of oil and grease in the bilges or in the exhaust from galley hoods.
- Keep quarters and workshops free of waste material.
- Properly stow bedding, unseasonable clothing, flammable liquids, paints, acids, gases, and chemicals.
- Be careful how you use open lights and electrical equipment where an explosive vapor might exist.

Fire Tetrahedron

Requirements for Combustion

Four things are required for combustion: *fuel* (to vaporize and burn), *oxygen* (to combine with fuel vapor), *heat* (to raise the temperature of the fuel vapor to its ignition temperature), and *chemical chain reaction*. A tetrahedron is a solid figure with four triangular faces. It is useful for illustrating the combustion process because it has room for the chain reaction and because each face touches the other three faces. The fire tetrahedron illustrates these requirements. It also illustrates two facts of importance in preventing and extinguishing fires:

- If any side of the fire tetrahedron is missing, a fire cannot start.
- If any side of the fire tetrahedron is removed, the fire will go out.

Solid fuels. The most obvious solid fuels are wood, paper, and cloth. These are found on board ship as insulation, canvas, packing, electrical cabling, wiping rags, and mattresses.

Types of liquid fuels. The flammable liquids most commonly found on board ship are JP-5, diesel fuel marine, lubricating oil, hydraulic oil, and oil-based paints and their solvents.

Gaseous fuels. Flammable gases found on board a ship include acetylene, propane, and hydrogen.

Oxygen. The oxygen side of the fire tetrahedron refers to the oxygen content of the surrounding air. Ordinarily, a minimum concentration of 16 percent oxygen in the air is needed to support flaming combustion. However, smoldering combustion can take place in an atmosphere with as little as 3 percent oxygen.

Heat. Heat is the third side of the fire tetrahedron. When sufficient heat, fuel, and oxygen are available, the tetrahedron is complete and fire can exist. Heat of ignition initiates the chemical reaction that is called combustion. It can come from the flame of a match, sparks, heat generated by friction, an oxyacetylene torch cutting or welding metal, an electrical short circuit, or an electrical arc between conductor and motor casing.

The fire triangle is a simple means of illustrating the three requirements for surface glowing or smoldering to exist. However, an additional requirement must be met for flames to exist. This requirement is an uninhibited combustion chain reaction.

Uninhibited combustion chain reaction. The fire tetrahedron includes the uninhibited chain reaction in the combustion process. The basic difference between the fire triangle and the fire tetrahedron is that the tetrahedron illustrates flaming combustion. Halon and dry-chemical fire-extinguishing agents directly attack and break down the chain reaction process. These are discussed later.

Theory of Extinguishment

A fire can be extinguished by cooling, oxygen dilution, or fuel removal. Another method of extinguishment is by chemical flame inhibition.

Removing the fuel. One way to remove the fuel from the fire is to physically drag it away. It is often possible to move nearby fuels away from the immediate vicinity of a fire, so that the fire does not extend to these fuels. Sometimes the supply of liquid or gaseous fuel can be cut off from a fire. When a fire is being fed by a fuel line, the fire can be extinguished by closing the proper valve.

Removing the oxygen. A fire can be extinguished by removing its oxygen or by reducing the oxygen level in the air to below 16 percent. Many extinguishing agents (for example, CO_2) extinguish fire with a smothering action that deprives the fire of oxygen.

Removing heat. Water, when applied in fog or a straight stream, is the most effective means of removing heat from ordinary combustible materials such as wood, paper, and cardboard. Cooling with water can ultimately stop the release of combustible vapors and gases associated with the burning of solid fuels.

Breaking the combustion chain reaction. Flaming combustion occurs in a complex series of chemical chain reactions. Once the chain reaction sequence is broken, a fire can be extinguished rapidly. The extinguishing agents commonly used to attack the chain reaction and inhibit combustion are dry chemicals and halons. These chemical agents directly attack the molecular structure of compounds formed during the chain reaction sequence. The breakdown of these compounds adversely affects the flame-producing capability of the fire. The attack is extremely rapid.

Classification of Fires

There are four classifications of fire: Class A, Class B, Class C, and Class D. The different classifications are briefly described in the following paragraphs.

Class A fires involve wood and wood products, cloth, textiles, fibrous materials, and paper and paper products, and are extinguished with water in straight or fog pattern.

Class B fires involve gasoline, jet fuels, oil, and other flammable liquids. These fires are extinguished with AFFF (aqueous film-forming foam), Halon 1211, Halon 1301, Purple-K Powder (PKP) and CO_2.

Class C fires are energized electrical fires that are attacked using non-conductive agents such as CO_2, Halon 1211, PKP, or water spray at prescribed distances. The most effective tactic is to first de-energize and then handle the fire as a Class A fire.

Class D fires involve combustible metals such as magnesium. Water in quantity, using fog patterns, is the recommended agent. When water is applied to burning Class D materials, there may be small explosions. The firefighter should apply water from a safe distance or from behind shelter.

Firefighting Agents

Many materials may be used as firefighting agents. The primary agents discussed in the following paragraphs are the most extensively used on board

Coast Guard ships. The primary firefighting agents are water, AFFF, CO_2, Halon, PKP, and aqueous potassium carbonate.

Water

Water is a cooling agent, and on board ship the sea provides an inexhaustible supply. For many years, the principal method of extinguishing a fire was the direct application of a solid water stream at the base of the fire. Although this method is still used, a more efficient method is to apply water in spray form.

Water will absorb heat until it changes to steam. The steam carries away the heat, which cools the surface temperature.

A secondary method of water extinguishment is steam smothering. When water changes into steam by absorbing heat, it expands 1,700 times in volume. The large quantity of steam displaces the air from the fuel, which smothers the fire.

Aqueous Film-forming Foam

AFFF is composed of synthetically produced materials similar to liquid detergents. These film-forming agents are capable of forming water-solution films on the surface of flammable liquids. The Coast Guard generally proportions AFFF by volume, six parts of AFFF concentrate mixed with ninety-four parts water. AFFF concentrate is a clear to slightly amber-colored liquid concentrate.

When proportioned with water, AFFF provides three fire-extinguishing advantages. First, an aqueous film is formed on the surface of the fuel, which prevents the escape of the fuel vapors. Second, the layer of foam effectively excludes oxygen from the fuel surface. Third, the water content of the foam provides a cooling effect. The principal use of foam is to extinguish burning flammable or combustible-liquid spill fires (Class B).

Carbon Dioxide

A method of extinguishing fires by smothering is the use of the inert gas CO_2, which is 1.5 times heavier than air. This makes CO_2 a suitable extinguishing agent because it tends to settle and blanket the fire.

General characteristics. CO_2 is a dry, noncorrosive gas that is inert when in contact with most substances and will not leave a residue and damage machinery or electrical equipment. In both the gaseous state and the finely divided solid (snow) state, it is a nonconductor of electricity regardless of voltage and can be safely used in fighting fires that would present the hazards of electrical shock.

Method of extinguishing. CO_2 extinguishes the fire by diluting and displacing its oxygen supply. If gaseous CO_2 is directed into a fire so that sufficient oxygen to support combustion is no longer available, the fire will die.

Halon

Halon is a halogenated hydrocarbon. The two types of halon used on board Coast Guard ships are Halon 1301 and 1211.

Halon 1301. For shipboard installation, Halon 1301 is superpressurized with nitrogen and stored in compressed-gas cylinders as a liquid. When released, it vaporizes to a colorless, odorless gas with a density five times that of air. Halon 1301 is installed and used in fixed flooding systems for extinguishing flammable liquid fires.

Halon 1211. This is used in portable fire extinguishers. The halons act by removing the active chemical species involved in the flame chain reaction.

Dry-Chemical Extinguishing Agent, Potassium Bicarbonate

Potassium bicarbonate (Purple-K Powder, or PKP) is a dry chemical used principally as a fire-extinguishing agent for flammable liquid fires. It is used in portable extinguishers.

Use of Purple-K Powder. PKP is used primarily to extinguish flammable liquid fires but can be used on electrical fires (Class C). PKP is highly effective in extinguishing both flammable liquid pool fires and oil-spray fires. Although PKP can be used on electrical fires, it will leave a residue that may be hard to clean. PKP can also be used extensively in the galley for such items as the hood, ducts, and cooking ranges. PKP is not effective on fires involving ordinary combustibles (Class A).

Method of extinguishing. When PKP is applied to the fire, the dry chemical extinguishes the flame by breaking the combustion chain. PKP does not have a cooling capability.

Aqueous Potassium Carbonate

Aqueous potassium carbonate (APC) is used on board Coast Guard ships for extinguishing burning cooking oil and grease in deep-fat fryers and galley-ventilation exhaust ducts. APC solution consists of potassium carbonate and water.

A technique often used in combating liquid-grease fires involving animal and vegetable oils and fats is the application of solutions such as APC, which, upon contact with the burning surface, generate a soap-like froth that excludes air from the surface of the grease or oil. Because the liquid grease cannot burn, the fire is extinguished.

Firefighting Equipment

There are many kinds of firefighting systems and equipment. Some of them are permanently installed and some are portable. These systems and devices must be maintained at maximum efficiency. All firefighting equipment is distributed in readily accessible locations. It is inspected frequently to ensure reliability and readiness for instant operation.

Fire-Main System

The fire-main system is made up of piping, pumps, fireplugs, valves, and controls. The system is designed to supply plenty of water for firefighting, sprinkling the magazines, washing the decks, or other purposes for which salt water is normally used.

The *piping* in a fire-main system is either a single line or a loop system. *Risers* are pipes leading from the fire main, which is usually located below the main deck, to *fireplugs* on the upper deck levels and to others located throughout the ship. The fire-main system is designed so that any damaged section of piping can be isolated. That is why there are so many cross connections and shut-off valves at various locations throughout the piping system.

Standard fireplugs usually have outlets either 2½ inches or 1½ inches in diameter. Some 2½-inch plugs are equipped with wye gates, which provide two hose outlets, each 1½ inches in size. In other cases a reducing fitting is used so that a single 1½-inch hose line can be attached to a 2½-inch outlet.

Forcible-Entry Tools

To do an effective job when firefighting, you must gain access to the fire. If the normal means of access are blocked, or locked, you will need to use forcible-entry tools to gain access to the area so that you can fight the fire.

The tools most frequently used for forcible entry are the sledge-hammer, ax, crowbar, wrecking bar, claw tool, hacksaw, bolt cutter, and exothermic cutting torch. The portable pack-type exothermic cutting torch may be used to cut holes in decks or bulkheads so that nozzles may be inserted to fight a fire. The outfit may also be used to cut away debris that would hinder firefighters or rescue workers. These cutting outfits are to be used for emergencies only, never for routine cutting jobs.

Smoke Curtains

Flame-resistant portable smoke curtains and blankets are stowed in repair stations. These curtains are made of fiberglass or other material. There are two kinds of smoke curtains. One is a two-panel curtain for use at doorways and in passageways to control the lateral spread of smoke; it is provided with five spring clamps. The other is a large blanket, 14 feet square, for use at open hatchways to control the vertical spread of smoke.

Dewatering Pump—Model CG-PIA

Designed for the purpose of dewatering boats in danger of sinking, this pump has a rated output of 140 gallons of water per minute.

Submersible Pumps

The portable submersible pump used on board naval ships is a centrifugal pump driven by a water-jacketed constant-speed AC or DC electric motor. The latest design is rated to deliver 140 gallons per minute against a maximum head of 70 feet. When the head is reduced to 50 feet, the output rises to 180 gallons per minute. Basket strainers are always used with submersible pumps when flood water is being pumped.

To dewater a compartment with a submersible pump, lower the pump into the water using the attached nylon handling line and lead the 2½-inch discharge hose to the nearest point of discharge. The amount of flooding water taken from a flooded space increases as the discharge head decreases. Therefore, dewatering is accomplished most efficiently if the water is discharged at the lowest practicable point and if the discharge hose is short and free from kinks.

Eductor

Eductors are jet-type pumps that contain no moving parts. An eductor moves liquid from one place to another by entraining the pumped liquid in a rapidly flowing stream of water (the Venturi effect). The eductor can perform low-head dewatering operations at a greater rate of discharge than can be obtained by straight pumping with available emergency pumps. Eductors are used to pump liquids that cannot be pumped by other portable pumps. Also, liquids that contain fairly small particles of foreign matter can be pumped by using an eductor.

Atmosphere Testing Devices

All closed or poorly ventilated compartments, particularly those in which a fire has just occurred, are potentially dangerous. The atmosphere may lack oxygen, contain poisonous gases, or present fire and explosion hazards.

Oxygen Analyzer

The oxygen analyzer is used to verify the percentage of oxygen within a space. The Navy's Bureau of Medicine requires a space to have an oxygen content of 20 percent before personnel can enter the space without a self-contained breathing apparatus or an air-line mask. The normal oxygen percentage level at sea level is 20.8 percent.

The oxygen analyzer provides a continuous reading of the oxygen content concentration from 0 percent to 25 percent oxygen in any environment. An oxygen level below 20 percent or above 22 percent should be considered as a potentially dangerous situation.

Combustible-Gas Indicators

The combustible-gas indicator is an instrument used to detect various flammable gases and vapors. The combustible-gas indicator does not identify combustible gas or gases; it merely indicates that one or more combustible gases are present in a certain proportion.

Drager Multigas Detector

The Drager multigas detector is used to test the atmosphere within a space for toxic gases such as carbon dioxide, carbon monoxide, nitrogen dioxide, and hydrogen sulfide. A separate detector bulb is used for each type of gas. For safety and accuracy, use only Drager detector tubes with Drager multigas detectors.

Firefighting Technique

Ships are in constant danger of sudden fire that can spread quickly and place the ship and its crew in the gravest danger. Fire on board ship must be handled promptly and effectively wherever it may occur. Every moment of delay increases the danger that the fire will spread.

Fighting fire on board ship is complicated. It involves the teamwork and efficiency of all shipboard personnel. All hands must learn their firefighting duties thoroughly. Then when fire strikes, they will be prepared to fight it promptly and efficiently.

A fire party follows these steps in fighting a fire:

- Locate and report the fire.
- Isolate the area.
- Fight the fire.
- Set a reflash watch.
- Overhaul the area.

Locating the Fire

Many fires break out in unattended spaces, such as storage areas, and gain considerable headway before they are discovered. The first sign of fire may be smoke pouring from a ventilating system outlet or seeping around a door or hatch cover. Smoke may have traveled a considerable distance from the fire, so locating the fire is usually the first job.

Repair-locker investigators examine all bulkheads, decks, and vents for abnormal heat. A compartment may be clear, but a hot bulkhead would indicate a fire in the next compartment. Follow up any odor or trace of smoke. Even when the main fire is located, there may be fires in other compartments, and they must all be fought at the same time. When the fire has been definitely located and its extent determined, a complete report of the situation is made to the repair locker.

Isolating the Fire

While the fire party fights the fire, it also sets up fire boundaries to isolate it. If a single compartment is involved, it is isolated from above and below as well as from all four sides. If a number of compartments are involved, the fire party treats them as a unit and protects all adjacent spaces to keep the fire from spreading.

Materials on the deck above are removed or thoroughly cooled to prevent the spread of fire. Usually an inch of water on the deck is enough to absorb the heat from the burning compartment below. Spaces below the fire must not be neglected or paint on the overhead may blister and ignite. Fog should be used to cool the overhead and prevent ignition of materials in such spaces. All ventilation systems in the fire area are secured to prevent heat and unburned vapors being carried to other compartments and to help starve the fire by cutting off its air supply.

If there is any doubt that the fire can be kept within the established boundary, a secondary or emergency boundary is set up. Additional hose lines are made ready for use at the secondary boundary.

Fighting the Fire

While the fire is being isolated, it should also be fought. The first step is to de-energize electric circuits in the compartments to protect personnel against shock. Always check fire-hose water pressure at the nozzle before entering any space. All members of the fire party will be dressed in firefighter ensemble. Heat and unburned gases build up pressure that should be vented by undogging the door slowly before it is opened completely for entry.

Reflash Watch

Fires that seem to be extinguished may start again from a smoldering fragment of material or through vapor ignition. The next step after extinguishing a fire on board ship is to set up a reflash watch for each compartment involved and to maintain it as long as the possibility of reflash exists.

Overhauling the Area

After all fire is extinguished, the area must be overhauled to prevent a reflash of the fire. To do this, members of the fire party break up all smoldering or charred material and saturate it. Whenever possible, materials are immersed in buckets of water.

Compartments are desmoked and dewatered. All compartments involved are checked for explosive vapors or liquids that may remain. At this time a full report is made to the commanding officer on fire and smoke damage and flooding.

Special Problems

In some spaces of the ship, firefighting is more hazardous than in others, and special equipment is needed. *Magazine fires* are extremely dangerous. In these spaces, sprinkling systems are installed, which are either automatically or manually operated to cool the ammunition and bulkheads until the fire can be extinguished by firefighters.

Dangerous *oil fires* may occur in machinery spaces. Therefore, a steam smothering system is usually installed in addition to standard firefighting equipment. Smothering is used first to smother the fire. If the fire spreads, water fog and foam must be used to extinguish it.

Bedding and other combustibles in living and berthing spaces are potential sources of fire. Some of these spaces contain fixed fog systems that are manually connected to fireplugs with hose lines. The fog will beat down the surface flames until fire hoses can be turned on the fire.

Caution

Some plastics and curtains used in living areas can burn easily and produce toxic gases. Fire parties must be aware of such dangers. The person in charge must know what is burning and how best to handle it in order to protect personnel.

Figure 29-1.
Coast Guard aircraft must be kept in first-class condition.

CHAPTER 29

Maintenance

Every Coast Guard item, no matter how simple or complicated, must be clean and operable for us to safely carry out our many missions. You may think that cleaning and painting are not important, but they are. Rusted or dirty equipment is not only harder to operate, but it eventually fails. That failure could come at a time when your life depends upon it.

This chapter describes the general procedures for compartment cleaning, basic paint ingredients, preparing surfaces for painting, the basic use and care of paints and paint applicators, safety precautions, and items you should not paint. If you ever have any doubt about the proper way to do any maintenance, ask your supervisor for help.

CLEANING

Maintaining clean conditions on board ship and ashore has always been an important part of Coast Guard activity. Cleaning is a job that involves practically every member, from the compartment cleaner to the inspecting officer. The close surroundings of Coast Guard life require each of us to have a personal interest in our living and working areas, not only for the sake of appearance but for our health and safety as well.

The Cleaning Bill

Each area of a ship or station is assigned to one of its various departments for upkeep. The Cleaning, Preservation, and Maintenance Bill describes these areas and outlines which department is responsible for them. This bill is carefully planned to ensure that all interior and exterior areas of a ship or station are assigned and that no areas overlap or are left out. Each division within a department assigns its personnel the duties of cleaning, preservation, and maintenance of the spaces for which it is responsible.

Compartment Cleaning

The term "compartment cleaner" generally applies to persons assigned to clean the living or berthing compartments or spaces such as passageways and heads. If you are assigned compartment cleaner duties, you will be responsible for keeping your spaces exceptionally clean, preserved, and in good order. Newly assigned personnel are closely supervised to make sure they understand what to clean and how to clean it. Items that may be unfamiliar to you, such as electrical and mechanical devices, are in almost every space on board ship. With this in mind, caution must be observed at all times. Ask your supervisor to point out hazardous items in your compartment, and observe all special cleaning instructions.

To reduce waste and provide for bulk storage, cleaning gear is stocked in and issued from the Bosun's Hold. Each division is periodically issued cleaning gear and is responsible for its proper storage and care. Because cleaning compounds and solvents are usually either flammable or toxic or both, they must never be left unattended or stored improperly. Always read warning labels and follow their directions carefully. Gear such as brooms and swabs must be cleaned after each use and placed in their storage racks. Gear adrift, such as rags, clothing, or personal gear, must be "policed up" immediately. Any items, if left adrift, could cause serious damage to the ship.

Sweepers

"Sweepers" is piped shortly after reveille, at the end of the regular working day, and at other times as scheduled. At these times, all persons assigned as sweepers draw their gear and sweep and swab down their assigned areas.

All trash and dirt should be picked up in a dustpan and placed in a trash receptacle. If dirt is swept over the side, the wind may blow it back on board or the dirt may stick to the side of the ship. In either case, additional work is necessary to clean the ship. At this time, you should empty all butt kits (ensure no butts are still burning) and trash receptacles as instructed. Never dump trash or garbage over the side of the ship without first obtaining prior permission. There are various times when all trash must be kept in a safe area on board the ship until it can be properly disposed.

Figure 29-2. Scrubbing and swabbing the deck in the Arctic looks so easy as demonstrated by these Coast Guardsmen.

The Cleaning Process

Dirt, soil, and contamination all describe the same thing: a foreign material on a surface where it is not wanted. Soil includes grease, oil, tarnish, rust, food residue, and stains. Most exposed surfaces that have been soiled may be cleaned by the proper use of cleaning agents.

Detergents are materials that have the ability to remove contamination and soil. There are other ways of cleaning besides the use of detergents or cleaning compounds. These include purely mechanical processes, such as removing rust from steel by sandblasting or cleaning decks by sweeping. For many cleaning problems, chipping, sweeping, sanding, or brushing may be needed. When detergent compounds are coupled with the mechanical action, however, a cleaner surface is usually produced with less time and labor.

The steps used in most detergent cleaning operations are as follows:

Wetting. The soil and the surface of the object being cleaned must be wetted. If the surface is not wetted properly, the cleaning results will be poor. Contrary to popular belief, water has very poor wetting properties. Its wetting ability, and therefore its cleaning ability, is improved greatly by the addition of materials such as soap or synthetic detergents, which cause the water to flow into tiny crevices and around small particles of soil.

Scrubbing. Dirt is loosened by the mechanical action of rubbing or scrubbing. Oil droplets, for example, are emulsified; that is, they are coated with a thin film of soap. The coating prevents them from recombining, and they rise to the surface. In a somewhat similar manner, solid particles are suspended in solution.

Rinsing. The rinse is very important. It removes loosened dirt from the surface along with the cleaning material.

Zone Inspection

Frequent inspections ensure that all spaces, machinery, and equipment are in a satisfactory state of operation, preservation, and cleanliness. One type of inspection, the zone inspection, divides the ship or station into zones.

Each zone is assigned to an inspection party or team. Usually the commanding officer will head one team while an officer or chief petty officer will head each of the remaining teams. If you are assigned to present a compartment, you present the space to the inspecting officer by saluting and greeting the inspector in the following manner: "Good morning (afternoon), sir/ma'am; Seaman Apprentice Jones (your rank and name) standing by compartment (name or number), for your inspection, sir/ma'am." You will stay with the inspecting officer during the inspection of your spaces to answer questions and provide assistance. Such things as storage cabinets, lockers, and drawers should be unlocked prior to the inspection for easy access. Usually the inspecting officer will give an overall grade to the space; for example; a grade of "outstanding" would indicate that no new discrepancies were noted and all previous discrepancies have been corrected. This is a grade of which you can be proud.

Field Day

Periodically, a field day is held. Field day is cleaning day. All hands "turn to" and thoroughly clean the ship inside and out, usually in preparation for an inspection. Fixtures and areas that sometimes are neglected during regular sweep downs (overhead cables, piping, corners, and spaces behind and under equipment) are cleaned; bulkheads, decks, ladders, and all other accessible areas are scrubbed; knife edges and door gaskets are checked, and any paint, oil, or other substances are removed; all bright work is shined; and clean linen is placed on each bunk. Field days improve the appearance and sanitary condition of the ship, aid in the preservation of the ship by extending paint life, and reduce the dirt taken into operating equipment. Dirt intake must be held to a minimum to prevent the overheating of electrical equipment and abrasive action in rotating machinery.

Because of weather conditions, there will be many days at sea when the ship's topside areas cannot be cleaned. At the first opportunity, all topside surfaces should be cleaned with fresh water and an inspection made for signs of rust and corrosion. If such signs are discovered, you should tend to the area immediately. A little work at that time will save you a lot of work later.

Deck coverings on board ship receive more wear than any other material. Unless proper care is given to them, costly replacement is required. There are several materials used for covering decks, but we will discuss only two: the resilient and the nonslip (nonskid paint) types.

Resilient deck coverings include vinyl tile and linoleum. These deck coverings do not need painting; however, daily sweeping and wiping away of spills as soon as possible is necessary. This type of deck covering is cleaned with a damp swab frequently, allowed to dry, and then buffed with an electric buffing machine. For more thorough cleaning, when the deck is unusually dirty, apply a solution of warm water and detergent with a stiff bristle brush or circular scrubbing machine and rinse with clean water to remove residual detergent. Stubborn dirt and black marks left by shoes can be removed by rubbing lightly with a scouring pad, fine steel wool, or a rag moistened with mineral spirits.

After the deck covering is washed and dried, it can be polished (with or without waxing) with a buffer, or it may be given a coat of self-polishing wax and allowed to dry without buffing. Deck coverings can be buffed several times before rewaxing.

Nonslip (nonskid paint) deck coverings contain pumice, which helps to provide a better footing. To clean a nonskid painted deck, use a cleaning solution consisting of one pint of detergent and five tablespoons of dishwashing compound. This is mixed with fresh water to make twenty gallons of cleaning solution. Apply the solution with a hand scrubber, let it soak for five minutes, and then rinse with fresh water. Nonskid deck coverings should not be waxed or painted because to do so will reduce their nonskid properties. If it becomes necessary to spruce up the appearance of a nonskid deck cover, it may be brushed with deck paint diluted with mineral spirits. The diluted paint should be as thin as possible so that the nonskid properties are not affected.

The Coast Guard uses paint primarily for the preservation of surfaces. It seals the pores of steel and other materials, prevents decay, and arrests rust and corrosion. Paint also serves several other purposes. It is valuable as an aid to cleanliness and sanitation because of its antiseptic properties and because it provides a smooth, washable surface.

Paint also is used to reflect, absorb, or redistribute light. For example, light-colored paint on a ship's interior distributes natural and artificial light to its best advantage.

Learning to paint properly requires the selection of suitable paints for the surfaces to be covered, the proper preparation of the surfaces before painting, and the correct methods of applying paint. Though the selection of suitable paints will not concern you at this time, you should know how to prepare the surface and how to apply paint with a brush and roller. Improper surface preparation and paint application, in that order, are the greatest reasons for paint failure.

Gaskets, Dogs, Air Ports

Rubber gaskets used for watertight, weather-tight, and airtight purposes should not be painted. The metal-bearing edges that come in contact with the gasket should be kept free from rust, grease, and paint. Emery should not be used. The dogs, dog bolts, nuts, hinges, and hinge pins or other parts of watertight doors, and the hatches, man-holes, and air ports upon which the watertight security of vessels depends must not be removed except for repair or adjustment. The removal of these parts for other purposes, such as cleaning or polishing, is prohibited, as this practice often results in the loss of the parts or their being replaced without correct adjustment.

PAINTING

Keeping cutters and boats in first-class condition means a constant battle against corrosion, and the only effective protection against corrosion is good paint properly applied to metal surfaces that have been carefully prepared for painting.

This section covers the basic procedures and safety precautions you should observe while painting. More detailed information can be found in the *Coatings and Colors Manual*.

Types of Paint

Anticorrosive paint is used on underwater hull areas to prevent salt water from acting on metal, and for boot topping. It is always used with antifouling paint and must go on first. It contains heavy pigments and must be stirred frequently. It dries very rapidly, so it must be applied in short, quick strokes.

Antifouling paint is applied to a ship's bottom to prevent marine growth—barnacles, worms, and plants. It contains copper oxide, which will pit steel plating wherever it touches, so it must always be used over anticorrosive paint.

Boot topping is a special paint applied to the hull at the waterline, covering the area between light-load and full-load drafts. Pretreatment coating is applied to all bare metal surfaces, if possible, before the primer coat is applied. It dries fast—within thirty minutes—and if not used within eight hours, must be discarded. It is highly flammable; observe all safety precautions. Modern paint systems, such as Formula 150, incorporate both pretreatment and primer in the same application.

Primers are base coats applied to wood and metal to make a smooth surface for final coats. Always apply two coats of a primer, and a third at all outside edges and corners.

Exterior topside paints consist of white or spar for vertical surfaces and deck gray for horizontal surfaces; spar is used on certain other areas.

Deck paints are green for officers' quarters and wardrooms, gray for other living and working spaces, and red for machinery spaces and shops.

Interior bulkhead and overhead paint is usually fire retardant. White is commonly used on both bulkheads and overheads; other colors are green for offices, radio rooms, pilothouses, and medical spaces; gray for electronic and flag offices; acid-resisting black for compartments in which acids are used; and various pastel shades in crew's living and messing compartments.

Machinery paint is usually gray enamel.

Surface Preparation

For paint to stick to a surface, all salt, dirt, oil, grease, rust, and loose paint must be removed completely, and the surface must be thoroughly dry. Salt and most dirt can be removed with soap or detergent and fresh water. Firmly imbedded dirt may require scrubbing with scouring powder. When oil and grease fail to yield to scrubbing, they may be removed with paint thinner or other approved solvents. After scrubbing or scouring, always rinse the surface with fresh water.

Equipment and Procedures

The removal of rust, scale, and loose paint requires the use of hand tools or power tools and paint and varnish removers. Hand tools usually are used for cleaning small areas; power tools are for larger areas and for completely cleaning decks, bulkheads, and overheads covered with too many coats of paint. Paint and varnish removers are used to remove paint from wood.

Hand Tools

The most commonly used hand tools are *sandpaper, wire brushes*, and *hand scrapers*. Sandpaper is used to clean corners and to feather paint—that is, to taper the edges of chipped areas down to the cleaned surface so that no rough edges remain. Paint will bond best to a clean surface that has been lightly sanded.

Sandpaper is graded according to the size of the abrasive grit on its surface. For example, it is graded from 12 to 600, the coarsest being 12 grit and the finest, 600 grit. Very fine emery (a natural abrasive) paper is sometimes used to polish unpainted surfaces.

A wire brush is a handy tool for light work on rust or on light coats of paint. Such a brush also is used for brushing weld spots and cleaning pitted surfaces.

Hand scrapers are made of tool steel. The most common type is L-shaped, with each end tapered to a cutting edge like a wood chisel. They are most useful for removing rust and paint from small areas and from plating less than one-fourth of an inch thick when it is impractical or impossible to use power tools.

Occasionally it is necessary to use a *chipping or sealing hammer*, but care must be taken to exert only enough force to remove the paint. Too much force dents the metal, resulting in high and low areas. The paint is naturally thinner on the high areas. Consequently, thin paint wears off quickly, leaving spots where rust will form and, in time, spread under the good paint.

Power Tool Safety Precautions

You must be properly trained and qualified before you operate portable power tools. The following safety precautions must be observed when working with electrical and pneumatic (air) tools.

- Wear eye and ear protection while chipping, grinding, sanding, or wire brushing. If dust is excessive, wear a respirator. Do not wear jewelry or loose-fitting clothing.
- Do not use defective tools. If you have any doubt about the condition of any tool, show it to your petty officer, who will have its condition determined.
- Make certain that electrical power tools are grounded properly. Every portable electrical power tool must be provided with a ground lead that connects the tool casing to the ship's structure.
- Give your full attention to your job.
- Do not operate power tools in areas where flammable vapors, gases, liquids, or exposed explosives are present.
- Do not allow power cords and air hoses to kink or come into contact with oil, grease, hot surfaces, or sharp objects.
- Power cords and air hoses must not be laid over ladders, steps, scaffolds, or walkways in such a manner as to cause a tripping hazard.

Aluminum Surfaces

Aluminum surfaces on board ship present a special problem because if they are not treated properly, considerable corrosion will result. Corrosion is greater when dissimilar metals (such as aluminum and steel) are in contact with each other and are exposed to an electrolyte (seawater, for example), which causes an electric current to flow between them, resulting in galvanic corrosion of the aluminum. The first sign of aluminum corrosion is a white, powdery residue in the area where the two dissimilar metals make contact, later by pitting and scarring of the aluminum surface, and finally by the complete deterioration of the aluminum in the area. Holes in aluminum plating enlarge, and the screws, bolts, or rivets pull out or may even disintegrate.

When painted aluminum surfaces are prepared for repainting, power sanders, if used, must be used with great care. It is best to use hand scrapers, hand and power wire brushes, or sandpaper of a very fine grit. *Never use scaling hammers.*

Steel Surfaces

The most important task in painting a steel surface is the preparation of the surface to be painted. Steel surfaces must be completely free of rust, loose paint, dirt, scale, oil, grease, salt deposits, and moisture. Old paint in good condition is an excellent base for repainting. When a surface is to be repainted and the old paint is not to be removed, the surface must be smoothed and thoroughly cleaned and dried before new paint is applied.

In touch-up painting, when only small areas or spots need repainting, it is essential that the removal of old paint be carried back around the edges of the spot or area until an area of completely intact paint, with no rust or blisters underneath, is reached. The edges of the remaining paint should be feathered.

When an old painted surface is to be completely redone, the old paint should be taken down to bare metal and a primer put on before painting. Never leave a bare metal surface exposed overnight. Always put on a primer coat before you quit for the day to prevent oxidation from starting.

Paint and Varnish Removers

Paint and varnish removers are used mostly on wood surfaces but may be used on metal surfaces that are too thin to be chipped or wire-brushed. The three types of removers in general use are *flammable, nonflammable,* and *water-based alkali.* All three are hazardous, and their safety precautions must be strictly observed. The removers should be used only in well-ventilated spaces. The alkali type should not be used on aluminum or zinc because of its corrosive properties.

Safety precautions. Paint and varnish remover must never be used around an open flame because some types are highly flammable. Removers must not be used in confined spaces because of their dangerous anesthetic or toxic properties that can kill or cause injury if you are exposed to them for long periods. Moreover, paint and varnish removers should not be used by persons who have open cuts or sores on their hands unless rubber gloves are worn. Avoid letting the remover touch your skin; watch out particularly for your face, eyes, and mouth. If paint or varnish remover should come in contact with the skin, wash it off immediately with cold water; if it gets into your eyes or mouth, seek medical attention as soon as possible. Read the product label for full details.

What Not to Paint

The following items must not be painted:

- Knife edges, rubber gaskets, dogs, drop bolts, wedges, and the operating gear of watertight doors, hatches, and scuttles
- Start-stop mechanisms of electrical safety devices and control switchboards on machinery elevators
- Bell pulls, sheaves, annunciation chains, and other mechanical communication devices
- Condenser heads and outside surfaces of condensers made of composition metal
- Sprinkler piping within magazines
- Glands, stems, yokes, toggle gear, and all machined external parts of valves
- Heat-exchange surfaces of heating or cooling equipment
- Identification plates
- Joint faces of gaskets and packing surfaces
- Lubricating gear, such as oil holes, oil or grease cups, grease fittings, lubricators, and surfaces in contact with lubricating oil
- Lubricating oil reservoirs
- Machined-metal surfaces (working surfaces) of reciprocating engines or pumps
- Metal lagging
- Rods, gears, universal joints, and couplings of valve-operating gear
- Rubber elements of isolation mounts

- Ground plates
- Springs
- Strainers
- Threaded parts
- Zincs
- Working surfaces
- Hose and applicator nozzles
- Electrical contact points and insulators
- Original enamel, lacquer, or crackle finish on all radio, electrical, and sound equipment, unless damage makes refinishing essential
- Decorative plastic, such as tabletops

Painting Safety Precautions

Painting can be dangerous if one is careless. Many paints are highly flammable, others are poisonous, and some are both flammable and poisonous. To increase your chances of remaining alive and healthy, observe the following precautions:

- When burning off paints, do not leave the burned-off paint on the deck where it may become powdered when stepped on. Do not play the torch on the paint long enough to produce smoke. The fumes and powders from paints are poisonous. Do not burn off paint in a closed compartment.
- Keep paint off your skin as much as possible. Wash your hands, arms, and face with soap and warm water before eating.
- Be sure you have adequate ventilation, and wear an approved paint/spray respirator whenever there is reason to believe the ventilation is inadequate in the place you are painting.
- Do not smoke, use an open flame, or use spark-producing tools in the vicinity of painting operations.
- Use only explosion-proof lights near painting operations.
- Do not wear nylon, Orlon, or plastic clothing or covering. These materials generate static electricity, which may spark and ignite paint vapors.
- Do not carry matches or cigarette lighters, or wear steel buckles or metal shoe plates. Too often one forgets and strikes a match or lights

a cigarette lighter in areas filled with explosive fumes. Steel buttons, buckles, and tabs can strike sparks that are invisible to your eyes, perhaps, but wholly capable of igniting paint vapors.

- When pouring solvents, make sure the containers are touching each other to prevent sparks.
- Never paint during electrical storms.
- Keep food and drink away from areas being painted.
- Do not use gasoline, turpentine, mineral spirits, or other solvents to remove paint from the skin, as the skin will absorb them.
- Follow carefully the instructions of your supervisor.

Paint Issue

Before paint, primers, or paint applicators are issued, your division petty officer will inspect the area to be painted to ensure that all preparations have been properly made. For example, are all items not to be painted properly identified or masked? Are all safety precautions understood and properly observed? Is the surface ready to be painted?

Remember, if you spill paint (or oil or grease, etc.) you are responsible for cleaning it up. At the end of working hours, all paint and brushes must be returned to the paint locker. Paint must be stored in its proper container, and all brushes and rollers cleaned as described later in this chapter.

Painting Methods

Three means of applying paint are used in the Coast Guard—*brush, roller,* and *spray.* The majority of personnel have no occasion to use paint sprayers; hence, their use will not be described here. Instead, this section concentrates on brushes and rollers, with which everyone in the Coast Guard should be familiar.

Paint Application by Brush

Smooth and even painting depends as much on good brushwork as on good paint. There is a brush for almost every purpose, so use the proper brush and keep it in the best condition.

The two most useful brushes are the *flat brush* and the *sash brush.* With a flat brush, a skillful painter can paint almost anything on board ship. Flat brushes are wide and thick, carry a large quantity of paint, and provide a maximum of brushing action.

Sash brushes are handy for painting small items, for cutting in at corners, and for hard-to-get-at spaces. The *fitch brush* also is useful for small surfaces. The *painter's dusting brush* is used for cleaning surfaces.

The following are some general hints to help you use a paintbrush properly:

Grip the brush firmly, but lightly. Do not put your fingers on the bristles below the metal band (ferrule). To hold it otherwise restricts your movements and causes undue fatigue.

When using a flat brush, do not paint with the narrow edge. This practice wears down the corners and spoils the shape and efficiency of the brush. When using an oval brush, if you revolve it too much, it soon wears to a pointed shape and becomes useless. Do not poke oversized brushes into corners and around moldings. Such a practice bends the bristles, eventually ruining a good brush. Use a smaller brush that fits into such odd spots.

Dip the brush into the paint, but not over halfway up the bristles. Remove the excess paint by patting the brush on the inside of the pot. Avoid overfilling the brush, otherwise paint will drip on the deck or other surfaces and will run down the handle.

Hold the brush at right angles to the surface being painted, with the ends of the bristles just touching the surface. Lift the brush clear of the surface when starting the return stroke. If the brush is not held correctly and is not lifted, the painted surface will be uneven, showing laps and spots and a daubed appearance. Also, a brush that is held at any angle other than a right angle will soon wear away at the ends.

For complete and even coverage, first lay on, then lay off. Laying on means applying the paint first in long strokes in one direction. Laying off means crossing your first strokes. Using the recommended method and crossing your strokes distributes the paint evenly over the surface, covering it completely and using a minimum amount of paint.

Always paint the overhead first, working from the corner that is farthest from the entrance to the compartment. By painting the overhead first, you can wipe drippings off the bulkhead without smearing the bulkhead paint.

When overhead surfaces are being painted, sections should normally be painted in a fore-and-aft direction; beams, in an athwartship direction. But where sections of the overhead contain many pipes running parallel with the beams, it is often difficult to lay off the paint in a fore-and-aft direction. In such situations, better results are obtained by laying off the paint parallel with the beams.

To avoid brush marks when finishing up an area you have painted, use strokes directed toward the last section finished, gradually lifting the brush near the end of the stroke while the brush still is in motion. Every time the brush touches the painted surface at the start of a stroke it leaves a mark. For this reason, never finish a section by brushing toward the unpainted area. Instead, always end up by brushing back toward the area already painted.

When painting pipes, stanchions, narrow straps, beams, and angles, lay the paint on diagonally. Lay off along the long dimension.

Always carry a rag for wiping up dripped or smeared paint. Carefully remove loose bristles sticking to the painted surface.

Cutting In

After you master the art of using a paint brush properly, you should learn to cut in. Cutting in is a simple procedure and anyone with a fairly steady hand can learn it quickly.

Suppose you have to cut in the angle between an overhead and a bulkhead. Start at one corner. Hold your brush at an angle of 75 degrees to 80 degrees from the bulkhead and 10 degrees from the overhead. Draw your brush along in fairly long, smooth strokes. This is one job where working slowly does not produce better results. The slower you stroke, the curvier your line.

Use of Rollers

Large areas, such as ships' decks and sides (free of rivets, bolts, cable, pipes, and so on), can be painted quickly by the roller method. The paint should be laid on and laid off in the same fashion as when brushes are used. A moderate amount of pressure must be applied to the roller to make sure that the paint is worked into the surface. If pressure is not applied, the paint does not stick and soon peels off. When the paint roller is properly used, it will apply a more even coat and use less paint than is possible with a brush.

Care of Brushes and Rollers

Unfortunately, far too many good paint brushes and rollers are ruined simply because painters have little or no idea how to care for them, or they are too lazy to clean them. Consequently, you should pay particular attention to the following hints, and heed them at all times. Treat applicators as though you paid for them yourself, and replace them when they are no longer usable.

Do not let a brush stand on its bristles in a pot of paint for more than a few minutes. The weight of the brush bends the bristles, making it almost impossible to do a good job. Never allow paint to dry on a brush. If you intend to leave a paint-filled brush for long periods, as over the noon hour, fold wax paper or other heavy paper around the bristles and ferrule in such a way that air is kept away from the bristles. Twist the paper around the handle and secure it with rope yarn or sail twine. Cover your pot of paint, and place both it and the brush in a safe place. Before starting to paint again, stir the paint thoroughly with a paddle—not with the brush. At the end of the day, before turning in your paint and brush to the paint locker, clean as much paint from the brush as possible by wiping it across the edge of the paint pot or mixing paddle.

Ordinarily, the person or persons working in the paint locker will clean and stow the brushes turned in. Occasionally, though, they require help, and you may be detailed to the job. If so, follow instructions carefully and do a thorough job of cleaning the brushes. Only brushes that can be cleaned with soap and water are to be cleaned. Others, unless they are to be used again right away, should be allowed to dry and then thrown away. The cost of disposing of thinners as hazardous waste precludes the reuse of these brushes.

Paint lockers usually have containers with divided compartments for stowing different types of brushes (that is, paint, varnish, shellac, and so on) for short periods of time. These containers normally have tight covers and are equipped for hanging brushes so that the entire length of the bristles and the lower part of the ferrule are covered by the solvent or cleaning oil kept in the container. Brushes are suspended so that the bristles do not touch the bottom, thus preventing them from becoming permanently misshapen.

Brushes to be used the following day should be cleaned in the proper cleaner and placed in the proper compartment of the container. Those not to be used again soon should be cleaned, washed in soap or detergent and water, and hung to dry. After drying, they should be wrapped in heavy paper and stowed flat. Do not leave a brush soaking in water. Water causes the bristles to separate into bunches, flare, and become bushy.

The proper cleaners for paint applicators are as follows:

Paint/Finish	Solvent/Cleaner
Natural and synthetic oil-base	Turpentine or mineral spirits paints and varnishes; chlorinated alkyd resin paints
Latex emulsion paints	Water
Chlorinated rubber paints	Synthetic enamel thinner xylene
Shellac	Alcohol (denatured)
Lacquer	Lacquer thinner

Paint rollers are cleaned in a different fashion. After use, the fabric cylinder should be stripped from the frame, washed in the cleaner recommended for the paint used, washed in soap and water, rinsed thoroughly in fresh water, and replaced on the frame to dry. Combing the pile of the fabric while it is damp prevents matting.

Figure 30-1.
Safety is crucial to many dangerous shipboard evolutions.

CHAPTER 30

Safety and
Environmental Health

We perform many tasks every day that are dangerous. If you do not follow three basic rules, you or someone else may become sick, injured, or killed. You are responsible for reporting hazardous conditions, using protective equipment, and observing safety precautions. The *Safety and Environmental Health Manual*, COMDTINST M5100.47, contains information about the programs the Coast Guard has developed to protect you.

This chapter gives you an overview of the Coast Guard Safety and Environmental Health Program. The program includes your rights and responsibilities, procedures for reporting hazardous conditions, and methods of protection against the hazards.

YOUR RIGHTS AND RESPONSIBILITIES UNDER THE COAST GUARD SAFETY AND ENVIRONMENTAL HEALTH PROGRAM

Current law provides rights and responsibilities for Coast Guard employees. As a member of Team Coast Guard, you have the right to a safe and healthful workplace. Your participation in the program is essential in providing such an environment. You have the right to

- *Have access* to copies of Coast Guard standards, procedures, and injury and illness statistics;
- *Comment* on proposed Coast Guard occupational safety and health standards;
- *Assist safety and health officials* in conducting inspections of Coast Guard facilities;

- *Request an inspection* of a work area alleged to possess unsafe or unhealthful conditions. You may make these requests either orally or in writing, and they must be recorded and maintained on file. You may remain anonymous if you wish; and

- *Appeal* if you disagree with the action taken on unsafe or unhealthful conditions. If you have exhausted all means available to achieve a satisfactory resolution of the conditions and you remain dissatisfied, you may write to the Office of Federal Agency Safety and Health Programs, Occupational Safety and Health Administration, Department of Labor, 3rd & Constitution Avenue NW, Washington, D.C. 20207.

You may exercise these rights without fear of reprisal, coercion, discrimination, restraint, or interference. If you think you are being treated unfairly because you exercised your rights, report it to the most immediate safety and health official.

The "Occupational Safety and Health Rights and Responsibilities" poster is required to be displayed in all workplaces. This poster spells out your rights and responsibilities and lists the safety representatives' names and phone numbers.

The *Hazard Communication Standard* expands on the previous rights to include a *right to know what hazards exist* in your workplace. All known hazards should be pointed out to you during your unit indoctrination period.

"Learn your job and the safety issues related to that job." You must comply with all occupational safety and health standards of the Department of Labor, the Department of Homeland Security, and the Coast Guard that apply to your own actions and conduct.

Identifying Hazards

Your ability to recognize hazards is directly related to your experience and familiarity with the task you are performing. Unfortunately, experience and job familiarity are often the cause of safety mishaps because people tend to be less careful in familiar situations. *Awareness of the ways people can be injured is a key element in recognizing hazards.*

Reporting Hazards

You must report safety and health hazards when you find them. Never leave a hazardous condition for the next person to find the "hard way" There are two ways to report a hazardous condition.

The first is a simple *oral report* to the person responsible for the area in which the hazard exists. Most hazards reported in this way can be corrected on the spot.

The second is a *written report* submitted on an Employee Hazard Report, Form CG-4903. Fill out Section One at a minimum. Instructions for completing the entire form are contained in COMDTINST M5100.47. The information from this form will be entered into a computerized information database so that other people can use it to correct similar hazards or document a widespread problem. Even if you make an oral report, a CG-4903 should be submitted for record purposes.

Hazard Abatement

Abatement is the action taken to reduce, control, or eliminate risk. All reported hazards must be abated. Requiring the use of protective equipment, clothing, and devices, and following safety procedures are examples of abatement methods.

PERSONAL PROTECTIVE EQUIPMENT (PPE)

Personal protective equipment is a form of guarding against hazards. PPE does not eliminate hazards; it only reduces the risks associated with the hazards. Examples of PPE are steel-toe boots, ear plugs, safety eyeglasses, hard hats, and rubber aprons. The following paragraphs look at some of the equipment you will use to do your jobs.

Eye Protection

Loss of eyesight is probably one of the most traumatic experiences a person can endure. The eyes are very delicate organs and must be protected. Eyes can be damaged from flying objects or debris, chemical splashes including acids and alkalis, and bright light, like that found in arc-welding or in the glare from sunlight.

The most common forms of eye protection are *safety eyeglasses*. They look like regular eyeglasses with shields along the sides to prevent objects from bouncing off the inside of the glasses into your eye. The lenses are made of special high-impact shatterproof plastic. They are available with prescription lenses; however, the Coast Guard does not usually provide prescription safety eyeglasses unless your job requires it. If you wear prescription glasses, you can use *safety goggles* designed to fit over the outside of them. They offer the same level of protection as safety eyeglasses.

Figure 30-2. Personal protective equipment must be worn by all responders in an effort to minimize injuries.

Face shields are worn on your head like a hat. They are usually built with the capability of being flipped up and down on a hinge, and they protect your face from flying debris or chemical splashes while offering additional protection over the eyeglasses or goggles. If a job calls for the use of other forms of eye protection, you should also wear a face shield.

Gas-welding goggles are used for gas welding or standing a fire watch. The lighter shade of the lenses in the goggles will protect your eyes from reflected, but not direct, ultraviolet rays. Never look at the arc of an arc-welding operation when wearing these goggles. Before using them, hold them up to a light source and check for cracks. Even small cracks will let harmful rays into your eyes and cause flash burn. Flash burn feels as if you have sand in your eyes.

You can best protect your eyes from the glare of reflected sunlight by wearing polarized sunglasses.

Hearing Protection

You should strive to protect your hearing with the same priority given to eyesight. Many jobs we do every day present high-noise-level hazards. There are two basic types of hearing protectors, ear plugs and aural protectors.

Ear plugs may be reusable or disposable. Disposable plugs are rolled between the thumb and forefinger and inserted into the ears. Once in the ear, they expand to form a sound-reducing barrier. They are maintenance free, inexpensive, and convenient and offer the same level of protection as the reusable type. Reusable plugs offer adequate protection but must be fitted to the person wearing them. They are available in various types and sizes.

When you first begin wearing ear plugs, there may be some discomfort. This is normal and usually goes away after a few minutes. Continued discomfort may be corrected by trying a different size or style. It is very important that you protect your hearing, so get a proper fit and wear them.

Respiratory Protection

Respiratory protection is required to prevent death or illness caused by harmful dusts, fumes, smoke, aerosols, mists, gases, vapors, and atmosphere without enough breathable oxygen. The Coast Guard strives to prevent contamination by confining the hazard, providing adequate ventilation, or substituting a less toxic material. Additional information on respirators can be found in the *Technical Guide: Practices for Respiratory Protection*, COMDTINST M6260.2.

If the above protection is not practical or not in place, you must wear a respirator appropriate to the task. When respirators are required, you are responsible for using them properly and reporting any malfunctions such as leaks, problems with fit, and air flow.

Your unit safety officer will assist your unit in selecting the right respirator for any given situation. There are three basic types of respiratory protection: air purifying, atmosphere providing, and atmosphere providing/ air purifying.

Air-purifying respirators. Air-purifying respirators in use in the Coast Guard fall into two basic categories: half-face and full-face. Their use is dependent upon several requirements.

- There *must* be adequate oxygen in the atmosphere.
- The contaminant(s) *must* be identified.
- There *must* be a distinct odor from the contaminant.
- The respirator *must* have the correct cartridge or filter.
- The contaminant *cannot* be absorbed through the skin.

Atmosphere-providing respirators. Supplied-air respirators provide breathable air. They *must* provide a positive pressure to the user. The Coast Guard uses the following types of supplied-air respirators:

- Self-contained breathing apparatus (SCBA)
- Airline

Atmosphere-providing/air-purifying respirators. This type of respiratory protection offers the benefits of an SCBA as a backup to the hose mask or airline respirators. Certain operations require this level of protection.

Approved Respirators

The respirator being used *must* be an approved type. The National Institute of Occupational Safety and Health and the Mining Enforcement and Safety Administration offer standards and specifications for approving respirators for various uses. *Approved respiratory protection will have an approval number beginning with the letters "TC."*

Protective Clothing and Equipment

So far we have talked about eye protection, hearing conservation, and respiratory protection. This section introduces you to protective clothing and equipment for the rest of your body.

Certain operations can result in injuries whether they are wounds, burns, or broken bones. Sickness and even death can result from absorbing a chemical through the skin. The choice of protective clothing and equipment depends on the hazards you expect to find when doing a job.

Some of the various forms of protective clothing and equipment you may use in the Coast Guard are listed below.

- Hard hats
- Combat helmets
- Hard hat/face shield combination
- Foul-weather clothing
- Personal flotation devices
- Body armor
- Aprons (leather, rubber, canvas)
- Gloves (leather, rubber, cotton)
- Boots (steel-toe, rubber)
- Full-body chemical-exposure suits
- Industrial safety harnesses

The job will dictate which clothing or equipment you need. Be sure you get the training you need to wear or use any protective equipment or clothing.

GENERAL SAFETY HAZARDS AND PRECAUTIONS

This section presents some general hazards and precautions you may find useful. Keep in mind, however, that there may be other hazards and appropriate precautions. They may be specific to your unit or the result of a new system or technology since this was written. Check with your supervisor or safety representative.

Confined Spaces

Closed compartments and unvented spaces may contain unexpected dangers. They may contain liquids or gases that are poisonous, explosive, or deficient in oxygen. The liquids or gases may be under high pressure. All confined spaces should be considered dangerous until proven otherwise. You should never enter any of these spaces on your own.

The spaces presenting this danger should be clearly marked, but this is not absolute. Enter *only* when directed and when you have taken these precautions:

- Break the seal on the entrance cover *before* removing hold-down bolts or other fastenings.
- Have the space certified as safe by a gas-free engineer.
- Wear appropriate respiratory protection as directed by the gas-free engineer.
- Wear a tending line when working in the space.
- Make sure you have a reliable safety tender standing by and a reliable method of communicating with each other. If you are the safety tender, and the person working in the space is overcome, *do not* enter the space without an SCBA. See the section on respiratory protection.

Boat Safety

Most people in the Coast Guard will be on board small boats at one time or another either as a passenger or a crew member. When you are boarding a boat or on the boat, follow these general guidelines:

- Obey all orders of the coxswain.
- *Do not* distract the boat crew.
- *Do not* engage in horseplay or move about, causing instability.
- *Do not* smoke in the boat.
- *Do not* sit on lifejackets . . . *wear them!*
- Keep all parts of your body in the boat and *do not* sit on gunwales.
- If the boat capsizes or gets swamped, *do not* panic. Stay with the boat and help others.

Deck Safety

A ship's deck presents numerous tripping hazards. These include cleats, bitts, pad eyes, and the larger objects like boat davits and winches. Learn where these hazards are before you need to go topside at night. When you are on deck, *do not* sit or lean on lifelines. Stand inboard of them and hang on at all times.

Larger ships have flight decks for helicopter operations. This deck has additional hazards. Rotating propellers are nearly invisible and will kill or maim you if you come into contact with them. Only personnel necessary to flight operations should be on the flight deck.

Lifting

The correct method of lifting certain loads is often neglected. If you do not lift loads properly, you may wind up with a hernia or injured back, or you may lose control of your load and drop it onto one of your crew mates.

When moving loads that are too heavy or bulky, get help or use a hand truck or dolly. There is no shame in doing a job safely and efficiently. If you are going to carry a load, follow these guidelines:

- Face the spot where you are going to put the object.
- Crouch down close to the object with feet spread to a comfortable width.
- Pull the object toward you.
- Get a good grip on the object.
- Make sure you can see over the top of the object.
- While keeping your back straight and nearly vertical, tuck in your chin and lift with your legs. Do not use a jerking motion.
- Pull the object in closer to center the weight.

- Do not twist your body when lifting or carrying. Twisting your body causes most back injuries.
- Do not change your grip while carrying the load. Set it down and change your grip.
- If placing a load on a bench or table, set it on the edge and push it away from you.

Handling Cargo

In addition to the lifting precautions, you should exercise the following precautions when handling cargo:

- Remove rings, watches, and bracelets.
- *Do not* throw or drop items from elevated places. Lower them on a line or carry them.
- Wear appropriate protective clothing and equipment such as steel-toe shoes, personal flotation devices, and hard hats.
- Secure hatch covers or doors to avoid interference with traffic flow.
- Use the nonworking side of the ship or an alternate path ashore to avoid traffic.
- *Never* stand in the bight of a line or near lines under strain.
- When using a hand truck, pull the load up, and lower (or push) the load down ramps.
- When steadying a load, *do not* stand between the load and a fixed object or under a suspended load.
- *Do not* ride a load in any direction, up, down, or horizontally.
- *Do not* engage in horseplay.

Lifelines

Lifelines are lines constructed around the edges of a ship's deck. They are protective barriers to keep people from being thrown or washed over the side. To ensure your safety, and the safety of your crew mates, you should observe these precautions:

- *Never* sit, stand, or lean on lifelines.
- *Never* hang or secure any weight on a lifeline.
- *Never* remove a lifeline without permission. If you do remove one, rig a temporary line in its place.

Ladders

Ladders are a common source of injury if precautions are not followed. In addition to shipboard ladders, the Coast Guard uses straight ladders, extension ladders, multiple-angle ladders, and stepladders. Some ladders are permanently installed, like fire escapes and on towers.

On board ship, most ladders are removable. If you have to remove one, be sure to secure hatches and gangways leading to the ladder. The steps of ladders seem to gather paint cans, tools, spilled liquids, and other hazards. Watch your step!

Stepladders are available in several heights. They are portable and are made of wood or aluminum. When using a stepladder, make sure that the legs are open as far as possible for maximum stability. If you are painting or using tools, set the paint can, pan, or unused tools on the shelf provided.

Straight and extension ladders are used for heights greater than those reached by most stepladders. They must be leaned against something to hold them up. If possible, secure the upper end of the ladder to the structure to prevent it from tipping over.

Some general rules for using ladders are summarized below:

- *Do not* leave anything on the steps of ladders.
- *Never* stand on the top two steps of a stepladder.
- *Never* use an aluminum ladder of any type when using electrical tools or working near electrical circuits.
- *Always* have someone hold the ladder steady while you are climbing.

Working Aloft

Working aloft can be quite hazardous. If safety precautions are not followed, you can be injured by falling or you can injure someone else by dropping something. Other hazards of working aloft include being poisoned by stack gases, electrocuted, or burned by radio transmissions. To avoid these hazards, follow these precautions:

- *Never* go aloft without permission from the OOD.
- *Ensure* that electrical and electronic equipment is turned off and tagged to stay off.
- *Always* use an industrial safety harness with safety line and working line.
- *Always* attach tools to the harness with preventer lines to keep them from being dropped.

- If working in the path of stack gases, *always* use an SCBA.
- *Always* have a ready-to-climb safety observer watching you.
- *Never* throw or drop anything from aloft. Lower the items with a hand line or suitable tackle.
- *Always* keep a safety line or snap-hook on a closed-loop hold. Radio and radar equipment on other ships create a charge you may feel when climbing. The shock will not hurt, but it may startle you and cause you to fall.
- *Do not* touch antennas, aloft or otherwise. You could be burned by the energy from your own ship, from nearby equipment ashore, or from another ship.
- Watch for rotating radar antennas. They could knock you to the deck.
- When finished with the work, report it to the OOD.
- Report any unsafe rigging aloft to the OOD.

Working over the Side

Working over the side can include painting or welding. You may work over the side when under way *only* with the permission of the commanding officer. While over the side, you must wear an industrial safety harness, an inherently buoyant life preserver, and a tending line. The line must be attached to the ship and tended by someone on the deck at all times.

Electrical and Electronic Equipment

All electrical and electronic equipment is hazardous. This does not mean you should be afraid it, but you should respect the potential danger and observe these safety precautions:

- *Always* inspect the equipment before using it. Look for damaged, cracked, or frayed power cords or defective parts. Return them to the electrician or electronics technician.
- *Always* remove from service equipment that causes a shock, regardless of how slight the shock may be.
- *Never* remove the third pin of a three-wire cord. This connection provides a ground that protects you in case the tool shorts inside. The ground is required on tools with metal cases.
- Use the plug when disconnecting a power cord.

- When using an extension cord, plug the equipment into the extension cord first, then plug the extension cord into the receptacle. Reverse the process when finished.

- *Never* remove or replace fuses. Only qualified personnel are allowed to make these repairs.

- *Never* attempt to make temporary repairs to electrical equipment. Take it to the electrical shop.

- *Do not* touch anyone who has been electrocuted—you will be electrocuted too. Secure the power and then remove the victim. Administer artificial respiration or cardiopulmonary resuscitation as needed. (You may pull a fire alarm, but *do not* leave the victim.)

- If you cannot secure the power when rescuing an electrical shock victim, use a wooden pole or cane or a dry piece of line or a belt to pull the victim free from the circuit.

Rotating Machinery

Knowing the correct operating procedures for machinery is vital to your safety. Many safety precautions for machinery are specific to the equipment, but the following general precautions apply to most equipment. *Do not* operate machinery you have not been trained to operate.

- *Never* place any part of your body into moving machinery.

- *Do not* ride machinery unless it is designed to carry you.

- *Do not* wear jewelry, neckties, or loose clothing when operating machinery.

- *Always* wear PPE appropriate to the operation (i.e., eye protection, hearing protection).

- *Always* remove from service any machinery requiring maintenance or adjustment. Tag it out if necessary.

- *Do not* use compressed air to clean parts of your body or clothing or to do cleanups. Use a vacuum cleaner or a broom. Compressed air can be used for cleaning machine parts as long as the air supply pressure is less than 29 psi.

- Ensure that all safeguards are in place when operating any equipment.

- Check safety devices, alarms, and sensors for proper operation, and report any problems.

- Clean up oil leaks and report them for repair.
- *Do not* allow tripping and fire hazards to accumulate. Keep all machinery spaces clean.

Power Tools

You will be using a variety of portable power tools. They may be electrical, hydraulic, or pneumatic. Regardless of the type, you must protect yourself against the hazards associated with the tool.

- *Always* wear protective equipment.
- *Always* use the installed guards supplied with the equipment.
- *Always* use tools that are in good repair. Return tools with defects for repair.
- *Do not* use tools that produce sparks near explosive or flammable vapors, gases, liquids, or materials.
- Make sure electrical tools are off before plugging them into a receptacle.

Tag-Out Procedures

Use DANGER and CAUTION tags for identifying equipment that is hazardous. However, your supervisor may prefer to tag out the equipment after you report the hazard. You must not remove these tags. They can be removed only by the person who placed them. If your job requires you to operate a tagged-out switch, valve, or control, *do not* operate it; report it to your supervisor. These tags are shown in figure 30-3.

Hazardous Materials

You may have to use, store, handle, or dispose of hazardous materials in your job. A standard labeling technique is used for identifying hazardous materials in three principal categories: *health, fire,* and *reactivity (instability)*. Each category is rated on a scale from zero to four to indicate the severity of the hazard. Four is most severe.

The sample label in figure 30-4 shows how the information is presented. The four diamonds are color-coded as follows: *Blue—Health Hazard*; *Red—Fire Hazard*; *Yellow—Reactivity*; and *White—Specific Hazard*. The specific hazard diamond is used to identify the type of material. Firefighters and health officials use this information to deal with spills or releases of the material. You must pay particular attention to the principal categories.

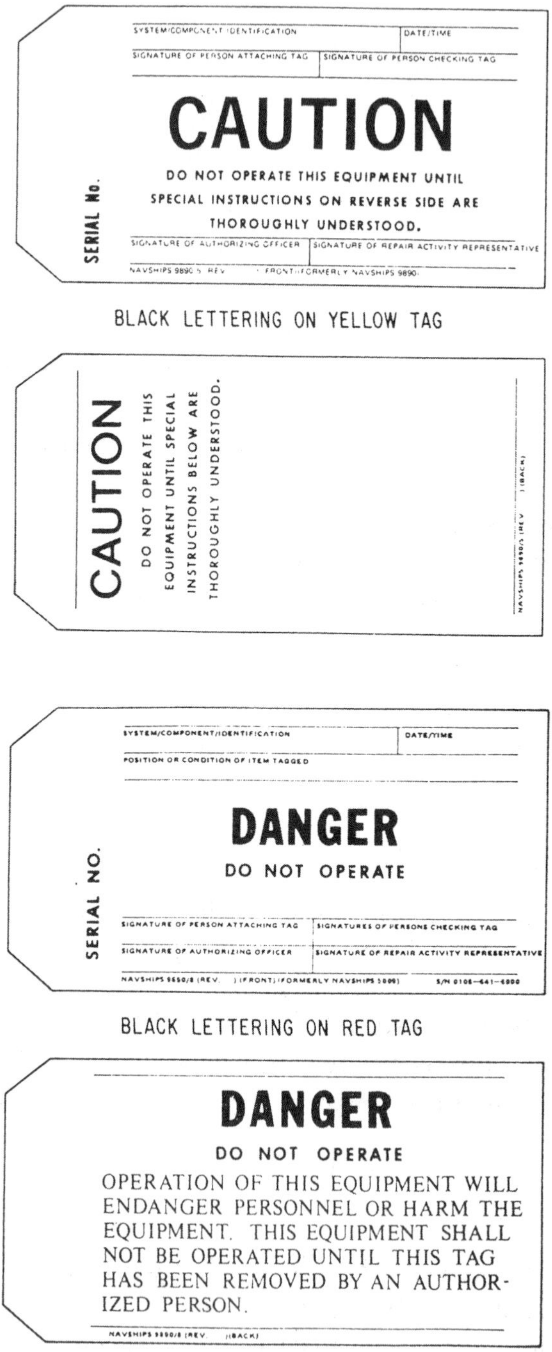

SYSTEM/COMPONENT IDENTIFICATION — **DATE/TIME**

SIGNATURE OF PERSON ATTACHING TAG — SIGNATURE OF PERSON CHECKING TAG

SERIAL No.

CAUTION

**DO NOT OPERATE THIS EQUIPMENT UNTIL
SPECIAL INSTRUCTIONS ON REVERSE SIDE ARE
THOROUGHLY UNDERSTOOD.**

SIGNATURE OF AUTHORIZING OFFICER — SIGNATURE OF REPAIR ACTIVITY REPRESENTATIVE

NAVSHIPS 9890 5 (REV) (FRONT) (FORMERLY NAVSHIPS 9890)

BLACK LETTERING ON YELLOW TAG

CAUTION
DO NOT OPERATE THIS EQUIPMENT UNTIL SPECIAL INSTRUCTIONS BELOW ARE THOROUGHLY UNDERSTOOD.

NAVSHIPS 9890 (REV) (BACK)

SYSTEM/COMPONENT/IDENTIFICATION — **DATE/TIME**

POSITION OR CONDITION OF ITEM TAGGED

SERIAL NO.

DANGER

DO NOT OPERATE

SIGNATURE OF PERSON ATTACHING TAG — SIGNATURES OF PERSONS CHECKING TAG

SIGNATURE OF AUTHORIZING OFFICER — SIGNATURE OF REPAIR ACTIVITY REPRESENTATIVE

NAVSHIPS 9890/8 (REV) (FRONT) (FORMERLY NAVSHIPS 3009) — S/N 0108—641—8000

BLACK LETTERING ON RED TAG

DANGER
DO NOT OPERATE
OPERATION OF THIS EQUIPMENT WILL
ENDANGER PERSONNEL OR HARM THE
EQUIPMENT. THIS EQUIPMENT SHALL
NOT BE OPERATED UNTIL THIS TAG
HAS BEEN REMOVED BY AN AUTHOR-
IZED PERSON.

NAVSHIPS 9890/8 (REV) (BACK)

Figure 30-3. Danger and caution tags.

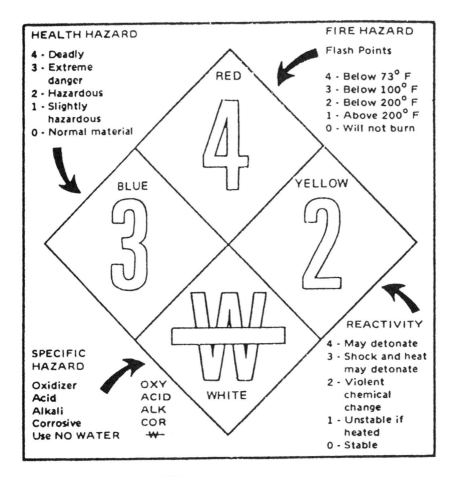

HEALTH HAZARD

4 - Deadly
3 - Extreme
 danger
2 - Hazardous
1 - Slightly
 hazardous
0 - Normal material

FIRE HAZARD

Flash Points

4 - Below 73° F
3 - Below 100° F
2 - Below 200° F
1 - Above 200° F
0 - Will not burn

RED
BLUE
YELLOW
WHITE

SPECIFIC
HAZARD

Oxidizer   OXY
Acid   ACID
Alkali   ALK
Corrosive   COR
Use NO WATER   -W-

REACTIVITY

4 - May detonate
3 - Shock and heat
 may detonate
2 - Violent
 chemical
 change
1 - Unstable if
 heated
0 - Stable

Figure 30-4. Hazardous material label.

All workplaces are required to maintain an inventory of all hazardous chemicals and a material safety data sheet (MSDS) for each chemical on board. The MSDSs *must* be available to you before using, handling, storing, or disposing of chemicals. An MSDS gives specific information on the type of hazard presented by the chemical, handling instructions, required and recommended safety precautions and protective equipment, and information on how to contact the manufacturer.

Motor Vehicles

You will probably operate some form of motor vehicle for the Coast Guard. Most people are able to drive an automobile. There are other forms of vehicles, however, including buses, trucks, forklifts, and tractors. Before operating these other types, you must be trained in their operation, the necessary safety precautions, and emergency procedures. Do not operate equipment for which you have had no training.

Regardless of the type of transportation you are using, Coast Guard regulations require you to wear seat belts, if available. Seat belts save lives. You have heard it all before, but the evidence is very clear.

In addition to seat belts, PPE is sometimes required when operating certain types of vehicles. For example, wear a hard hat when operating a forklift and hearing protection when driving a tractor.

Recreation and Sports

Many people enjoy participating in some kind of sports and recreation. Unfortunately, off-duty injuries do occur when engaging in these activities. To reduce the risk of injury, follow these rules:

- Get a physical examination. Talk with a physician about the activity in which you are going to engage.

- Condition yourself before engaging in it. The physical ensures that you are healthy. Conditioning ensures that you are fit. Health and fitness are not the same.

- Before exercising or playing a game, be sure you warm up and stretch your muscles out.

- Use protective equipment appropriate to the activity.

- After strenuous exercise like aerobics, take the time to cool down. This will help reduce muscle soreness.

Aircraft Operations

No matter what career field you choose in the Coast Guard, you may at some time fly on Coast Guard aircraft. Many personnel not normally involved in aviation take flights occasionally, when they need to look at something from the air or when flying is the fastest way to get needed people somewhere. Oil spills requiring monitoring teams to scan large areas quickly or situations where law enforcement teams must be put in place rapidly are two such cases.

Non-aviation personnel need to be aware of the hazards around operating aircraft. Helicopter rotor blades and tail rotors and transport aircraft propellers are deadly when turning. Jet engines can create enough suction at high power to pull people into them if they get too close, and the exhaust is extremely hot and dangerous. Antennas on the outside of the aircraft carry hazardous electrical charges and must not be touched.

If you have the opportunity to fly on a Coast Guard aircraft, one of the crew members should brief you on all the hazards and the procedures for avoiding them. If you have any questions, ask a crew member.

There are some general safety precautions you should always follow when near aircraft:

- *Never* approach an aircraft unless the crew knows you are there and has signaled you to approach.
- *Always* approach helicopters from the side, never from the front or back.
- *Always* wear hearing protection around operating aircraft.
- *Never* touch anything painted red or yellow without the crew's permission. These colors indicate hazards or emergency equipment.
- Aircraft emergency exits are indicated by yellow markings on the inside of the aircraft. Make sure you know where they are and how to operate them. Ask the crew during your briefing.
- In case of emergency, obey the crew members' orders.
- *Do not* touch helicopters or hoisted loads before discharging the static charge.

Figure 31-1.
A Coast Guard team member stands over the MV *Cape Orlando* (T-AKR-2044) at the Port of Anchorage in 7° weather. The team was tasked with providing security for the vessel while in port.

CHAPTER 31

Other Vessels and Mariners

To fulfill the many missions of the Coast Guard, our personnel have always had to deal directly with other mariners and the marine industry. The Coast Guard has always done well in dealing with these people. Although this relationship has not always been one of affection and admiration, it has usually been marked with a mutual respect. A general understanding of the types of vessels and the personnel who operate them will help acquaint new Coast Guard personnel with the marine industry.

Vessels are characterized as fitting into one of three general categories: commercial vessels, private vessels, and public vessels.

COMMERCIAL VESSELS

Before beginning a discussion of different vessel types, it is important to understand that the basic underlying purpose of commercial vessels is to make money. Unlike other vessels with which you may be familiar (pleasure craft, Coast Guard cutters, etc.), commercial vessels are conceived, designed, and built with this one ultimate goal in mind.

Cargo Vessels

Cargo vessels are designed to carry a variety of dry cargoes in various forms of packaging. They are generally classified as either bulk, break-bulk, or container vessels.

Bulk Carriers

Certain types of dry cargo do not lend themselves to packaging. Materials such as metal ores, grain, and minerals are more efficiently carried in bulk. Bulk carriers are commonly employed on the Great Lakes to carry iron ore from the mines to ports in the area of the steel mills. Other bulk carriers carry grain from the United States to foreign countries. The cargo holds in such vessels are designed to facilitate off-loading through the use of slightly sloping sides so that cargo settles to the middle of the hold. The cargo is usually off-loaded using cranes with large scoops aided by bull-dozers lowered into the holds. Bulk carriers carrying either dry-bulk cargo or liquid-bulk cargo usually transport cargo on only one leg of a voyage.

OBOs (Ore/Bulk/Oil)

OBO carriers can carry either ore, dry-bulk cargo such as grain, or oil. With this cargo-carrying capability, the vessel becomes more "fully employed."

Break-Bulk Vessels

This type of cargo vessel corresponds with the general concept of a tra-ditional "freighter." These vessels are sometimes called "general cargo" vessels. Cargo for this type of vessel is usually in bales, bags, or small crates that are sometimes palletized to speed loading and unloading operations in port. Additionally, these vessels can carry cars, trucks, or heavy machin-ery. Because of their ability to carry a wide range of cargoes and their self-contained cargo-handling capability, these vessels usually service ports with limited shore-side pier facilities.

Roll On/Roll Off (RO/RO) Vessels

RO/ROs are designed to carry cargo on multiple decks, usually in trail-ers, or as individual vehicles that are themselves cargo. Access from one deck to another is by way of ramps or elevators.

The "cargo" is driven on board through a stern ramp or side ramp through watertight closures in the side of the vessel. The advantage of this type of vessel is that cargo can be quickly loaded or unloaded at a port without relying on elaborate shore equipment. More and more RO/RO vessels are being used by the military for such applications as support vessels for the Afloat Prepositioning and Ready Reserve Forces, carry-ing jeeps, tanks, armored personnel carriers, and other rolling military hardware.

Refrigerator Ships

Refrigerator ships are outfitted with refrigerating equipment to preserve a cargo. The number of "reefer" spaces varies on a given vessel depending upon its trade. Usually reefer ships will carry other general cargo as well.

Offshore Supply Vessels (OSV)

OSVs are special-purpose freight vessels designed to carry drilling supplies from the shore to drilling units operating offshore. They usually have a large deck area for carrying drill pipe and miscellaneous supplies and have internal tanks for carrying dry or liquid drilling mud. Some OSVs are fitted with large winches on the after-deck for the purpose of anchor handling.

Container Vessels

Containerization of cargo was started in the late 1950s and early 1960s to assist in the intermodal transportation of cargoes. Cargo is placed in a container at the point of manufacture; the container is placed on a truck, transferred to a rail car, transferred to another truck, and finally placed on board a container vessel. Thus, the cargo itself is touched only two times—once during initial container loading and once at the destination where the container is unloaded. The containers themselves are constructed of fiberglass, aluminum, or steel and are in standard lengths of either twenty or forty feet.

Certain types of specialized container vessels, known as LASH or Seabee vessels, are specifically designed to carry, load, and unload barges filled with cargo instead of containers. Once unloaded, these barges are towed to discharge ports.

Lighter Aboard Ship (LASH). The typical LASH vessel has a crane that lifts the barges out of a cargo hold, carries them to the stern of the vessel, and lowers them to the water for launching and towing by a tugboat. The reverse process takes place during the loading of the vessel. The barges are relatively small and, when loaded to capacity, draw only eight feet of water. Thus, LASH vessels can transport cargo to and from ports that have no deep water or crane facilities. In addition to the savings accrued through the containerization concept, additional savings are realized from reduced or eliminated port fees, pilot fees, and tug charges.

Seabee. The Seabee is very similar to the LASH type except that, instead of using a crane on the ship, a stern elevator is used. The barges are moved along tracks to the stern of the vessel and placed on the elevator. The elevator is lowered to the water to launch the barge. The procedure is reversed during loading.

Bulk Liquid Vessels

Many types of liquids are transported by ships. Although some liquids are packaged in drums or other small containers and carried on various types of cargo vessels, many liquids are more efficiently transported in bulk form. Vessels designed for this purpose are called tankers. Virtually the entire vessel has tank spaces for cargo, fuel oil, or ballast.

Tankers. Tankers are essentially floating cargo tanks. Over the years tank ships have increased in size in the interest of fuel economy and the growing consumer need for oil and chemical products. Tankers are usually subdivided into the following categories:

- Ultra large crude carriers (ULCCs): 500,000 or more barrel cargo capacity; over 1,000 feet in length; carry crude oil
- Very large crude carriers (VLCCs): 300,000- to 500,000-barrel cargo capacity; 900 to 1,000 feet in length; carry crude oil
- Product carriers: various capacities and lengths; carry refined cargoes such as gasoline, aviation gas, and diesel
- Chemical carriers (commonly called "drugstore" tankers): specially designed tankers built with coated tanks and piping systems; carry a variety of corrosive or hazardous chemical cargoes

Liquefied gas carriers for LNG/LPG. For years the oil fields of the world produced natural gas along with various types of crude oil. If a ready market for the gas existed in the vicinity of the wells, pipelines were constructed and the gas was sold. The energy crisis during the early 1970s made liquefied natural gas/liquefied petroleum gas (LNG/LPG) transport by ship economically feasible, so a number of ships were built. By cooling these gases to −265 degrees Fahrenheit, or by pressurization, they become liquefied and occupy only $\frac{1}{600}$ of their original volume. Many ship designs are in use today, but they all rely on very special materials and construction techniques to withstand unusual vessel-loading conditions and the extremely low temperatures. Most of these vessels use a portion of the liquefied gas that "boils off" during voyages for fuel in the vessel's engines.

Tank Barges

Tank barges are distinguished from tankers in that they are not self-propelled. Barges comprise a very large group of vessels that carry many types of cargo and serve a variety of functions, including lightering of cargo and the fueling (bunkering) of larger vessels. In fact, there are just about as many types of barges as there are cargoes. Most barges are unstaffed, but some carry a small crew. The following is a list of the more common barge types:

- Tank barges
- Hopper barges
- LNG/LPG barges
- Chemical carriers

Passenger Vessels

Passenger vessels, as the name implies, are those vessels that are designed solely to carry passengers. They are diverse in their designs, and it is virtually impossible to address all the possible configurations. If they could be categorized, the categories might look like this:

- Ferries: Vessels of various sizes designed for carrying both passengers and vehicles.
- Head boats: Small passenger vessels used for taking people on fishing trips, for which they pay "by the head."
- Passenger liners: Large ships designed to carry passengers for pleasure purposes.
- Excursion boats: Small passenger vessels designed to take people on sightseeing tours or dinner/dance cruises.
- Crew boats: Small passenger vessels designed to transport workers involved in the offshore oil industry.

Towing Vessels

There are a number of types of commercial towing vessels common to most U.S. waters. They are similar in that they are designed to move or maneuver other floating objects so their design is oriented toward power and maneuverability. Typical towing vessels include harbor tugs, large seagoing tugs, and tow boats or "pushers" that push barges on the Western Rivers and Intracoastal Waterway.

Fishing Vessels

Commercial fishing vessels make up the most numerous single category of commercial vessels. Fishing vessels likely to be encountered by Coast Guard personnel include lobster boats, gillnetters, shrimpers, oyster boats, crabbers, stern trawlers, and clammers. There is no such thing as a typical fishing vessel. Fishing vessels are at the top of the list for the number of lives lost on all commercial vessels.

Figure 31-2. USCGC *Liberty* (WPB-1334) and a twenty-five-foot response safety boat from Station Juneau escort a cruise ship.

Mobile Offshore Drilling Units (MODUs)

MODUs are very special types of vessels. They are really movable platforms from which oil wells can be drilled. Actually, they are vessels only when they are moving from one site to another; at other times they are considered offshore facilities. There are three basic types of MODUs:

- *Semisubmersibles* are used in deep-water drilling. Upon arrival on site they are held in place using various multiple anchor systems.

- *Jack-up rigs* are self-elevating units used in shallow-water drilling. They are towed into place, at which time the legs are lowered to the ocean floor and the rig is "jacked-up" above the water level.

- *Drill ships*, built in the shape of regular ships but with the addition of a derrick, move to sites under their own power and are anchored in place; they are typically used for exploratory drilling.

MERCHANT VESSEL ORGANIZATION

Although the primary function of most merchant vessels is to make money by transporting cargo, a vessel is more than just a complex of structural members, fittings, and machinery. A vessel is also the people that crew it. History has proven that a ship is no better than its crew. The crew directs the vessel from point to point and, by its skill and experience, enables a vessel to survive in the often-fierce environment.

For operational and administrative purposes, all merchant vessels are divided into several departments. The two most important of these are the *deck department* and the *engine department*.

The single most important deck department responsibility is that of supervising the handling and storage of cargo. The administrative head of the deck department is the vessel's master. Since the master is also in command of the entire vessel, many of the deck functions are taken over by the chief mate.

Unlike the Navy and Coast Guard, where officers are usually given only one responsibility at a time in a ship, the merchant vessel deck officer must constantly assume responsibility for such diverse operations as navigation and piloting, the operation of navigation equipment, ship's maintenance, the use of lifesaving and firefighting equipment, vessel handling, and calculations for cargo loading that may affect the vessel's trim and stability.

Within the engine department, the chief engineer is the highest-ranking officer. Additional personnel required within the engineering department include "qualified members of the engine department" and wipers, who perform watch-standing duties and maintenance under the supervision of licensed officers.

The third major department on merchant vessels is the *steward's department*. The chief steward and the chief steward's cooks are responsible for the feeding of the passengers, if applicable, and the officers and crew.

Passenger vessels have a fourth department called the *staff department*. This department is made up of officers who hold a U.S. Coast Guard certificate of registry as chief purser, purser, pharmacist's mate, surgeon, professional nurse, or medical doctor. These officers have nothing to do with the navigation of the vessel but provide services and medical care for passengers. As mentioned, all of the above staff officers are typically found only on board passenger vessels, with the exception of the purser, who may also serve in certain cargo and container vessels.

The Coast Guard is responsible for specifying the minimum staffing of United States merchant vessels on the basis of the numerous statutes and the regulations that clarify the application of these statutes. In developing these minimum staffing standards for a particular vessel, the Coast Guard is concerned primarily with the safety of life at sea. In other words, the Coast Guard ensures that the vessel has a sufficient number of qualified personnel to be capable of coping with the normal hazards of the sea. The following licensed and unlicensed mariners are likely to be part of a commercial vessel's crew.

Merchant Vessel Personnel

Master

The master has the overall responsibility for the vessel. The master is not considered a member of the crew but rather the onboard owner's representative. The master normally does not stand a watch but must maintain a valid Coast Guard license that certifies competence to navigate and administer the operation of the vessel.

Chief Mate

The chief mate (the "mate") is responsible for cargo loading, regardless of the type of vessel, for maintaining the vessel's stability, and for the general operations of the ship other than in the engineering department. The chief mate is also the deck department administrator for all paperwork and documents.

Second Mate

The second mate works for the chief mate and is normally designated the vessel's navigation officer. In port the second mate may be responsible for cargo transfer operations.

Third Mate

The third mate assists the chief mate and is responsible for the maintenance of the lifesaving equipment.

Chief Engineer

The chief engineer (the "chief") has the overall responsibility for the engineering department as well as for the maintenance and repair of the vessel's deck machinery and the hydraulic and electrical systems. By holding the highest level of license within the engineering department, the chief engineer is considered to be on an equal status with the master. However, the chief engineer is a member of the crew and answers to the master but does not stand watches.

First Assistant Engineer

The first assistant engineer is responsible for the overall supervision of the engine room and for the major maintenance and repair of machinery and various system components.

Second Assistant Engineer

The second assistant engineer is responsible for the main propulsion machinery, which includes the boilers (if applicable to the plant).

Third Assistant Engineer

The third assistant engineer is responsible for the auxiliary machinery and the general upkeep of the engineering department.

Radio Officer

The radio officer operates and maintains the vessel's radio equipment. Because of the equipment installed on board modern vessels, the radio officer does not normally stand a watch.

Other Crew Members

Able-Bodied Seaman (AB)

An AB acts as the helmsman and also can serve as the vessel's lookout. The AB is qualified to operate the vessel's lifeboats. At least 65 percent of the deck crew on board a major vessel, exclusive of the licensed officers, are required to be rated "able-bodied seamen."

Boatswain

The boatswain is the AB who heads the deck gang. The boatswain is responsible for maintaining the cargo gear and maintaining the general upkeep of the vessel within the deck department. The boatswain reports directly to the chief mate.

Ordinary Seaman

The ordinary seaman is an entry rating in the deck department. For the most part, ordinary seamen carry out the vessel's cleaning and upkeep functions and attend to minor repairs. The ordinary seaman rating equates to a nonrated individual in the Coast Guard.

Qualified Member of the Engine Department

Individuals with the following special qualifications make up these ratings: refrigeration engineer, deck engineer, fireman/water tender, junior engineer, electrician, machinist, pump man, deck engine mechanic, engineman.

Oiler

The oiler is responsible for engine room upkeep and minor maintenance of machinery. The oiler is a watch stander.

Wiper

The wiper is an entry rating for the engine department. The wiper performs a similar function in the engineering spaces as the ordinary seaman does on deck.

Tankerman

A tankerman is any person holding a certificate issued by the Coast Guard attesting to competency in the handling of certain bulk liquid cargoes.

Lifeboatmen

Lifeboatmen must possess the knowledge and ability to launch and recover the vessel's lifeboats or life rafts, to handle the various types of lifeboats/life rafts in a seaway, and to operate and be familiar with all of the lifesaving equipment and provisions on board the lifeboats/life rafts. The number of certified lifeboatmen varies depending upon the lifeboat and the life-raft requirements for each vessel.

PRIVATE VESSELS

Private vessels are pleasure boats, yachts, research craft, nonfederal training vessels, exploration vessels, and certain supply and support vessels. Because a vessel is private does not exempt it from federal safety and environmental protection requirements. Service in "passengers for hire" can bring a private vessel into an "inspected vessel" category. Passengers for hire are people on board who contribute some consideration toward their passage, which can include buying a ticket or reimbursing expenses.

For the most part, when we speak of private vessels we are thinking of the millions of state-numbered pleasure craft in the United States. Pleasure craft and other private vessels operating on U.S. navigable waters are subject to Coast Guard and state jurisdiction. The fact that a vessel is state numbered does not exempt that vessel from federal requirements. The applicability of federal law is a matter of the craft's service and economic circumstances.

PUBLIC VESSELS

The term "public vessel" generally means any vessel owned and operated by a department or agency of the federal government or of a state or a subdivision thereof, and which is engaged exclusively in official, noncommercial operations. This term also applies to foreign vessels owned and operated by a foreign government recognized by the United States. Public vessels include vessels of the Coast Guard, Navy, Army, Air Force, National Oceanic and Atmospheric Administration, Army Corps of Engineers dredges, and support vessels crewed by civil-service personnel.

Figure 32-1.
Coast Guard Base Boston is located on Boston's historic waterfront.

CHAPTER 32

Duty Ashore

C oast Guard shore units support our fleets of cutters and aircraft or operate in specific program areas of their own. An assignment ashore may present opportunities to serve in small-boat search and rescue, respond to an environmental hazard, or keep the flow of supplies and personnel "moving to the mission." The *Staffing Standards Manual* (COMDTINST M5312.11) provides a detailed description of each type of Coast Guard field unit, including manning and specific watchstanding responsibilities.

UNIT ORIENTATION

After reporting to a unit ashore, you may expect your initial period to be taken up with the *check-in* process. Often this procedure simply consists of touring the unit with your sponsor, completing the paperwork, interviews, required reading, and area familiarization necessary to settle you in at the unit. As discussed later in this chapter, you will begin the required qualification process necessary for the responsibilities you will be assigned in the unit watch organization.

SHORE UNIT ORGANIZATION

You learned about basic shipboard organization in this manual. Shore units, too, are organized for order and efficiency to the greatest degree possible and are headed by a commanding officer or *officer-in-charge*; as with ships, an executive officer or *executive petty officer* serves as the second-in-command and administrator. Similarly, the shoreside executive officer is assisted by unit personnel assigned to collateral duties, boards, and

Figure 32-2. Coast Guard Air Station Port Angeles, Washington.

committees. But whereas departments prevail as the major subdivision for most units ashore, there is little further resemblance to a ship's organizational structure.

Field units ashore, such as sectors and small-boat stations, may differ significantly from one another in their organizational arrangement because of the variance in personnel allowances, tenant/landlord command relationships, and assigned functions. Conversely, larger, more complex activities, such as district staffs, are identically structured in accordance with the *Coast Guard Organization Manual* (COMDTINST M5400.7 series).

The key to understanding the shore unit to which you are assigned is your familiarity with the specific unit's organizational rules and regulations. These documents, sometimes published as *Standard Operating Procedures*, a unit's *Organization and Regulations Manual*, or a series of individual directives, parallel the system of shipboard bills previously discussed in-asmuch as they govern by force of order the operation and administration of the unit and its personnel.

STANDARD ROUTINE

The routine of the day for Coast Guard personnel assigned ashore may, as you might expect, differ considerably for each type of activity. Personnel assigned to units within metropolitan office spaces, such as recruiters and staff, often work under conditions and scheduling that closely resemble those of their civilian counterparts. This is certainly not the case for the crew of a small-boat or air station during the peak of search-and-rescue season, or for the Coast Guard's thousands of watchstanding personnel serving at a variety of units.

Shore units having berthing and messing facilities will publish and enforce as regulations specific policies for these accommodations, including such specifics as meal hours, reveille and taps, smoking regulations, and material inspection schedules and standards. Additional timely information about the unit's routine may be found in the plan of the week.

PLAN OF THE WEEK

The XO or senior administrative officer will prepare and issue a plan of the week, which serves as a unit calendar of official events to all hands for the following seven days. The plan of the week includes:

- The schedule of the normal routine, and any variations or additions;
- The orders of the day, detailing the training schedule, duty section, liberty sections, personnel and material inspections, and so forth;
- Notices of matters that should be brought to the attention of all concerned; and
- Reprints of regulations and orders that are to be brought to the attention of the station's company.

Copies of the plan of the week are widely distributed throughout the unit and publicly displayed. Everyone is responsible for knowing what is in the plan of the week because any instruction or direction it provides constitutes an official order.

WATCH ORGANIZATION AND QUALIFICATIONS

As on board ship, nearly all Coast Guard field units ashore must employ a watch organization to meet round-the-clock mission responsibilities. Such organizations will vary in personnel numbers and complexity in relation to the size and functions of the unit.

Ashore, periods of military responsibility that may encompass one or more watches, in addition to boat or flight crew, maintenance, emergency response, or other tasks, are known as *duty*; personnel in the section of the command's watch organization remaining on board the unit to perform this rotating responsibility are said to "have duty." The frequency, or *rotation*, of your particular duty will also vary according to its nature and may be affected by the seniority of your rate or rank.

Periods of duty are normally of twenty-four hours in duration, although periods of up to forty-eight hours are sometimes authorized at small units, such as small-boat stations, to increase the consecutive length of the follow-on liberty period.

Soon after reporting on board, you will likely begin the training process, sometimes called a *break-in*, which will lead to your qualification for the type of duty you will *stand*.

DUTY DAYS AND WATCHSTANDING

Depending upon the unit mission, organization, and your rate and rating, or rank, watchstanding might be either a duty-day responsibility or your primary responsibility at the unit. The latter case is true, for example, for operations specialists assigned to command centers and vessel traffic services.

If you are nonrated, your duty day and accompanying watchstanding responsibilities will vary with the type of unit to which you are assigned. At a small-boat station, you may take a turn at the radio watch and work at boat maintenance or grounds keeping when off-watch. At a training center, you might serve a period of telephone watch or perhaps in a standby capacity as the duty vehicle driver. A roving security watch might be a responsibility of your duty day at a base in addition to housekeeping responsibilities.

Figure 33-1.
Santa Claus steps off a Jawhawk helicopter in a remote Alaskan village.

CHAPTER 33

Coast Guard Community Relations

Just by the very nature of what we do, the Coast Guard and our local communities need each other. In the Coast Guard, we believe we should reach beyond simply carrying out our missions. We strive to cultivate productive relationships within our communities. We want to get to know you, our neighbors, and encourage you to get to know us as well. Beyond our local involvement, many national Coast Guard programs give you the opportunity to learn more about who we are and offer us the opportunity to listen to your concerns. Our community relations efforts help introduce you to our role in America's homeland defense and explain our relevance in your daily lives.

COAST GUARD ART PROGRAM

The Coast Guard Art Program (COGAP) uses fine art as an outreach tool for educating diverse audiences about the U.S. Coast Guard. Through displays at museums, galleries, libraries, and patriotic events, Coast Guard art tells the story of the service's missions, heroes, and history to the American public. Art is also displayed in offices of members of Congress, senior officials of the executive branch of government, and other military services and Coast Guard locations throughout the country. Coast Guard artists are volunteer professional artists who donate their time and talents to help COGAP fulfill its missions.

Today the collection comprises nearly 2,000 works that recount the Coast Guard's history from the early beginnings of our nation to the present day. Artists have also captured the daily missions the 40,000+ men and women on active duty in the Coast Guard perform, including homeland security, search and rescue, marine environmental protection, drug interdiction, military readiness, and natural resource management. The program is a partnership between the Coast Guard and the Salmagundi Club, which has been a New York City artistic and cultural center for over 130 years.

COAST GUARD BAND

Since 1945 the U.S. Coast Guard Band has been one of the five premier service bands in the United States. The Coast Guard Band routinely tours throughout the United States and has performed in the former Soviet Union and in England. Additionally, the band regularly furnishes programs to National Public Radio for broadcast throughout the country. Concerts have been broadcast in Australia, Japan, and Europe.

Each year the band conducts tours with free public concert appearances. Daytime school visits are often included in these tours. If you would like your community to be considered for a future tour, you will need to

Figure 33-2. Members of the U.S. Coast Guard Pipe Band march up Fifth Avenue in the 250th St. Patrick's Day Parade, Manhattan, New York City, 17 March 2010.

submit your suggestion one year in advance. The band also fulfills a limited number of public requests for patriotic programs at national conventions and meetings of nationally recognized civic, patriotic, and veterans organizations. Sponsors for these events must fund all expenses such as transportation, meals, lodging, and promotion efforts. For information on the Coast Guard Band's tour schedule or to submit a request for a special program appearance, consult http://www.uscg.mil/band/.

COAST GUARD CITIES

The very nature of the Coast Guard's mission creates a need for understanding between the Coast Guard and the local community. Coast Guard commands everywhere are urged to develop the kinds of relationships that enable unit commanders to sense public attitudes and interests. In turn, many cities have made special efforts to acknowledge the professional work of the Coast Guard men and women assigned to their area. Making Coast Guard men and women and their families feel at home in their home away from home is an invaluable contribution to morale and service excellence.

The Coast Guard is pleased to recognize Coast Guard Cities—those cities that have extended so many considerations to the Coast Guard family and their dependents. To date, twenty-one cities have been designated by Congress as Coast Guard Cities, including Grand Haven, Michigan; Eureka, California; Mobile, Alabama; Wilmington, North Carolina; Newport, Oregon; Alameda, California; Kodiak, Alaska; Rockland, Maine; Portsmouth, Virginia; Traverse City, Michigan; Astoria, Oregon; Sitka, Alaska; Clearwater, Florida; Newburyport, Massachusetts; Sturgeon Bay, Wisconsin; Camden County, Georgia; Cape May, New Jersey; Elizabeth City, North Carolina; New London, Connecticut; Carteret County, North Carolina; and San Diego, California.

AIRCRAFT EXHIBITS AND FLYOVERS

Coast Guard aircraft participate in appropriate public events that help contribute to public knowledge of the Coast Guard. Participation may be a flyover, demonstration, or static display. Appropriate events include airport dedications, air shows, expositions, and fairs. Static displays must be held at airfields or heliports.

Civilian sponsors must agree to provide or reimburse transportation, meals, and lodging costs of Coast Guard participants. Sponsors also must provide suitable aircraft fuel at military contract prices. Sponsors are required to pay all costs over military contract prices, including any transportation and handling charges, if fuel is not available at such prices. Aerial demonstrations must be within Federal Aviation Administration guidelines and must be over open water or suitable open areas of land, where spectators will be safe.

The Coast Guard may support flyovers for civic-sponsored, public ceremonies such as Armed Forces Day, Memorial Day, Independence Day, Veterans Day, and for similar local holidays overseas; memorial services for dignitaries of the armed forces or federal government; national conventions of veterans' organizations; and occasions of more than local interest designed primarily to encourage the advancement of aviation.

CEREMONIAL HONOR GUARD

The Coast Guard Ceremonial Honor Guard represents the commandant, the Military District of Washington, and the U.S. Coast Guard through ceremonial operations held before world leaders and dignitaries. Ceremonies can include parades, funerals, White House dignitary arrivals as well as presenting colors at local and official functions. Honor Guard members participate in joint service activities as well as Coast Guard functions. The Honor Guard performs in excess of 1,600 ceremonies annually.

The Honor Guard is composed of 73 members, with a lieutenant (O-3) serving as the Honor Guard Company Commander, two junior officers (usually O-2) serving as operations/weapons officer and supply/training officer, a chief petty officer (E-7) as the Honor Guard chief, and four petty officers (ranging from E-4 to E-6). The remaining members of the Honor Guard are "first-tour" nonrated personnel (E-3) coming directly from Training Center Cape May. The officers and nonrates serve a two-year tour of duty in the Honor Guard while the chief petty officer and petty officers serve four year tours.

VII

PERSONAL FITNESS AND SURVIVAL SKILLS

Figure 34-1.
Regular exercise will keep you healthy and fit for duty.

CHAPTER 34

Wellness

Wellness is an important part of Coast Guard life. It is more than simply the absence of illness. Wellness refers to your lifestyle habits, how you live your life. More than half of the diseases affecting humans are caused by poor habits: tobacco, diet, lack of exercise, unmanaged stress, and so forth. To reach and maintain an optimal state of good health, you must actively pursue a healthy lifestyle. If you do, you are guaranteed to look better, feel better, play better, think better, work better, and live longer!

What does it mean to lead a healthy lifestyle? Most important is good nutrition or eating right. Then there is achieving and maintaining a high level of physical fitness. Avoiding tobacco products is an essential part of a healthy lifestyle, as is moderation in drinking alcoholic beverages or not drinking at all. Next is managing stressful situations and successfully keeping your cool. Lowering your risks for heart disease, cancer, and other long-term problems is also part of a healthy lifestyle. This means keeping your cholesterol and blood pressure under control and periodically visiting your medical and dental department for screening examinations. And finally, a healthy lifestyle means paying attention to simple but critical safety precautions—always wearing your seat belt in cars and always wearing a helmet when riding a bicycle or motorcycle.

In thinking about wellness, it is sometimes helpful to consider your body as a machine. It is an incredibly complicated and highly efficient machine, and only you are responsible for caring for it. Just as a finely tuned, expensive sports car requires the correct fuel, oil, tires, and maintenance procedures to run properly, your body needs the correct food, exercise, rest, and care to stay healthy. Just as you would avoid putting sand in your gas tank, you should avoid poisoning your body with tobacco, excess alcohol, drugs, or other toxic substances. And just as you would carefully follow the maintenance procedures in your car's owner's manual, you should have periodic physical and dental exams to ensure your overall health.

NUTRITION

Nutrition is the foundation of wellness. Since you literally are what you eat, a healthy lifestyle starts with your diet. There are seven steps to eating right: (1) eat a variety of foods; (2) maintain a healthy weight; (3) limit your intake of total fat, animal fats, and cholesterol; (4) eat plenty of vegetables, fruits, and grain products; (5) use sugars in moderation; (6) use salt and sodium only in moderation; and (7) if you drink alcoholic beverages, do so in moderation. (It is okay not to drink!)

PHYSICAL FITNESS

There are five essential parts to being physically fit: aerobic capacity, muscle strength, muscle endurance, flexibility, and maintaining a healthy level of body fat. Each of these is important, and you cannot neglect any of them if you want to be fit. Physical fitness helps make you a more productive and capable member of the Coast Guard and a healthier and happier person in your private life.

Aerobic capacity is a measure of your body's endurance or stamina. Good endurance enables you to maintain activities (either work or play) for extended periods of time. Aerobic capacity depends on the efficiency of your heart, lungs, and muscles to get and use oxygen. To increase your aerobic capacity, you must exercise with your large muscles (your legs and arms) for at least thirty minutes at a time, at least three times per week.

Aerobic exercises include running, jogging, cycling, stair-climbing, rowing, swimming, jumping rope, cross-country skiing, and fast walking. Aerobic exercise is also an effective way to burn calories, which is important if you want to lower your level of body fat.

Muscle strength is a measure of the force you can produce, allowing you to lift, push, pull, and carry. These are all important in your Coast Guard duties, but they are also important in your recreation and home activities. The only way to build muscle strength is to exercise against resistance, either by lifting weights or by lifting your body-weight against gravity

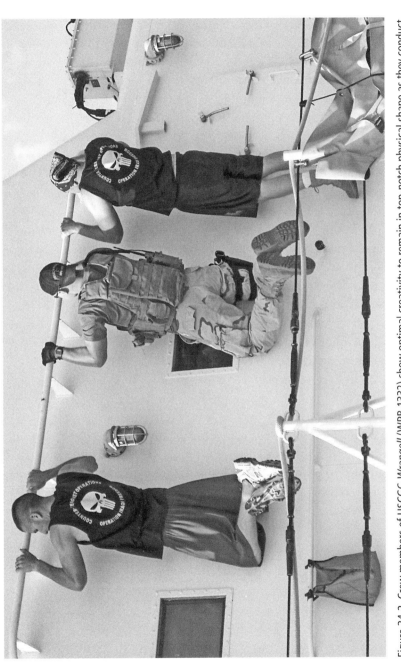

Figure 34-2. Crew members of USCGC *Wrangell* (WPB-1332) show optimal creativity to remain in top-notch physical shape as they conduct pull-ups from a bar on the superstructure of the cutter.

(push-ups, pull-ups, calisthenics, etc.). Strength-training exercises not only help you build muscle but also help you lose fat. The more muscle mass you have, the higher your rate of metabolism and the faster you burn calories, even when you are asleep!

Muscle endurance is your ability to perform hard work over a prolonged period of time. Muscle endurance is important in your Coast Guard duties for both emergency situations (e.g., firefighting, damage control, cardiopulmonary resuscitation) and for operational requirements (e.g., boat crew or cutter duties in rough seas). The best muscle endurance exercises are pull-ups, curl-ups, push-ups, parallel bar dips, and calisthenics. You can also build muscle endurance by lifting weights. But instead of using heavy weights and a small number of repetitions (five to ten), as you would do to build muscle strength, you should use light weights and a high number of repetitions (fifteen to twenty) to build endurance.

Flexibility describes your ability to move your joints through a full range of motion. You need good flexibility to develop muscle strength; to prevent injuries to your muscles, tendons, ligaments, and joints; and to protect against back injuries. Proper stretching, both before and after exercise, is the most effective way to develop good flexibility.

Body composition refers to the relative amounts of fat and lean tissue in your body. The more fat you have, the harder it is on your joints and the harder your heart has to work. Trying to carry around an extra ten to twenty pounds of body fat is like walking with a ten- to twenty-pound pack on your back. Your goals should be to maximize your muscle mass and minimize your body fat. An ideal body composition for men is a body fat of 15 to 17 percent, while for women it is 18 to 22 percent. Try not to think of losing "weight" if you are too heavy; what you really want to lose is fat. The best way to lose fat and preserve muscle is to maintain a low-fat, healthy diet, perform aerobic exercise for at least thirty minutes three times per week, and perform strength-training and muscle endurance exercises.

TOBACCO

Tobacco is one of the most harmful and addictive substances available in our society. It is responsible for more deaths and more disease than any other single agent. It is as addictive as heroin, so once you start, it is hard to get free of its hold. Cigarette smoking not only leads to heart disease and cancer but it also causes you to have wrinkled skin, so you look pre-maturely old. Smokeless tobacco (dip, chew, etc.) can give you mouth and throat cancer while you are still young. Young men and women have died in their teens and twenties from using dip! Do not let this happen to you. Tobacco is definitely not a part of a healthy lifestyle.

ALCOHOL

It is okay not to drink. Alcohol has been a part of our society for a very long time, and people have had problems with it for just as long. If you drink, do so responsibly. "It is okay not to drink" means not only absti-nence but also knowing your limits. It means it is okay to stop after one or two drinks and not to feel pressure from your friends to drink more.

It also means that you, in turn, should not pressure your friends to drink more if they want to stop. The Coast Guard can provide you with profes-sional help if you develop a problem with alcohol, but the best medicine is preventive medicine. Drink responsibly or not at all, and remember, it really is okay not to drink.

Figure 35-1.
An aviation survival technician is lowered into the water.

CHAPTER 35

Swimming and Lifesaving

Anyone going to sea, whether on a high-endurance Coast Guard cutter or merely across a harbor in a response boat, should know how to swim. Many emergencies—shipwreck, flood relief, and other types of rescue work—require the use of expert swimmers. Just being able to swim is often not enough; you have to be a good swimmer. The following information will help, but the only way to improve is to practice.

CONTROL OF POSITION AND BUOYANCY

Jellyfish or Tuck Float

Double up, hugging legs, with face submerged. Float like a ball with only part of your back above the water. This is a good exercise for beginners; it acquaints them with body buoyancy and develops their ability to hold their breath. From this position, beginners learn to open their eyes under water and familiarize themselves with floating. From this position, too, you can slowly but firmly massage cramps in legs or toes. At times, it may serve as a rest for muscles tired from the stretch of swimming.

Back Float

Lie on your back with your head thrown back, chin and nose above water, arms and legs relaxed. Some people can float well this way; others keep sinking as they exhale.

Try to keep your lungs full of air. Do not become panicky if your face does sink. Gently work your hands so as to raise your chin enough for a quick breath, and then let yourself sink. Your face will soon rise above the surface. Gulp a breath while drifting, rest and relax while sinking. This is the preparatory position for learning the backstroke and an excellent way of resting and floating during long periods in the water.

Dead Man's Float and Glide

Lean forward, face down in the water, and relax completely. Play "dead man" all over. This float leads to the crawl, or speed stroke, and to the breast, or froglike, stroke. It teaches relaxation in the water, how to swim under water with eyes open, and how to get the most out of a glide.

Breathing

In general, breathing for swimming is the same as for any other sport. Try to keep it regular, do not hold your breath too long, and do not pant. Breathe through your mouth when your face is out of water; exhale by blowing out through your mouth and nose to keep them clear of water. Try to take air in fast; blow it out slowly.

Treading Water

Being able to tread water is a vital skill and should be practiced until you can do it for minutes at a time, not just a few seconds. It is required for every situation in which you need to raise your head to look around, or call for help, or use your hands.

There are several methods of treading. Practice all of them; by changing from one to the other, you will be able to stay afloat longer. The most commonly used methods are mentioned here.

Modified frog kick for treading. Stand upright in the water. Draw legs up together with the knees spread out to the sides. Separate and extend legs. Finally press the legs together until the feet are ten to twenty inches apart. The movement of drawing, spreading, and pressing them is then repeated. Pushing the legs out and pressing them together gives an upward thrust to the body.

This kick can be practiced while hanging on to the edge of a pool. Its advantage is that the slow movement is not as tiring as some of the other treading kicks and it is adjustable; you can tread low in the water or thrust up high by this same kick.

Scissors kick for treading. This is similar to walking and is easy to learn. It is best practiced at first on land. Lie on your side with feet together and legs extended, and then draw them slowly up in a semi-tucked-up position. Separate the legs, one moving forward, the other backward, then squeeze them together. The movement should be slow, steady, and continuous. Practice for a while, holding on to the edge of the pool or a firm object. The scissors kick is easy to learn and is a natural movement, effective but not tiring. It also leads to the sidestroke, which is important in distance swimming and lifesaving.

Sculling

Sculling with the hands enables people to control their body in the water in a number of positions, with or without the help of their feet. Sculling and treading combined will save more energy than using the feet or hands alone. Sculling has the advantage of teaching a sense of balance in the water. It develops the muscles of the hands and arms needed for pushing against the water, and it teaches people how to get a "grip" on the water.

Vertical sculling. One method of sculling while holding oneself upright in the water is to weave the hands vigorously back and forth in front of and to the sides of the body in a figure-eight movement. This forces the water downward and keeps the body up.

Another method is to drop the hands down near the hips in the water; then turn them to about a 45-degree angle away from the body, thumbs downward. With the palms of the hands, push the water away from the body until the hands are about two feet out from the sides of the body. Immediately rotate the hands so that the thumbs are up, and then return to the starting position, pulling water toward the sides of the body. When the hands reach the sides, they are again rotated until the thumbs are down, and the movements are repeated. The emphasis is on the push-away part of the stroke.

A third method is to place the hands palms down on the water, press down on the water several feet in depth, and recover again.

All these methods, or combinations of them, will be useful at different times.

Flutter-back sculling. This is one of the best sustaining strokes for staying afloat for a long time or for moving slowly and easily in some direction. Lie on your back and kick slowly up and down, with legs and feet in a knock-kneed and pigeon-toed position, and make a slow sculling movement at the hips with your hands.

SWIMMING STROKES

In general, there is no need for speed swimming. There may, however, be occasions where speed is needed, such as in swimming away from a sinking ship, catching up with a boat or raft, or swimming against a rapid current.

In the crawl, or freestyle stroke, it is necessary to lift your arms out of the water, and water-soaked clothing can make this very difficult and tiring. At the same time, a lifejacket and other gear can hamper the freedom required for the crawl strokes. The best strokes to use in emergencies are those with underwater arm recoveries, as discussed in the sections that follow.

Dog Paddle or Human Stroke

The dog paddle is a valuable sustaining stroke for people fully clothed or wearing lifejackets. It is particularly useful in crowded or narrow places because most of the movements are under the body, and the position of the head enables you to look about.

In the dog paddle the legs execute a slow up-and-down kick, the knees are bent, and the feet are extended in a pigeon-toed manner. The arms are moved alternately. Each arm is pushed ahead in the water until extended, then brought downward and backward, held straight the whole time, and a strong grip is taken on the water. At the end of the pull the arm is bent, drawn up, and shoved forward again. Avoid the tendency to hold your breath and to stroke too fast.

Sidestroke

The sidestroke is a well-rounded stroke for long-distance swimming and lifesaving rescues.

Leg Action. Lie on your side, whichever side is the most natural. Start with your feet together, bend your knees and move your heels toward your hips. When your legs are up as far as is comfortable, separate your feet, making your top leg move forward and the bottom one backward. After they are separated as far as possible, straighten your legs and snap them together from this spread position like closing a pair of scissors. Make sure to stop your feet as they come together. This is the "scissor kick." It is the same as used in treading water.

Arm action. Lying on your side with one ear in the water, reach the lower arm ahead of you along the surface of the water and the top arm back alongside the top leg. When beginning the stroke, cup the lower hand a little, as if holding a handful of water, and sweep the hand downward toward the front of the chest. Here it meets the other hand, which (at the same time) has been slowly brought up in front of the chest, with the hand knifing edgewise in the water to lessen resistance. At the meeting position, the hand that has just come up in front of the chest is cupped and pushed down along the body to the top of the upper leg. While this motion is being carried out, the lower arm is returning to its extended position, knifing edgewise through the water. This ending and starting position is then held during the glide portion of the stroke.

Combined stroke. Push or glide into a sideways position, the lower ear in the water, feet together, with the lower arm reaching ahead and the upper arm resting on the top leg. Start moving the arms and legs at the same time, bringing up the heel while pulling with the lower arm. When the hands meet in front of the chest, the feet should begin separating. In a continuous motion, keep moving the arms and snap the feet out and together using the scissor kick motion. Glide for a short distance and repeat the stroke.

Breaststroke

The breaststroke is invaluable for long-distance swimming, swimming through heavy surf, or swimming through debris-filled water. It is frequently used as an approach stroke for water rescues.

The kick. Lie face down with legs together and fully extended. Bring your heels up toward your hips and spread your knees about the width of your hips, with heels slightly wider than the knees. When the knees are brought up to just beneath the hips, turn your feet outward and move them to the sides until they are past the width of the hips. To finish the kick, push feet back down, making a circle until they return to their starting position.

Arm action. Start face down, and with arms extended, hands together, and palms down. Push palms out and back, pushing downward slightly until the hands are in line with shoulders. Bring the elbows up to the sides and the forearms and hands just under the chest and neck. Not stopping, extend hands forward under the water's surface back to the starting position. Hold this position for the glide.

Combined stroke. Start with arms extended forward and legs straight. The hands start pulling to begin the stroke. Just after the hands have pulled a few inches, the legs start their recovery and the head begins to lift for a breath. As the legs move through the propulsive part of the kick, the arms are moved forward to the extended glide position. Take a breath when the mouth lifts out of the water, with hands pressing down and back. The head should be moved back into the water with the neck straight as the power is exerted by the kick. If you want to chat with your shipmate or keep an eye on something ahead, you can keep your head up as you swim, but this reduces your glide and makes the stroke harder.

Elementary Backstroke

This stroke will enable you to swim on your back for great distances without becoming tired; it is outstanding for survival swimming. You will also find it to be the easiest stroke to learn and accomplish.

Leg action. Lie on your back, extend your legs, and keep your legs together. Keep your thighs straight and bend your knees so that your heels drop down and move toward your hips. When your heels are directly underneath your knees, turn your feet outward and spread your feet and legs. Without stopping, push your feet down and back making a circle with your heels until they come together. This is the "breaststroke kick" adapted for the back stroke. You may also hear it called the "whip kick."

Arm action. Perform a back glide to get into position, with arms down along the sides of your body. Slowly bring the hands up underwater, over the front of the body to about mid-chest, with fingertips brushing your ribs. Then stretch your arms straight outward at shoulder height. Continuing, sweep your slightly cupped hands even with the surface of the water in a wide arc until they are back at the starting point, along your sides. Remember to keep your hands underwater at all times and to put your chin to your chest during the arm stroke to prevent getting water in your face.

Combined stroke. The combined movements of the stroke are very easy to learn. Start with feet together and hands at your sides. Move your hands up at the same time as you drop your heels. As you extend your arms, turn your toes out. Now sweep your hands to your sides as you bring your feet together in a circular motion. Make your movements smooth and continuous, and rest only at the end of the stroke to allow a long glide. Keep your head fairly well back, with your ears in the water. Take your time and relax.

SPECIAL INSTRUCTIONS

Cramps

Cramps are rarely as dangerous as people believe. Take a deep breath and float quietly for a while. If the cramp is in an arm or leg muscle, massage it slowly and firmly.

Removal of Clothes

If it is necessary to remove your clothes in the water, the heaviest articles should be removed first. To remove trousers, assume a back–float position, unzip them, slide them down, and flutter kick out of them.

Remember that shoes are a real protection against sharks and sunburn, and that clothing in general is a protection against exposure. Remove only clothing that interferes with your keeping afloat.

Using Clothing to Keep Afloat

Shirts may be inflated by tying the cuffs and collar, blowing air into the opening, and then holding the garment under water.

Trousers make better floats than shirts. After the trousers are removed, they should be floated on the surface with the fly turned up. A single knot should be tied in each leg, then one side of the waist of the trousers should

Figure 35-2. A muscle cramp in the foot can be worked out by massage.

be grasped with each hand, and the garment should be worked around on the surface until the legs are at the back of the head and neck. When this position is reached, the trousers should be flipped over the head, and the waist brought down smartly on the surface, trapping a good pocket of air in each leg. The waist should then be gathered under the water and held in one hand.

Mattress covers, sea bags, laundry bags, pillow cases, and sacks may be used for support by capping the openings on the surface.

SWIMMING IN DANGEROUS WATERS

There are several emergency techniques that should be studied and practiced.

Underwater Swimming

The ability to swim underwater is very important in avoiding surface hazards such as floating debris, oil, or flaming oil or gasoline. It has lifesaving value, for it enables a swimmer to rescue a shipmate who has gone under. There are occasions, too, when it is necessary to recover articles lost overboard in shallow water.

Almost any swimming stroke can be used underwater. The breaststroke and the sidestroke lend themselves particularly well to underwater swimming. The main thing is to learn breath control. Practice holding your breath and keeping your eyes open as much as possible.

Swimming in Shark-Infested Waters

A lot of research has been done on shark attacks in recent years. During World War II, people in the water were told to splash wildly to drive away sharks, but more recently it has been discovered that sharks seem attracted to frantic movements. Sharks are extremely dangerous and unpredictable predators and have extraordinary sensory receptors. Species of sharks vary in their feeding habits, with the result that some are considered more dangerous than others. Sharks seem to sense fright exhibited by creatures in the water. They also seem to be attracted to bright colors, but most of all, sharks are attracted by blood. Some recent experiments indicate that sharks exhibit a characteristic "S" twist before striking, and immediately before biting they roll their eyes upward losing sight of their prey. These experiments indicate that during the last phase of the attack, the sharks are guided by sensors located in their snouts. For some reason, in these

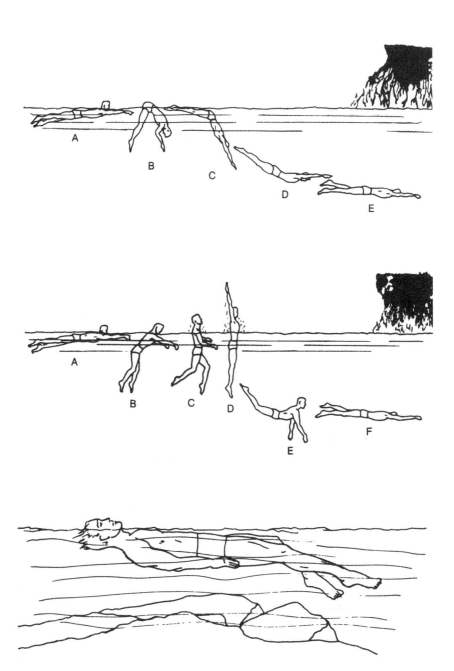

Figure 35-3. Swimming in dangerous water: (top) procedure for making surface dive; (center) alternate procedure for submerging; (bottom) in shallow rapids, swim feet first on your back, "fin" the hands at hip level to protect against rocks.

Figure 35-4. Use side stroke or breast stroke in surf to preserve strength. Ride small waves; surface dive to end the ride before the wave breaks.

experiments, the presence of metal objects caused the shark to divert that last phase of his attack toward the metal object and away from the original object of the attack.

Although there is no good advice on how a swimmer might avoid shark attacks, except by avoiding sharks, the information that is currently available indicates that the following actions may reduce the risk of shark attack.

- Avoid bleeding into the water.
- Wear shoes and other protective clothing, particularly clothing of dark, drab shades; save the bright colors to show above the surface to attract rescuers.
- Remain calm and remember all species of sharks are not the same, nor do all sharks have the same level of aggressiveness.
- Swim slowly and quietly to minimize attracting the attention of sharks.
- In the presence of nonattacking sharks, do not strike at them because that may provoke an attack that might not otherwise occur.
- If the water is clear enough to see sharks swimming beneath the surface, watch for a telltale "S"-like twisting by a shark, which may indicate that an attack is imminent.
- If a shark is about to attack, its jaws may possibly be diverted at the last moment by thrusting an object such as a piece of driftwood toward the shark's mouth.
- Do not confuse sharks with dolphins and porpoises, both of which are mammals, known in fact and legend to be friendly and helpful to sailors in the sea.

Figure 35-5. In shark-infested waters, swim as quietly as possible. Always wear shoes. Try to hold onto some floating object, especially in cold water. Keep moving to maintain circulation. In a current, either swim parallel to the shore or diagonally to it in order to get out of the current. In seaweed or underwater plants, move calmly; do not thrash about.

Swimming in Surf

Along coastlines and on reefs, waves steepen and break, turning into what are called "breakers." Surf consists of more than one of these breakers. Coast Guard boats often cross bars in leaving and entering inlets and harbors, and often encounter broken waves at sea, on bars, and in inlet channels; seldom, however, do Coast Guard boats go deliberately inshore into the surf, except during special emergency operations.

You are most likely to encounter surf after the foundering of a Coast Guard boat or if you are swept overboard from a boat near the beach. Surf conditions in these cases are probably going to be very rough.

You *should* be wearing a personal flotation device (PFD). The PFD-equipped swimmer's best chance of avoiding injury is to make sure that the PFD is properly secured, and then to assume a rolled-forward position to protect the neck, back, and limbs. Take deep breaths whenever possible, and keep the eyes open underwater to help you reorient to the surface.

A non-PFD-equipped swimmer must avoid being caught on the surface by a large, plunging breaker. Such a swimmer must dive under approaching crested waves, and then surface soon enough after the wave has passed to get another breath before diving under the next wave. Dives should be as deep as possible to avoid getting caught in the surface water as it curls over from the back face of the wave. Waves can be flattened by onshore winds yet retain all the force of much higher seas. The swimmer in such surf must watch for approaching seas and rise with them or dive as necessary to conserve air and strength. A swimmer must move quickly onto the beach to avoid the back rush of water after each wave.

Strong inshore feeder currents will set parallel to the shoreline, usually in the direction of the wind and surf. These currents are formed by waves that have been pushed onto the beach and that must find an outlet to return to the level of the sea. These outlets are marked by lower wave heights, slower wave speeds, confused current swirls, and foam. When these feeder currents turn to sea, they are called "rip currents." They flow seaward at angles from 45 to 80 degrees from the inshore feeder currents.

Swimmers must avoid these rip currents. If the swimmer seaward of the breaker line can determine the direction of the wind, or the angle at which the surf is approaching the beach, it is possible to pick the area where a rip current is least likely. That area will be just downwind of the place in the surf line where the waves seem lower and break farther to sea.

Should a swimmer get caught in a rip current, the best strategy is to swim gently at right angles to the direction of the current, never against it. Since most rip currents angle away from the beach, swimming at right angles to the current, down the beach and in the general direction that the wind and surf is moving, will enable the swimmer to make slow progress toward the shore without becoming exhausted. Once clear of the current, the swimmer should head directly toward the beach. Because of their unique profession, Coast Guard personnel should learn to swim comfortably in surf, studying its movements and rhythms. Under emergency conditions, experience and knowledge of the surf may make survival possible.

LIFESAVING

Effective lifesaving in the water depends upon the rescuer's presence of mind, knowledge of the methods that may have to be used, and strength and skill in carrying them out. Drowning people usually panic. Their one idea is to keep their heads above water. The whole technique of lifesaving is centered upon that fact. If a drowning person grabs or tries to grab some part of the rescuer's body or clothing, the rescuer should sink, taking the victim under water. Under water, the victim usually will let go. Under no circumstances should a rescuer strike a drowning person. A drowning victim's system has already had sufficient shock to cause severe physical reactions, and a severe blow may cause heart failure.

Lifesaving Approaches

There are several ways to approach a drowning person; practice all of them.

Frontal approach. The rescuer should swim slowly toward the victim and attempt to calm and reassure the victim by talking. The victim should be told exactly what the rescuer is going to do and must be instructed to follow orders. Then the rescuer should grasp with the right hand the victim's left wrist, turn the victim's body slowly, and use one of the carries described below.

Rear approach. If the victim is too excited to pay attention to directions, the rescuer should swim behind the victim, grasp the chin with the right hand, apply pressure to the back with the left hand, and use one of the carries.

Underwater approach. The underwater approach is by far the safest because the drowning person does not have a chance to get a grip on the rescuer. The rescuer should swim within ten feet of the victim and surface-dive to a depth at which the victim's legs can be easily reached. If the victim is facing the rescuer, the rescuer should turn the victim in the opposite direction by pressure on their upper legs. The rescuer should slide the right hand up the drowning person's back and grasp the victim by the chin, applying pressure to the back with the left hand. The victim should be brought to the surface as quickly as possible and one of the carries used.

Lifesaving Carries

All of these methods should be practiced (see figure 35-6).

Hair carry. This is the easiest of the carries because it allows the most freedom of movement on the part of the rescuer. The rescuer turns on the left side, slides the right hand up the back of the victim's head to the top and grasps the hair tightly, using the left arm and legs for swimming sidestroke. The rescuer may swim on either side, changing hands when necessary in order to rest.

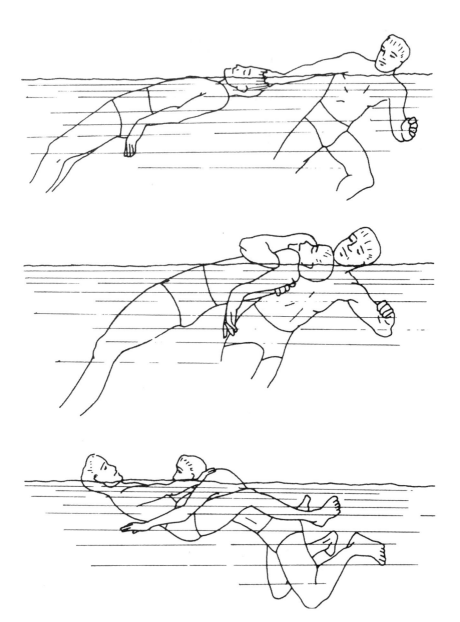

Figure 35-6. Approved methods of carrying a person in the water are (top) hair carry; (center) cross-chest carry; and (bottom) tired swimmer's carry.

Head carry. The rescuer swims on the back, holding the victim's head above the water with both hands meeting under the victim's chin.

Cross-chest carry. The rescuer turns on the left side, places an arm over the victim's right shoulder, across the chest and under the left arm. The victim's body should be supported on the rescuer's right hip, and the rescuer should swim sidestroke, using the free left arm and both legs. This carry may be done from either side but should not be used unless the victim is in bad condition, nor should it be used for long distances because it is the most tiring.

Tired swimmer's carry. If the victim has enough control to obey orders, the victim should be ordered to turn on the back, face the rescuer, spread the legs, and place both hands on the rescuer's shoulders with the arms stiff. The rescuer assumes the position for the breaststroke and swims, pushing the victim ahead. The breaststroke is useful because it leaves the arms and legs unhampered for swimming and is only slightly more tiring than ordinary swimming.

Carries from the water. The most common carry is the *fireman's carry.* On reaching shallow water, the rescuer stands the victim up temporarily, places the right arm between the victim's legs, and throws the helpless person's body over the right shoulder. In this position, the victim's right wrist is grasped in the right hand of the rescuer, whose left arm is free.

A convenient carry for short distances is the *saddle back.* After bringing the victim into waist-deep water, the rescuer moves the victim's body around behind, with the head to the right. The rescuer grasps the victim with both arms around the body, hoists the victim across the back, leans forward, and carries the victim ashore.

SNORKEL AND SCUBA DIVING

Both of these techniques are being used more and more in military operations and as a form of recreation. Each offers many dangers to an untrained swimmer, and no one should attempt to snorkel or dive with SCUBA gear without training. Certification courses are available and will greatly benefit any swimmer. Any diving done under orders for the Coast Guard must be performed only by trained divers.

Figure 36-1.
Survival at sea requires being prepared at all times.

CHAPTER 36

Survival

Normally, the Coast Guard helps those who get into trouble, and you will be trained to handle such emergencies. There may come a time, however, when your own ship, boat, or aircraft is in trouble, and the problem of doing something about it may be up to you. This chapter gives basic information on survival if you should find yourself in distress. Special instructions will be given to members of rescue boats or aircraft crews. No matter where you are, you must familiarize yourself with such other instructions as are available. Perform drills diligently, learn the use of survival equipment, keep your lifejacket and other lifesaving gear in good condition, and regularly review instructions so that you can manage in an emergency. Refer to the *Coast Guard Rescue and Survival Systems Manual*, COMDTINST M10470.10 (series), for detailed information.

PERSONAL FLOTATION DEVICE

Personal flotation devices (PFDs) come in a number of forms. Shipboard military personnel are usually provided with specially designed PFDs that are inherently buoyant, meaning they are filled with a substance such as fibrous glass or kapok that provides flotation. Other military PFDs include less cumbersome, inflatable, and formed plastic types. Commercial-type PFDs are used by Coast Guardsmen in certain situations.

These are referred to as Type I or III and are assigned a Coast Guard approval number. Type I commercial PFDs provide similar flotation to military PFDs that are inherently buoyant. PFDs are often termed "life preservers," "lifejackets," or "life vests." The official definitions for these terms are somewhat tricky, so if you become involved with this

Figure 36-2. A boatswain's mate demonstrates the proper method of wearing a Type III lifejacket.

terminology in a legal sense or in a written report, be sure to use the correct term. Generally speaking, the term "life preserver" is correct for PFDs provided on board Coast Guard vessels.

Offshore Lifejacket Type I PFD

The typical Coast Guard life preserver is the Offshore Life Jacket Type I PFD. It is an inherently buoyant PFD. This PFD is designed to provide superior flotation characteristics and reliability. It will support the head of an unconscious person.

The Type I PFD is designed somewhat like a regular jacket. It has a back, a collar, and a right and left chest pad. The cover or outer shell is made of nylon and is bright orange in color. The collar and right and left chest pads have flotation material in them. There are three black nylon straps. These are used to secure and adjust the PFD:

- The chest-strap
- The waist-belt
- The right and left suspender webbing

The PFD has four three-by-five-inch pieces of retro-reflective material. They are individually located on the right and left collar and chest pads.

Characteristics of the Type I PFD

The PFD provides superior flotation in the open ocean, rough seas, or remote waters where rescue may be slow in coming. It is designed to keep the head of an unconscious person above the water and will turn *most* unconscious wearers faceup in the water. It is approved as the abandon-ship lifejacket for military, commercial, and all other vessels carrying passengers. It provides a minimum adult buoyancy of twenty-two pounds.

Signaling Gear on the Type I PFD

Personnel Marker Light (PML). A chemical light developed for use on PFDs to provide an artificial light for nighttime spotting. There is one PML attached to the PFD on the left chest pad suspender webbing.

Signal Whistle. A sound-producing device constructed of plastic that has a lanyard attached for easy access and to prevent loss. It is located in the pocket on the left chest pad.

Proper Care

If the PFD is wet or damp, the nylon cover material and straps will mildew and could break down, causing premature failure. The PFD must be air-dried. *Do not* place any PFD in a dryer; this will damage it.

Donning the Type I PFD

1. Slip into the PFD.

2. Close chest-strap buckle.

3. Secure waist-belt snap-hook; grasp free end of belt and pull snug.

4. Grasp free end of suspender webbing and pull downward until snug.

Removing the Type I PFD

To remove the PFD just reverse the steps:

1. Lift up on the suspender webbing tension hardware. This will release the tension on these straps.

2. Pull outward on the waist-belt snap-hook to release the tension. When tension is released, unsnap the snap-hook from the "D" ring.

3. Push down on the exposed center of the male end of the chest-strap until it disconnects.

4. Take the PFD off.

Stowing the Type I PFD

Follow these steps to properly prepare the PFD for storage:

1. Lay the PFD on a flat surface with the manufacture's label facing up toward you. All of the straps should be disconnected or loose. The flotation collar should naturally fold itself under the PFD.

2. Connect the chest-strap buckle.

3. Connect the waist-belt snap-hook.

4. Take the slack in the waist-belt and loosely wrap it around the back of the PFD and weave it around itself on the "D" ring side of the belt.

Navy Vest-type PFD without Collar, Work Type (Inherently Buoyant)

The Navy vest-type PFD is lightweight, formed, plastic-filled, and provides 17½ pounds of buoyancy. The unit is composed of three cotton drill-covered sections that are assembled through a series of straps to form the completed PFD.

Each section has a formed or molded pad two inches thick, notched for flexibility, allowing the PFD to conform more closely to the shape of the body. Because this PFD-type vest is light in weight, the wearer is able to work in comparative comfort.

The work-type PFD is buoyant enough to keep the wearer afloat, but it has no self-righting capability and will not keep an unconscious wearer's head out of the water while awaiting rescue.

Coast Guard–Approved Type III PFD (Inherently Buoyant)

The main advantages of Type III PFDs are wearability, ease of donning, simple construction, and neat appearance. The disadvantages are poor flotation characteristics (no righting movement), and minimum buoyancy (15 pounds). Therefore, type IIIs should be worn only where greater freedom of movement is required and the mission and environment are less hostile.

Navy Inflatable Yoke-type PFD

The Navy inflatable yoke-type PFD is a lightweight preserver that can be blown up either by a CO_2 cylinder or by mouth. It is fastened around your waist by a web belt. When blown up, it holds itself closely around your neck. An extra piece of webbing provides a handhold for assisting you out of the water. A length of line with a wood T-toggle enables you to attach yourself to a boat, to float lines, or to other survivors. When not needed, the inflatable preserver is carried in a pouch, normally worn at your back.

When you need the preserver, pull the pouch around to the front, remove the preserver, slip it over your head and inflate it. To inflate the preserver, grasp the lanyard attached to the inflater and jerk downward as far as possible. This will release the carbon dioxide gas into the air chamber. If the preserver has two inflaters, pull both lanyards. For more buoyancy, push in the mouthpiece or the oral inflation valve and blow up the preserver yourself. To deflate the preserver, open the oral valve.

ABANDONING SHIP

Preparations

There are two important things to do if time permits: find out the distance and direction to the nearest land, and see that your equipment is in good condition. What clothing and gear to take with you will depend on whether you are in a hot or cold climate and whether you will take to a boat or raft, or if you will depend upon your lifejacket to keep you afloat. Clothing is especially important in cold weather, but even in hot weather you need protection from the sun.

In cold weather, you need

- Long woolen underwear and woolen socks to keep your body warm;
- A windproof jacket to protect your shoulders and arms;
- A helmet or cap to keep your head warm and water out of your ears;
- Gloves to warm hands and to prevent burning them on lines; and
- An anti-exposure suit, if available.

Equipment

The more equipment you have, the better. Take these as a minimum:

- Flashlight, knife, whistle
- Line, at least six feet long, tied under your arms with snap-on free end to make fast to a line on a raft, boat, or rescue ship
- Sunglasses
- Wallet, money belt

Actions before Leaving Ship

The actions you will take before leaving the ship depend on the amount of time you have. Remember to

- Keep calm.
- Test lifejacket valves and inflating tubes; do not leave the ship with a leaking jacket.
- If it is available, drink hot tea or coffee to reduce the effects of cold.

Abandon-Ship Procedure

The following procedure is a guide to be followed whenever practical.

Leaving the Ship

For survival, it is best that you abandon ship fully clothed. If possible, get away from the ship in a lifeboat or life raft. If it is impossible to leave the ship on a life raft, lower yourself into the water using a firmly anchored hose or line. When you have a choice, leave the ship from the windward side and from whichever end of the ship is lower in the water. If it is necessary to jump into the water, hold the legs together and the body erect. Injured personnel should always have their leg straps adjusted before they are lowered into the water. Inherently buoyant-type PFDs must be securely fastened and kept close to the body by folding the arms across the chest and gripping the jacket with the fingers. This procedure will prevent buoyant PFDs from riding up and striking your chin or neck when you hit the water. If you are wearing an inflatable PFD, it must not be inflated until you are in the water. The same procedure is followed for jumping with a noninflated PFD as with the inherently buoyant PFD. The PFD should be inflated as soon as you are in the water and clear of flames.

In the water. When in the water, swim away from the ship as rapidly as possible and climb into a lifeboat if one is available.

Ship surrounded by flames. When the ship is surrounded by burning oil and abandonment is essential, jump feet first through the flames and swim to windward under the surface of the water for as long as possible. When the air in your lungs is exhausted, spring above the water in a vertical position, push the flames away with a circular motion of your hands, quickly take a deep breath, and turn around to swim to windward, submerge feet first in a vertical position and swim under the surface again. This procedure should be repeated until you are well clear of flames. Any buoyant articles of clothing and shoes should be discarded. Only the inflatable PFD should be worn, if possible during this abandon-ship procedure, and the preserver should be inflated only after you are clear of the flames. Inherently buoyant PFDs will not permit the wearer to swim beneath the surface.

Survival in Cold Water

When forced to abandon ship in cold waters, put on as much clothing as possible (including hat and gloves) or an anti-exposure suit, if available, and be sure your lifejacket is properly adjusted. Once in the lifeboat or raft, huddle up close to your shipmates to conserve body heat. If there is water in the lifeboat, immersion foot can be prevented by untying shoe laces, elevating legs, and exercising ankles and toes at regular intervals. However, if you develop frostbite or immersion foot, do not rub the affected parts.

If you must enter the water, make sure you are heavily dressed and wearing a lifejacket. Enter the water as slowly as possible in order to reduce the shock of initial entry. Once in the water, remember that any movement will increase the rate of body-heat loss and reduce your chance for survival. Swim only to a nearby boat, raft, or piece of debris that you may be able to use to escape from the water. If you must remain in the water, cross your legs, bring your knees up toward your chest and hold your arms around your chest (heat-escape-lessening posture). This posture will help reduce body-heat loss.

If you should find yourself in the water, especially cold water, in a survival situation, follow these tips:

- Keep as much of your body out of the water as possible. If you can, climb onto an overturned hull or piece of debris immediately.
- Wear a PFD. This will eliminate the need for physical activity to stay afloat, thereby saving body heat.
- If you have time, put on extra clothing and be sure to cover your head.
- Protect yourself against the wind by staying to leeward.

If forced to remain in the water, keep your head above water and as dry as possible. Swimming will produce a faster cooling rate; therefore, once flotation is ensured and essential survival actions have been taken, stop swimming and assume the heat-escape-lessening posture. If there are other people in the water, assume a huddle position with them. All survivors face inward, with sides and chests close together and arms around each other. Either of these two actions will increase survival time by up to 50 percent.

Hypothermia is a killer. Its danger is not restricted to freezing the body to death. Prolonged exposure to cold renders the body incapable of coordinated movement and the mind incapable of rational thought. Long before victims are unconscious, they may be unable to take the steps necessary to save their own lives.

Survival in a Hot Climate

Wear shirt, trousers, hat, and shoes. They may be a hindrance in the water, but if you are adrift for any time, they are necessary to protect against the effects of sun, wind, and salt water. Take the same equipment as for cold weather, and take the same actions before leaving the ship.

Avoid panic; keep calm. If you panic, others may as well, and all of you may be in more trouble. Nonswimmers are most likely to lose their heads after abandoning ship, so encourage them, help them, and keep talking to them. Keep your wits about you, keep control of the situation, and you will come through.

ADRIFT IN A BOAT OR RAFT

Ingenuity and foresight are required to make the most of all available equipment. In a powerboat, use the fuel for making the most distance, not speed. The best mileage will be made at the slowest speed that can be held with the clutch fully engaged. If the boat has sails, use them and save the fuel for an emergency. If there are no sails, try to jury-rig a mast and sail with oars, boat hooks, pieces of wreckage, clothing, and tarpaulins.

The most serious situation is to be adrift in a life raft with no way to rig a sail. In that case, make your rations and water last as long as possible. The longer you hold out, the better your chances for rescue or a safe landing. People have endured for more than a month in a raft with practically no rations.

Rescue units will start looking for you at your last-known position. Do not leave that area unless you can make it to shore or to a well-traveled shipping lane.

Organizing for Survival

The first thing to do is to secure all gear to the raft or boat. Anything not lashed down is liable to go overboard and be lost.

- Rig a tarp or sail for protection from the sun, and get under it. The water will reflect enough sun to burn your skin, so stay out of the sun as much as possible. Remember that the same tarp or sail will catch drinking water when there is rain.

- Inventory the provisions on board and store them where they will be safe, even against high seas. Plan for rationing food and water. No one should eat or drink during the first twenty-four hours. However, in cold climates, eating every two hours will help you to stay warm.

- Do not get excited and do not rush. Do things deliberately to conserve energy and to avoid perspiring.

- Save your clothes. Although it is warm during the day, you may need something to wear at night.
- Arrange watches and cooperate in standing them. Stow your signaling gear—mirror, marking dye, flares, and smoke signal—where they can be reached in a hurry.

First Aid

In general, the first-aid procedures afloat are the same as those ashore. There are some special considerations, however, that apply to survivors at sea.

Fuel oil from a sinking ship will float on the water and accumulate on the skin of survivors. If swallowed, it may cause vomiting. In the eyes, it will cause redness and smarting. Ordinarily these symptoms will disappear in a day or two. Wipe oil from the eyelids with a clean cloth or bandage.

Petrolatum (Vaseline) may be used to ease chapped skin and cracked lips, but it is better to prevent sunburn by wearing or rigging whatever protection is available against the sun. Even a dark tan is not a sure protection against sunburn. Reflected light from the ocean may burn the eyeballs as well as the skin. Guard against the sun when it is directly or almost directly overhead. Boils often appear on the skin of people who have been in lifeboats or rafts for several days. Unless skilled medical assistance is at hand, no treatment should be attempted. Simply cover them with protective bandages and do not try to open them. Resist exhaustion and conserve strength by sleeping and resting as much as possible.

Coast Guard–Approved Rafts

The standard features of Coast Guard–approved rafts are

- Constructed of neoprene-coated nylon
- CO_2 overpressurized inflation system with relief valves that provide for cold inflation
- Two independent buoyancy tubes
- Self-erecting canopy
- Boarding ladder or towing bridle
- Survival equipment (drinking water, signal flares, etc.)
- Inflatable floor
- Exterior and interior lights

- Stabilizing pockets (water)
- Rainwater catchment assembly
- Rescue line (throwable)
- External and internal lifeline
- Sea anchor

Crew survival rafts can be released and inflated either manually or automatically by use of the hydrostatic release. The life rafts can be deployed manually by releasing the raft from the stowage rack and pulling the operating painter from the raft container. The container should preferably be placed in the water prior to inflation to allow ample room for expansion of the raft. If practical, personnel should board the raft directly from their vessel to avoid unnecessary boarding problems and exposure to the water, especially in cold oceans. Immediately upon boarding the raft, complete these steps:

- Get clear of sinking vessel in case of explosion or surface fire.
- Search for survivors.
- Salvage floating equipment that may be useful; stow and secure all items.
- Ensure that the sea anchor is properly deployed.
- Check the raft for proper inflation, leaks, and points of possible chafing. Bail out water that may have entered the raft. Be careful not to snag the raft with shoes or sharp objects.
- In cold oceans, inflate the raft floor immediately and put on exposure suit, if available. Rig an entrance cover. If with others, huddle together for warmth.
- Check the physical condition of all on board. Give first aid as necessary. Take seasickness pills if available. Wash oil or gasoline from clothing and body.
- If there is more than one raft, keep close together to expedite rescue operations. If possible, rafts should be tied together about fifty feet apart bow to stern, from towing bridle to towing bridle.
- Make a calm estimate of the situation and plan a course of action carefully.
- Ration water and food; assign duties.

- Keep a log. Record time of entry into water, names, and physical condition of survivors, ration schedule, winds, weather, direction of swells, times of sunrise and sunset, and other navigation data. Inventory all equipment. While in the raft, encourage, help, and keep talking to your shipmates.

Rescue and Assistance

Although shipboard rafts are intended for crew survival, these rafts may be used, at the discretion of the commanding officer, for other purposes such as rescue and assistance. If the raft is used as a rescue platform for recovering survivors, the canopy fabric may be cut away.

Four/Six-Man Rescue and Survival Raft

The four/six-man rescue and survival raft was modified to provide search-and-rescue boats with a multipurpose raft. This raft is a basic Coast Guard–approved raft modified to incorporate a stowable canopy and a heavy, water-ballast system for stability.

The raft is packed with the canopy stowed. The exterior and interior lights and the rainwater catchment assembly have been removed from the canopy. A heavy, water-ballast system has been added for increased stability in heavy weather. Because of this system, the raft will not be as affected by wind as conventional rafts would be.

Survival Items

Signal Whistle

A signal whistle is used to attract the attention of rescue ships or personnel in foggy weather or at night. It is made of plastic with a lanyard attached for easy access and to prevent loss. The whistle's range is one thousand yards.

Survival Knife

The survival knife is a multipurpose survival tool stowed in its sheath when not in use. It is a hunting knife with a five-inch steel blade. One side of the blade is honed and the other side is serrated.

The survival knife is the most valuable general-purpose survival tool. It can be used for cutting wood and material and for opening cans, and it is a hunting knife as well as a weapon. At all times, the survival knife should be kept clean and sharpened and should be returned to the sheath when not in use.

Signal Kit MK-79 MOD-0

The MK-79 MOD-0 signal illumination kit is a distress signaling device. It is small and lightweight so that it can be carried in flight suits or life rafts.

The MK-79 signal kit consists of one MK-31 projector and seven MK-80 screw-cartridge flares. Included with the launcher and flares is a bandolier, which stores the flares until use as well as protective caps for the flares and an instruction sheet.

The projector aims and fires the flares. Each flare consists of a red pyrotechnic star. Upon activation, this star is propelled to heights of from 250 to 650 feet. The star burns for a minimum of 4½ seconds and has 12,000 candlepower.

The kit is operated by pulling the thumb trigger into the safety catch, screwing the cartridge into the projector, pointing the cartridge in a safe, upward direction, and releasing the thumb trigger. Full safety and operation instructions are in the *Coast Guard Ordnance Manual* M80002 and NAVORD OP 2213.

Smoke and Illumination Signal Marine MK-124 MOD-0

The MK-124 MOD-0 is used for day or night signaling with smoke or flare. This marine smoke and illumination signal consists of a metal cylinder 5 inches long and 1½ inches in diameter, each end of which is fitted with a protective rubber cap. On the flame (night) end there are two raised ridges encircling the cylinder. These ridges positively identify this end, by the sense of touch, for nighttime use. A label adhered to the outer surface around the whole body of the signal further identifies the smoke (day) and flame (night) end and provides precise instructions for use. The MK-124 MOD-0 illumination signal is intended for either day or night signaling on land or sea. Because of its weight (6½ ounces) and size, the MK-124 MOD-0 may be carried in a life vest, flight suit, or life raft. It emits orange smoke for day use and a red flare for night use. Burning time is 20 seconds. To activate the MK-124 Mod-0 slide the plastic lever out away from the cylinder and pull down on it with your thumb. This will ignite the flare mixture (night) or the smoke mixture (day), depending upon which type of display is desired.

The MK-124 Mod-0 signal must be operated, stored, and handled in accordance with procedures found in the *Coast Guard Ordnance Manual*, M80002, and NAVORD OP 2213.

Emergency Signaling Mirror

The emergency signaling mirror is used to attract rescue ships or aircraft. It measures 2 inches by 3 inches and replaces the 3×5-inch mirror. There is a hole in the corner through which a lanyard passes so that the mirror may be looped around the neck.

The emergency signaling mirror is intended to reflect sunlight at ships, aircraft, or rescue parties. Instructions for using the mirror are printed on its back.

Distress Signal Lights

CG-1

The CG-1 is a battery-operated strobe light used to signal rescue parties. The light is a lightweight, compact, battery-operated portable unit with all circuitry encapsulated within the case. The battery entry and the case are completely watertight when the battery is installed. The case is provided with a slide-up ON/slide-down OFF switch to permit one-handed operation.

A three-foot nylon cord attaches the light to the equipment case by either of two attachment loops on the light case. The CG-1 is intended for equipping aircrew members and shipboard personnel with a high-intensity, visual distress signal in the event of abandonment of aircraft or a man overboard.

Personnel Marker Lights

The personnel marker light (PML) is a chemical light developed for use on PFDs. It is used to attract the attention of search-and-rescue aircraft, ships, or ground parties. Once activated, it provides light for eight hours. The PML is equipped with a pin-type clip and should be attached in the same location on the PFD as the one-cell flashlight.

To activate the PML, slide the protective sleeve from the PML, breaking the sealing band. Firmly squeeze the lever against the light tube until the ampoule inside the light breaks.

Equipment in Rafts and Boats

The following items should be in all rafts or lifeboats:

- Air pump
- Canned water
- Dye marker

- First-aid kit
- Flashlight (waterproof)
- Food packet
- Knife
- Paddles
- Sea anchor
- Signal gear
- Signaling mirror
- Sponge
- Water desalting kit
- Whistle

Other Survival Kit Items

Concentrated food. The Navy emergency rations for life rafts meet minimum nutrition requirements for personnel shipwrecked with limited supplies of water. Each can contains a total of fifteen sugar-type tablets, eight malted milk tablets, two multivitamin tablets, and two packs of chewing gum. The daily ration would normally be one of these cans per person.

Dye marker. The dye marker consists of a can of fluorescein (a material that glows in very little light) packed in a waterproof container. On the water, it forms a yellow-green patch that usually is visible for two hours for a relatively long distance. At night, it can be seen only in bright moonlight, and rough water quickly disperses it.

First-aid kit. The camouflaged kit for life rafts contains an assortment of bandages and some tubes of petrolatum (for burns). The bandages consist of compresses to cover open wounds, long strips of gauze to hold the compresses in place, and triangular all-purpose gauze bandages, which may be used to hold other bandages in place, or as slings, head bandages, and so on.

Fishing kit. The fishing kit contains gloves, a knife that will float, sinkers, pork rind for bait, an assortment of hooks and fishing rigs, a dip net, a bib with pockets to hold the equipment, and directions printed on waterproof paper. The kit is enclosed in a can with a key opener.

Flashlight. One of the most commonly used night signals as well as a most useful piece of equipment is the flashlight. A waterproof model is standard, but any flashlight should be kept out of the water to ensure the best service and longest life. The equipment containers are provided with spare batteries; nevertheless, flashlights should be used only when necessary.

Food and Water in the Environment

Fish caught in the open sea are healthful and nourishing, cooked or raw, and will supplement and even replace canned rations as long as sufficient drinking water is available. When fish are caught near shore, avoid the poisonous kinds. Poisonous fish can be used as bait but must not be eaten.

Drying Fish

Well-dried fish will stay unspoiled for several days. To dry large fish, clean them, cut them in thin narrow strips, and hang them in the sun. Small fish, a foot long or less, should also be cleaned before drying, the backbone removed, and slits cut across the inside about a quarter of an inch apart. They should then be hung in the sun. Fish not cleaned will spoil in a half day. Do not eat dried fish unless you have plenty of drinking water.

Fishing Tips

Never tie a fishing line to finger, hands, feet, or boat. A big fish can cut or break your line or carry off your tackle. Do not pay out all your line but leave slack in the boat. Have someone hold the end of the slack line. Do not lean over the side of the boat when a fish is hooked. A boat can be capsized in this fashion.

- Try to catch small fish rather than big ones.
- Keep the bait moving to make it look alive.
- Keep parts of any bird or fish you catch and use them for bait.
- Clean hooks and lines of fish and fish slime, and dry them in the sun.
- Do not get your lines tangled or let hooked fish tangle them. Two people can fish at the same time, but they must be careful.
- In a rubber boat, be careful that your knife and fishing hook never get a chance to prick, stick, or make a hole in your boat.
- Make a fair division of your catch.
- Keep fishing. You never know when fish may bite.
- Fish will be attracted by a light at night. Often a bright moon or flashlight shining on a cloth will draw fish.

Solar (Sun) Stills

Solar stills are sometimes provided for creating potable water. They can distill two pints of water each day. Usually, a still is provided for each person the boat is designed to accommodate. The stills operate most efficiently in direct sunlight. They will function to a certain degree during cloudy or overcast days, but they cannot distill water at night, on dark days, in polar waters, or during inclement weather. The solar still is one of the most reliable sources of potable water at sea. The water is pure but may appear somewhat milky because of talc sprinkled on the inside of the plastic cover to keep it from sticking to itself.

Eating on Land

Most tropical fruits are good to eat. It is safe for humans to eat the same things that monkeys eat. Unknown roots, fine shoots, and other herbs must be experimented with cautiously. Eat any unfamiliar plants in very small quantities until their effect is ascertained.

Turtles and their eggs may be eaten cooked or raw. All animals are safe to eat. Snakes, even poisonous ones, are edible, but their heads should be removed. Grubs and grasshoppers can be eaten if the wings and legs are picked off before cooking. Caterpillars should not be eaten. All birds are good to eat, cooked or raw. Their blood and liver are edible. The feathers, meat, guts, and even the toes make good lures or bait for fish. Birds can sometimes be caught in nets; the larger ones will often take a bit of fish on a hook dragged on top of the water. Some birds will go after a baited hook tossed into the air.

Water Discipline

If possible, take a drink of water before abandoning ship. After abandoning ship, conserve body water; try to keep cool and avoid sweating. Unless you become very thirsty, drink no water for the first twenty-four hours on the raft. Thereafter, drink a pint (sixteen ounces) of water a day if your supply is limited. Drink a pint and a half (twenty-four ounces) a day, or more if necessary, if you have an abundant supply of rain water and the one-pint allowance does not satisfy you. It is better to keep up the rate of sixteen ounces a day (four ounces—half a cup—four times a day) until only ten ounces are left, rather than to use smaller amounts for a longer time. When only ten ounces are left, use it only to moisten the mouth. (Sea water and urine should never be drunk. Sea water will cause vomiting, diarrhea, and delirium.)

When the supply of water is limited, bird and fish flesh should be consumed sparingly. The moisture in fish flesh is more than offset by the increase in urine resulting from the body's assimilation of the fish protein. Use fish for food, not water.

On land, water can be obtained from a hole dug at low tide just below the high-water mark. The water in the hole may be salty and discolored, but it can be drunk in small quantities. Large amounts will cause sickness.

Mental Attitude

The most important factor in survival after shipwreck in any except the coldest waters is the element of mental and emotional stability. How well people adjust to the psychological stress of being cast adrift may mean the difference between their rescue and loss. Superior equipment is useless when people are emotionally and mentally disorganized.

Cold-Weather Protection

Many cutters and aircraft operate in cold climates where exposure to cold weather and frigid water can cause hypothermia and death. To protect crews from these dangers, cutters and aircraft have hypothermia protective devices.

The *anti-exposure coverall*, or "work suit," is designed for routine work on deck when operating in cold environments. In addition, the anti-exposure coverall provides protection against hypothermia in the water. Similar to insulated coveralls, the anti-exposure coverall is water resistant and buoyant but still provides the mobility necessary for working and climbing.

The *survival suit* is provided as an abandon-ship garment on board cutters operating in cold climates. The suit is a one-piece garment with high flotation characteristics that completely covers the body except for the face. The survival suit is designed so that one size will fit all. The thermal qualities of the $3/16$-inch nylon-lined neoprene foam will keep a person warm, wet or dry.

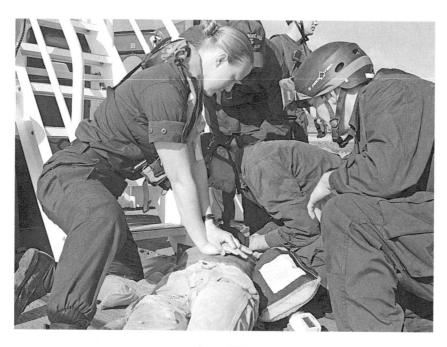

Figure 37-1.
After recovering the victim, Coast Guard members begin
to administer first aid.

CHAPTER 37

First Aid

Firirst aid is emergency treatment for injured or wounded people. It consists only of immediate, temporary assistance to save life, prevent further injury, and preserve the victim's vitality and resistance to infection. The way to remember your first-aid priorities is to simply go by the ABCs. First, ensure the victim has an adequate *airway*. Then check for *breathing*. Next check *circulation* by checking for signs of life and obvious bleeding. When these are under control, treat for or prevent *shock*. The ABCs are the serious life threats and should be corrected first before things such as splinting fractures.

TAKE IMMEDIATE ACTION

Know what to do and do it. When coming upon a victim, take care to protect *yourself* by surveying the scene to be sure there are no hazards such as live electrical wires or loose gear that may have caused the injury and may be dangerous to you. Never enter a space, whether on a cutter or a shipping container, where a shipmate is unresponsive. There may be deadly gases present that contributed to his or her collapse. Secure any hazards before starting treatment. Next, check to see if he or she is awake—a victim who is conscious and speaking obviously has an adequate airway, can breathe, and has a pulse, so the next priority is to check for hemorrhage, control it, then treat for shock. Do not move victims unless absolutely necessary to save them from fire, gas, drowning, or gunfire. A fractured bone may cut an artery or nerve; a broken neck or back may result in spinal cord injury and paralysis, or death. Make an injured person comfortable and cover to keep warm. If a victim must be moved, know how to do it.

CONTROL THE SITUATION

Ask for medical assistance if possible. Use a cell phone, radio, messenger, or send someone in a car. Have someone keep bystanders clear. Loosen clothing about the victim's neck, chest, and abdomen. Examine the victim to determine the extent of injuries. Look for bleeding, wounds, fractures, or burns. Notice the color of the skin. Determine the state of consciousness by asking the victim questions. Bleeding from the nose and ears is often a sign of a fractured skull. Bloody froth coming from the mouth or nose often indicates damaged lungs. Check the pulse rate and strength. Check dog tags, if present, for blood type; see if the victim carries anything (bracelet, tag, or card) about drugs or medicines required or forbidden.

Loss of blood is frequently not as serious as it may look, but lack of breathing is very dangerous. Breathing must be restored first. For that reason, artificial respiration is covered first in this chapter. After that, all other first-aid treatments follow in alphabetical order.

RESCUE BREATHING AND CPR

Signs

Absence of or insufficient breathing, or cyanosis (blueness of the skin, lips, fingernails, tongue, or mucous membranes of the mouth) are signs that indicate the need for rescue breathing. If these signs are present and the victim has no pulse, the victim is in cardiac arrest and cardiopulmonary resuscitation (CPR) is needed.

Securing the Scene

Before you can recognize or treat a victim's condition, you must first secure the scene. This is done by removing the cause or removing the victim from the cause. For example, if your victim was electrocuted, secure the power.

Recognition and Treatment

After you have secured the scene, perform your ABCs:

- *Airway*: open then look, listen, and feel for . . .
- *Breathing*: if victim is not breathing, call for help, then give two full breaths. Next, check for . . .
- *Circulation* (pulse): if there is a pulse, continue artificial respirations (rescue breathing). If there is no pulse, start external cardiac compressions along with artificial respiration (CPR).

Rescue breathing or CPR must be continued until

- Normal breathing and pulse have resumed;
- The victim is pronounced dead by a doctor;
- You are no longer able to continue; or
- You are relieved by someone trained in CPR.

Frequently, after recovery of respiration, the victim will again stop breathing. Closely monitor your victim. If natural breathing stops, artificial respiration should be resumed at once.

Rescue Breathing

This is always your first step of treatment, followed by a pulse check in infants and children, which will determine the necessity for external cardiac massage (see figure 37-2). Follow these steps:

- Place the victim on their back and kneel beside the head (a).
- Shake and shout to establish the victim's level of consciousness (b).
- Place one hand under the chin and the other hand on the forehead, so that your thumb and forefinger can pinch the nose closed (c).
- Tilt the head back gently by lifting the chin and pushing the head down. This will open the airway in the majority of cases (c). Note: If neck or back injuries are suspected, use an alternate method for opening the victim's airway called the modified jaw thrust (k).
- Look, listen, and feel. Bend over the victim so that your ear is a few inches from the mouth and you are looking at the chest. Look for the chest to rise while listening and feeling with your ear for the movement of air (d).
- Take a deep breath (about twice the normal), open your mouth widely, and place your mouth over the victim's mouth and blow, giving two full breaths (e).
- Watch for the victim's chest to rise. As soon as this happens, remove your mouth from the victim's and allow the air to escape naturally after each breath. Give two breaths initially then repeat.

If the chest does not rise, one or more of the following conditions exist and must be corrected:

Air leak. Make sure you have an airtight seal between your mouth and the victim's.

(a) Kneel beside victim

(b) Shake and shout

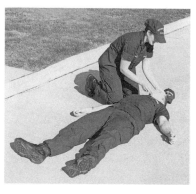

(c) Open the airway

(d) Look, listen, and feel

(e) Give two full breaths

(f) Check the pulse

Figure 37-2. Rescue breathing.

(g) Locate correct hand position

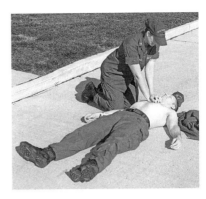

(h) Compress breastbone

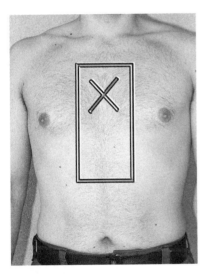

(i) Breastbone

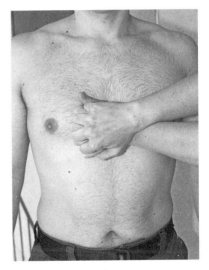

(j) Correct hand position

(k) Modified jaw thrust

(l) Clearing the mouth

Airway obstruction (more likely). Reposition the head to attempt to open the airway and try again. If the chest still fails to rise, look into the victim's mouth and remove any foreign bodies (false teeth, etc.), vomit, or blood clots, then try another breath. If the chest still will not rise, the choking sequence should be followed.

Children and infants require less air to sufficiently ventilate. Infants only require the amount of air you can hold in your cheeks. However, as long as you watch the chest when you blow and remove your mouth when the chest begins to rise, no damage will occur.

Mouth-to-nose breathing may be carried out using the same techniques as mouth-to-mouth, except, of course, the victim's mouth is held closed while your mouth is placed over the victim's nose.

Airways and tubes should be used only by personnel trained in their use or by medical officers. Not only are they dangerous when used by untrained personnel but they usually are not available when an emergency arises.

External cardiac compressions should be started and combined with rescue breathing to give CPR only *after* mouth-to-mouth breathing has been started and after it has been determined that the heart has stopped. Many cases will be encountered in which the person has stopped breathing but the heart is still beating, and these cases require only rescue breathing.

CPR must be started without delay. You have only four to six minutes in which to initiate this rescue technique. After that time, irreparable damage begins to occur in the victim's brain. Therefore, do not leave the victim to summon aid. Send another person for help, and start resuscitation immediately. Do not waste time moving the victim except when absolutely necessary.

External Cardiac Compressions

After mouth-to-mouth breathing has been instituted with two full breaths, check to see if external cardiac compressions should be started. They are needed only if the heart has stopped beating. Perform external cardiac compressions as follows:

- Check for signs of life.

- Find proper hand placement. Locate the notches at the upper and lower ends of the breastbone (figure 37-2-g). Place the heel of one hand over the lower third of the breastbone—point "X" in (i)—and the other hand on top of the first (j).

- Compress breastbone toward the backbone by exerting downward pressure on hands (one and a half to two inches for adults) with the weight of your upper body. Keep elbows straight. Pressure is then released smoothly (h). This cycle is repeated at a rate of one hundred times per minute in adults and children. Children and infants' breastbones are not as strong; therefore, external cardiac compressions should be done with one hand for children and two fingers for infants. Compress the breastbone one to one and a half inches for children and half to one inch for infants at a rate of one hundred times per minute.

Timing

Timing is one of the most critical aspects of CPR. Without correct timing, CPR is useless. Remember the following while performing CPR as a rescuer performing chest compression or rescue breathing; it could be the difference between life and death for your victim.

- The rescuer performing chest compressions (compressor) must maintain a steady tempo and rate. A helpful way to maintain this tempo is count aloud as you compress the chest (one, two, three, four, etc.). During two-rescuer CPR, the compressor must pause after the thirtieth compression to allow the rescuer performing the rescue breathing (ventilator) to administer a breath.
- During one-rescuer CPR, the rescuer must administer both the breaths and chest compressions. This is performed as a cycle of thirty compressions, then two breaths. If your timing is right, you should be completing four cycles per minute.

While performing CPR, check the following. *Hand placement:* fingers should be kept away from the ribs to avoid fractures. *Pulse:* should be checked after two minutes in infants and children.

While performing CPR, the victim's stomach may become distended with air. This is especially true if the airway is not clear. It is not dangerous and should be left alone. Often during CPR the victim vomits, so you must be ready to turn the head to one side and clean out the mouth with fingers or a cloth (figure 37-2-l).

Summary

CPR should be started at once in any case where breathing and the heart have stopped. If only breathing has stopped, use rescue breathing. Do not

waste time seeking help or equipment. Begin treatment at once by following these simple instructions:

- Put victim on a firm surface, faceup.
- Start rescue breathing.
- Check for signs of life in adults and a pulse in infants and children.
- If signs of life or pulse are absent, start CPR.
- Continue CPR until someone of higher authority takes over, until
  victim recovers, until scene becomes unsafe, until an automatic
  external defibrillator is ready to be used, until someone of equal
  or higher training takes over, or until you are physically unable to
  continue.

GENERAL FIRST-AID TREATMENT

Animal Bites

The dangers in an animal bite are rabies and infection. If the bite is by a
pet, the pet must be kept under observation to determine if rabies treatment is required. Bites by wild animals or unknown domestic animals
should be checked by a doctor to determine if rabies treatment is necessary.

Rinse the wound under running water and then wash with soap and
water. Apply antiseptic. Treat for bleeding and shock as necessary.

Bandages

Bandages are used to hold dressings (compresses) to the wound. Dressings
should be sterile, but bandages can be of any material that will sufficiently
keep the dressing in place. The bandage may be applied in turns: *circular,*
spiral, or *figure eight. Triangular* bandages may be tied on the head or face,
shoulder or hip, chest or back, foot or hand. *Cravat* bandages are employed
for wounds of the head, neck, eye, temple, cheek, ear, elbow, knee, arm,
or leg. *Roller* bandages are wrapped on the hand and wrist, forearm or leg.

Bleeding

An average human body contains twelve pints of blood. One pint can
be lost without harmful effect. A loss of two pints will usually produce
shock. If half the blood is lost, death almost always results. Bleeding must
be stopped quickly.

In *arterial bleeding,* bright red blood spurts out; this bleeding is very serious. In *venous bleeding,* dark red blood flows steadily; this also can be very

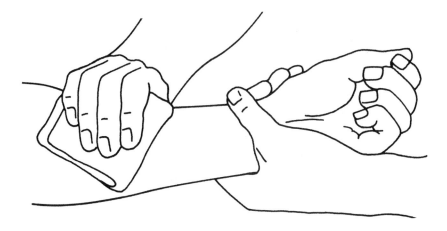

Figure 37-3. Control of bleeding by applying direct pressure on the wound.

serious. *Capillary bleeding*—usually not serious—comes from a cut or small abrasion. Practically all bleeding can be stopped if *pressure* is applied to the wound. If direct pressure does not stop the bleeding, pressure should be applied at the correct pressure point (see below). Where severe bleeding cannot be controlled by either of these methods, pressure by means of a tourniquet should be applied.

Direct Pressure

Use a sterile dressing or the cleanest cloth available—a freshly laundered handkerchief, a towel, or an article of clothing. Fold it to form a pad, place it directly over the wound, and fasten it in position with a bandage. If the bleeding does not stop, try applying direct pressure by hand steadily, for five or six minutes, over the pad (see figure 37-3).

In cases of severe bleeding, do not worry about infection—stop the flow of blood. If nothing else is available, jam a shirt or other large material into the wound. Remember, direct pressure is the first method to use to control bleeding.

Pressure Points

Bleeding from a cut artery or vein may often be controlled by applying pressure to the correct pressure point and elevating the part if it is a limb. A pressure point is a place where a main artery lies near the skin surface and over a bone. Pressure there compresses the artery against the bone and

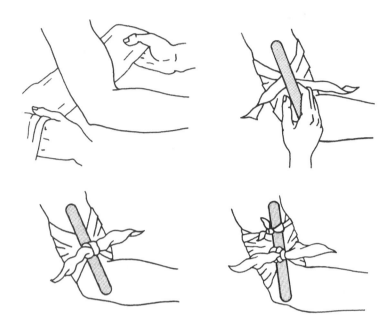

Figure 37-4. Applying a tourniquet: tie bandage with overhand knot, tie square knot over stick, twist to tighten the tourniquet.

shuts off the flow of blood to the wound. Remember to apply pressure at the point nearest the wound and between the wound and the heart.

It is very tiring to apply finger pressure, and it can seldom be maintained for more than fifteen minutes. As soon as possible, use a compress held securely over the wound by a bandage.

Tourniquet

If the bleeding is still severe, you may have to apply a tourniquet. Remember, this may result in gangrene when left on too long.

A tourniquet consists of a pad, a band, and a device for tightening the band so that the blood vessels will be compressed. The type found in many first-aid kits consists of a web band two inches wide by five feet long, with a buckle for fastening. Wrap it once about the limb, then run the free end through the slit in the felt pad and through the buckle. Draw the band tight enough to stop the flow of blood.

An improvised tourniquet should be a band approximately two inches wide. Do not use rope, wire, string, or very narrow pieces of cloth; they will cut into the flesh. A short stick may be used to twist the band and tighten the tourniquet.

The band should be placed two inches above the wound. A tourniquet is never used unless bleeding cannot be controlled any other way. Once a tourniquet has been applied, only medical personnel should release it. By the time the tourniquet is put on, the victim has already lost a considerable amount of blood. The additional loss resulting from loosening the tourniquet may easily cause death. If a tourniquet is used, mark the patient's forehead with a "T" and the time. Improper use may produce gangrene. Do not use a tourniquet except as a last resort.

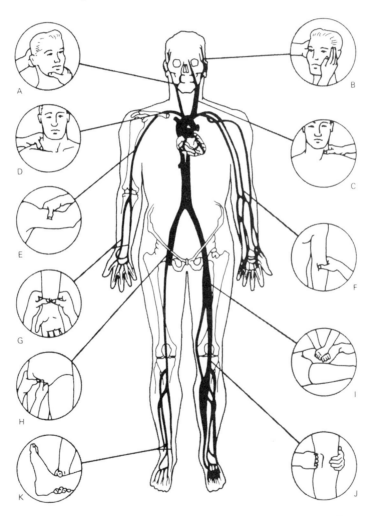

Figure 37-5. Control of bleeding by application of pressure at points where artery lies near the skin and over a bone.

Bleeding Internally

If a patient is bleeding internally, he or she may be thirsty, restless, fearful, and in shock. Nothing should be given by mouth, even if the victim is thirsty, for if anesthesia has to be given, the patient might vomit and aspirate material into the lungs.

Bleeding in the stomach may be indicated by a wound in that area and bloody vomiting (often resembling coffee grounds) or bloody stools. The treatment to be administered for internal bleeding is the same as for shock. Immediate medical care should be sought since internal bleeding may be rapidly fatal or may cause death by slower bleeding over a longer period of time.

Blisters

The skin covering a blister is better protection than any bandage and should therefore be left alone. If the blister is where it might be broken, open it with a sharp knife or needle that has been heated in flame. Press out fluid with a bit of sterile gauze and apply a sterile bandage.

Burns and Scalds

Burns and scalds are caused by exposure to intense heat (fire, bomb flash, sunlight, hot metal solids, hot gases, or hot liquids). Contact with electric current may cause severe burns.

Burns are classified according to the depth of injury to the tissues. A burn that reddens the skin is called a *first-degree* burn. A burn that raises a blister is a *second-degree* burn. A burn that destroys the skin and actually chars or cooks the tissues is a *third-degree* burn. In some third-degree burns the destruction extends down to muscles and bones.

The size of the burned area may be far more important than the depth of the burn. Third-degree burns frequently appear as blanched areas surrounded by reddened or blistered skin. A first-degree burn that covers a very large area of the body may be more serious than a small third-degree burn.

The main dangers from burns are shock and infection. First-aid treatment must be directed toward relieving pain, combating shock, and preventing infection.

Minor Burns

Cool the burns immediately. Cool-water treatment has proved both comforting and effective. Submerge the burned part in cool water, but do not freeze. Continue until no pain is felt when the burned part is withdrawn from the water. Be careful not to overcool. If the victim begins to shiver, stop cooling and dress the burns. Do not apply ointments or other medicines to the burned area. Cover the burn with a sterile bandage to prevent infection. The pain will be greatly lessened if the bandage is airtight and fairly firm. Commercially available clear plastic food wrap makes an excellent airtight burn dressing; just apply the wrap and cover with a bandage.

Serious Burns

Immediate care is to cool the burns. Tissues continue to burn after removed from the heat source. Keep the victim's head slightly lower than the feet, and see that the victim is warm enough. Do not remove clothing immediately. Cover the victim with a blanket. Do not overheat. Remember that exposure to cold will increase shock. Seriously burned victims should be given nothing by mouth except on direct order of a physician or medical officer.

Chemical Burns

Chemicals in contact with the skin or other body membranes may cause chemical burns by direct chemical destruction of the body tissue. This kind of injury can be caused by acids (nitric acid, sulfuric acid, hydrochloric acid, etc.) and caustic alkalis. Here are some guidelines for treating chemical burns:

- Wash off the chemical immediately with large amounts of clean, fresh, cool water. If it is not possible to put the victim under running water, immerse the affected areas in water, or pour great quantities of water over the burns. If the chemical is a dry powder, do not apply water but remove the agent by "dusting" and removing the clothes.

- Dry gently with sterile gauze. Do not break the skin or open any blisters.

- Treat for shock. See also "Eye Injuries."

Information on chemicals can be obtained through the *Chemical Hazards Response Information System* (COMDTINST M16465.12C series).

Choking

Choking occurs when an object obstructs the airway. The obstruction may even be the tongue in an unconscious victim. This is relieved by the airway-opening methods explained in the CPR section. Most commonly choking is caused by a piece of food in adults or small toys or food in children. The universal choking sign is one or both hands clutching the throat. If the obstruction is not removed, the eventual outcome is death; therefore, *first-aid action must be immediate.*

Ask the victim, "Are you choking?" If the victim can speak and is getting enough air, encourage him or her to relax and try to cough the object up. Do not attempt to remove the object yourself. Get help if the obstruction is not cleared in a short time.

In complete obstruction, or in partial obstruction where the victim is not getting enough air, back blows and abdominal thrusts should be performed. The choking sequence follows:

- Call for assistance.
- Stand behind the victim.
- Tilt the victim slightly forward and give five back blows in an upward motion between the shoulder blades, then wrap your arms around his or her waist.
- Make a fist and place the thumb side of your fist just above the navel but below the lower margin of the ribs.
- Grasp your fist with your other hand and press into the victim's abdomen in an inward, upward motion five times.
- Repeat until the object is coughed up or until the victim goes unconscious.

For unconscious victims, perform the airway and breathing sequence as in CPR. If you are unable to get air into the lungs, reposition the head and try again. If there is still no air exchange, chest compressions are needed.

- Place the heels of both hands on the center of the sternum as in CPR.
- Give thirty chest compressions.
- Open the mouth and with the index finger attempt to remove the object.
- Reposition the head and attempt artificial respiration.

If no air exchange, repeat above steps until obstruction is cleared or until you are relieved by another rescuer.

Electrocution

A person knocked unconscious by electrical current may need artificial respiration or CPR at once. If the victim is still in contact with any

First Aid For Choking

1

- **ASK: Are you choking?**

- If victim cannot breathe, cough, or speak...

2

- **Give the Heimlich Maneuver.**

- Stand behind the victim.

- Wrap your arms around the victim's waist.

- Make a fist with one hand. PLACE your FIST (thumbside) against the victim's stomach in the midline just ABOVE THE NAVEL AND WELL BELOW THE RIB MARGIN.

- Grasp your fist with your other hand.

- PRESS INTO STOMACH WITH A QUICK UPWARD THRUST.

3

- **Repeat thrust if necessary.**

4

- **If a victim has become unconscious:**

  - Sweep the mouth.

American Red Cross

5

- Attempt rescue breathing.

6

- Give 6-10 abdominal thrusts.
- Repeat Steps 4, 5, and 6 as necessary.

Figure 37-6. First aid for choking. *Courtesy of American Red Cross*

object that might carry current, cut the contact. Pull the switch if you know where it is, but do not waste time looking for it. Pull or push the victim away from contact with a dry rope, piece of clothing, pole, oar, or board.

Rubber gloves are good if you have them, but they will not protect against very high voltage. Most work gloves will be damp and sweaty; do not use them. Do not touch the victim until he or she is clear of contact. Commence artificial respiration or CPR. Send for medical assistance.

Exposure to Extreme Cold

This can result in loss of body heat, frostbite, and immersion foot. In the treatment of all injuries from exposure to extreme cold, it is essential to get the victim into a warm place as soon as possible. It is almost impossible to give effective first-aid treatment while the victim is still exposed to the cold. Remember, however, that there may be other life threats that must be treated before the victim can be moved into a warmer place.

General Loss of Body Heat (Hypothermia)

This abnormally low body temperature usually results from total immersion in cold water or exposure to cold temperatures without adequate clothing and can result in loss of consciousness. The victim will appear pale and unconscious and may be taken for dead. Breathing is slow and shallow and the pulse faint or unobtainable. Body tissues feel semirigid, and the arms and legs may be stiff.

First-aid treatment consists mainly of bringing the body temperature to normal. The victim should be wrapped in warm blankets in a warm room. Take care to warm the body and not the arms and legs as cold blood from the extremities can cause cardiac arrest if circulated to the body too fast. If the victim has been immersed in cold water, remove the wet clothes. Be alert for further complications such as respiratory or cardiac arrest. It is very important to handle the victim very gently and move the victim only if necessary. Get the victim to a medical officer as soon as possible.

Frostbite

Exposure to dry cold causes frostbite, especially in the cheeks, nose, chin, ears, forehead, wrist, hands, and feet. The skin turns white or gray, then bright pink.

Frostbite may also be caused by contact with certain chemicals that have rapid freezing action, such as industrial gases. Injuries caused by these substances are often referred to as chemical burns, but the body tissue is actually frozen rather than burned.

When the frostbitten area is warmed up, it immediately becomes red and swollen, and large blisters develop. Severe frostbite causes gangrene, in which body tissues and sometimes even bone are permanently destroyed. If deep tissue has been destroyed, the injured part may have to be amputated.

The first-aid treatment for frostbite is rewarming the frozen tissue, but only if there is no danger of the part refreezing. Get the victim into the warmest available place as soon as possible. Undress the victim. If feet or legs are frostbitten, do not allow the victim to walk. Do not handle the frostbitten area unnecessarily, and do not exert any pressure against it.

Place the victim in bed and covered. Do not let anything touch the frostbitten parts. Keep sterile gauze or cloth pads between the toes and fingers, and keep frostbitten parts slightly elevated. Obtain medical assistance as soon as possible. *Caution: Never rub or massage frostbite with ice or snow. Do not apply cold water. Do not expose a frostbitten area to cold air.*

Immersion Foot (Trench Foot)

Immersion foot results from prolonged exposure to a combination of moisture and cold. People on life rafts or in unprotected lifeboats are most likely to get immersion foot from exposure to near-freezing seawater, but cases have occurred as a result of longer immersion in warmer water. This condition sometimes affects other parts of the body, such as knees, hands, and buttocks.

A person who remains in a cold, wet place, standing or crouching in one position for a long time, is also very likely to develop this condition. Immersion foot causes a feeling of heaviness or numbness. All sensitivity may be lost. The affected areas become swollen. The skin is first red, then waxy white, then yellowish, and finally a mottled blue or black. The injured parts remain cold, swollen, discolored, and numb; or they thaw, the swelling increases, and the skin becomes hot, dry, red, and blistered. In severe cases, there may be gangrene.

The following first-aid treatment should be given while medical aid is being summoned:

- Get the victim off his or her feet as soon as possible; keep the victim as warm as possible.
- Expose the injured part to warm, dry air.

- If the skin is not broken or loose, the injured part may be left exposed; however, if it is necessary to move the victim, cover the injured part with loosely wrapped fluff bandages of sterile gauze.
- Do not apply salves or ointments. Be careful not to rupture blisters.

Eye Injuries

For wounds of the eye, apply thick, dry sterile bandages. If the eyeball is injured, do not allow any pressure of the bandage against the eyeball. If soft tissue around the eye is injured, apply a pressure bandage. Keep the victim lying down while being transported and until medical aid is available.

Objects

To remove an object in the eye, such as a floating speck of dirt, have the victim look upward, and remove the object with the corner of a clean handkerchief or with a cotton swab moistened with clean water. Either gently pull the lower lid down with your finger or turn the upper lid back over a match stick or cotton swab applicator by grasping the lashes of the upper lid between the fingers and averting it. If these methods fail, cover the eye by bandaging and obtain medical help. *Never* attempt to remove an object that is *imbedded* in the eye.

Chemical Burns

Flush the eye immediately with large quantities of fresh, clean water. Hold the victim's head over a drinking fountain so that the water flows from the inside corner of the eye toward the outside corner; or have the victim lie down with the head turned slightly to one side, then pour water into the inside corner of the eye and let it flow gently across the eyeball to the out-side corner. If the victim is unable to open his or her eyes, hold the eyelids apart so that the water can flow across the eyeball. Do not use anything except water. Another way to wash chemical substances from the eye is to have the victim open and close the eyes several times while the face is immersed in a pan of fresh water.

Cover the eye with a small, thick compress; fasten the compress in place with a bandage or an eye shield, and get medical care as soon as possible. Refer to the *Chemical Hazards Response Information System*, if the manual is available.

Fainting

Fainting is a reaction of the nervous system that slows down the flow of blood to the brain. To prevent fainting, a person should lie flat for ten minutes with the head lower than the rest of the body. If a person who appears to have fainted does not recover consciousness almost immediately, that is not simple fainting. Unconsciousness may be caused by asphyxia, deep shock, poisoning, head injury, heat stroke, heart attack, stroke, or epilepsy. Get medical help at once.

Fish Hooks

If the barb of a fish hook is buried in the flesh, bandage the hook so that the flesh is protected from further injury and seek medical help.

Fish Stings

Many sea creatures (Portuguese men-of-war, jellyfish, stingrays, scorpion fish, toadfish, etc.) are poisonous. The symptoms of their stings range from burning, stinging, reddening of the skin, hives, pus sores, abdominal cramps, numbness, dizziness, pain in the groin and armpits, nausea, muscular pain, breathing difficulty, and constriction of the chest, to prostration and shock.

For Portuguese men-of-war and jellyfish stings, remove the tentacles immediately and wash the skin with rubbing alcohol. Apply ammonia or a soothing lotion. If available, meat tenderizer is an excellent remedy. Make a paste with it and apply to the sting area. Treat for shock. Care must be taken that you do not come in contact with the tentacles. Use a stick or some other means to remove them.

A stingray wound should be washed immediately with cold salt water. Much of the toxin will wash out, and the cold water causes blood vessels to constrict, slowing circulation and acting as a mild painkilling agent. *Do not* rinse with *fresh* water. The wound should then be immersed in hot sea water for thirty minutes to an hour, with the temperature as high as the victim can stand without injury. (Hot compresses can be applied for wounds in areas that cannot be immersed.) A sterile dressing should then be applied.

Fractures

A closed fracture is one in which a bone is broken but there is no break in the skin. An open fracture is one in which the skin is broken. All fractures require careful handling; send for medical aid as soon as possible. Do not move a fracture victim until the fracture has been splinted, unless necessary to save life or prevent further injury. Do not attempt to "set" a broken bone. Treat for shock. Stop the bleeding in an open fracture. Use direct pressure or the pressure point system.

Apply splints over clothing if the victim is to be moved a short distance or if a doctor will see the victim soon. Otherwise, apply well-padded splints after the clothing has been cut away. Be careful when you handle any fracture to avoid additional shock or injury. Splints vary according to the area and nature of the fracture. Follow these rules:

- For a bone fracture, place the splint(s) so that the joint above and the joint below the fracture cannot move.

- For a joint fracture, place the splint so that the bone above and the bone below the fracture are immobilized.

Collarbone

On the injured side, place the forearm across the chest, palm turned in, thumb up, with the hand four inches above the elbow, and support the arm in this position with a sling. Fasten the arm to the body with several turns of bandage around the body and down over the hand to keep the arm close against the body.

Skull

It is not necessary to determine whether or not the skull is fractured when a person has had a head injury. The primary aim is to prevent brain damage. Do not let the victim move. Try not to move the victim more than necessary. Do not let the victim get cold, and do not give anything to drink. Stop any bleeding and get immediate medical assistance. Do not plug the ears and nose to stop bleeding or leaking of fluid from these areas.

Spine

Pain, shock, and partial paralysis result from damage to the spine. Severe pain in the back or neck after injury should be treated as a fractured spine. Treat for shock. Keep the victim flat, and do not move the head.

If the victim cannot move his or her legs, feet, or toes, the fracture is probably in the back; if the victim cannot move his fingers, it is probably in the neck. If transportation is necessary, carry the victim faceup on a backboard, rigid stretcher, door, or wide frame. Never try to lift such a victim with fewer than four people; pick up the victim by the clothing and slide the stretcher underneath.

Pelvis

Treat a victim with a pelvis injury for shock but do not move unless absolutely necessary. If a victim must be moved, handle the same as for a victim of a fractured spine. Bandage legs together at ankles and knees; place a pillow at each hip and fasten them in place. Fasten the victim securely to the stretcher.

Heart Attack

Heart disease is the number one killer in the United States today. Heart attacks are caused by a lack of blood supply to the heart muscle. Symptoms of a heart attack include chest pain that may radiate to the arm, jaw, or neck and cause shortness of breath, nausea, sweating, dizziness, anxiety, and weakness. Fortunately, new treatments are reducing the number of deaths due to heart attack. These new treatments must be started early, making recognition and prompt transport of victims of prime importance.

Treatment includes making the victim comfortable by loosening clothing. Keep the victim calm and quiet; do not allow a heart attack victim to walk around. Give the victim oxygen if you are trained in its use. Get the victim to a medical officer as soon as possible. Monitor the pulse and breathing—be ready to do CPR if necessary.

Heatstroke

Heatstroke and heat exhaustion are caused by excessive exposure to heat, direct rays of the sun, or heat of machinery spaces, foundries, bakeries, and so forth. Under similar circumstances, one person may develop *heatstroke* and another *heat exhaustion*. There are important differences between the two afflictions. Each represents a different bodily reaction to excessive heat, and for this reason the symptoms and treatment are different.

Heatstroke results from a failure of the heat-regulating mechanism of the body. The body becomes overheated and the temperature rises to between 105 and 110 degrees Fahrenheit, but there is no sweating or cooling of the body. The victim's skin is hot, dry, and red. The victim may have preliminary symptoms such as headache, nausea, dizziness, or weakness, but very often the first signs are sudden collapse and loss of consciousness. Breathing is likely to be deep and rapid. The pulse is strong and fast. Convulsions may occur. Heatstroke may cause death or permanent disability; at best, recovery is likely to be slow and complicated by relapses.

The longer the victim remains overheated, the greater the danger of death. Follow these first aid measures to lower body temperature:

- Move victim to a cool place, remove the clothing, place victim on back with head and shoulders slightly raised.

- Sponge or spray the body with cold water, and fan victim so the water will evaporate rapidly.

- Get victim to a medical facility as soon as possible. Keep victim cold while being transported.

- Do not give the victim anything to drink unless directed to do so by a doctor.

Heat Exhaustion and Heat Cramps

In heat exhaustion, there is no failure of the heat-regulating mechanism, but there is a serious disturbance of the blood flow, similar to the circulatory disturbance of shock. Through prolonged sweating, the body loses large quantities of salt and water.

Heat exhaustion may begin with a headache, dizziness, nausea, weakness, and profuse sweating. The victim may collapse and lose consciousness but can usually be aroused rather easily. The temperature is usually normal or even below normal; sometimes it may go down as low as 97 degrees Fahrenheit. The pupils of the eyes are usually dilated, the pulse is weak and rapid, and the skin is pale, cool, and sweaty. Sometimes there may be severe cramps in the abdomen, legs, and arms. Follow these first aid measures:

- Move the victim to a cool place but not to where victim will be exposed to strong drafts or become chilled.

- Loosen clothing and make the victim as comfortable as possible.

- Keep the victim quiet and lying down, with the feet and legs somewhat elevated. Be sure that the victim is kept comfortably warm. You may have to cover the victim with blankets, even if the air is quite warm.

- The pain of severe cramps may be relieved by the application of heat to the affected muscles or by stretching the affected muscle group.

- If the victim is conscious and able to swallow, give water to drink. If available, a commercially prepared electrolyte beverage is helpful. Replacement of the salt and water that has been lost by sweating is probably the most important part of the treatment and often brings rapid recovery.

- If recovery is not prompt, get medical attention as soon as possible.

Insect Bites

Wash insect bites with soap and water. Itching can be relieved by using cold-water compresses. Ice, applied to bee stings, will prevent pain and swelling. Some people experience severe allergic reactions to bee and other insect stings. The person stung should be carefully observed and given immediate medical attention if allergic reactions are either observed or are part of that person's medical history.

Ticks can be dislodged by applying mineral oil to the tick; be careful not to crush the insect or leave its jaws embedded in the flesh. Certain types of ticks carry Rocky Mountain spotted fever or Lyme disease, which can be fatal. Inflamed bites require medical assistance.

Poisoning

First-aid treatment for poisoning depends partly upon whether the poison enters the body by *ingestion* (taking the poison into the stomach by way of the mouth), by *inhalation* (breathing poisonous gases), by *skin contact*, or by *injection*. (Poisoning by injection is covered under "Snakebite" and "Spiders and Scorpions.")

Symptoms

Intense pain frequently follows poisoning; nausea and vomiting may occur. The victim can become delirious or may collapse and become unconscious, and almost always is likely to have trouble breathing. Some poisons characteristically cause paralysis; others cause convulsions. Shock is always present in acute poisoning.

Emergency treatment is usually required because poisons act very rapidly. Treatment by a physician should be obtained as soon as possible.

The first treatment for ingested poisoning is to remove any remaining poisonous material and then contact the Poison Help line at 1-800-222-1222. The Poison Help line offers free, confidential medical advice twenty-four hours a day, seven days a week. You may be directed to induce vomiting in the victim. Vomiting should not be induced if the victim has ingested a strong acid or alkali or if the victim is unconscious. Vomiting may be brought on by tickling the throat, or by administering a nauseating fluid such as salt dissolved in water.

Activated charcoal can absorb poisons in the body. It is given as a mixture of two to four tablespoons dissolved in a glass of water after the victim stops vomiting. Always save empty containers of the offending agent and send to the hospital with the victim.

Inhaled poisons can come from refrigeration machinery, firefighting equipment, paints and solvents, photographic materials, and many other types of shipboard equipment that contain volatile and sometimes poisonous chemicals. Fuel oil and gasoline vapors constitute special hazards.

Carbon monoxide is present in all exhaust gases of internal combustion engines. First-aid treatment for carbon monoxide, and for all gases, follows:

- Get the victim out of the toxic atmosphere into a well-ventilated space.
- Remove contaminated clothing.
- Watch the breathing. Give artificial respiration if necessary. Give oxygen if it is available and you know how to use it.
- Keep the victim lying down and quiet. Treat for shock. Call a medical officer as soon as possible.

Poisoning by skin contact is not usually a first-aid problem. Such poisoning can be fatal; some types build up over a long period of time. There is no real cure. Know the substances that can poison by contact and be careful in handling them. They include gasoline, benzene, naphtha, lead compounds, mercury, arsenic, carbon tetrachloride, and TNT.

Shock

Shock is a term used to describe events that take place when one of the components of the circulatory system fails. It can be *pump failure* (the heart), *pipe failure* (blood vessels), or *fluid failure* (the blood). The result is a lack of blood supply to the brain and vital organs, which can lead to death. Shock can be caused either by injury to the nervous system, causing dilated blood vessels; injury to the heart, causing loss of pump function; or most commonly because of severe bleeding (not enough blood to go around). In an attempt to compensate for the lack of blood supply to the vital organs, the heart rate increases. The pulse, however, is weak and the blood flow to the brain is reduced. Unconsciousness or confusion may result.

Most injuries will cause shock, which may be slight (lasting only a few seconds) or serious enough to be fatal. Shock may begin immediately, or it may be delayed as much as several hours.

Shock is recognized by the following symptoms: The pulse is weak and rapid; breathing may be shallow, rapid, and irregular; the skin feels cold to the touch and may be covered with sweat; the pupils of the eyes are usually dilated. A shock victim may complain of thirst and may feel weak, faint or dizzy, and nauseated. The victim may be very restless and feel frightened and anxious.

As shock deepens, these signs gradually disappear and the victim becomes less and less responsive, even to pain. Shock victims may insist they feel fine and then pass out.

Keep the victim flat on the back with feet higher than the body. Keep the victim warm by using blankets. Do not use artificial heat. Do not give anything by mouth. A shock victim is often thirsty; moisten lips with cool water but do not give anything to drink.

Snakebite

Poisonous snakes in the United States include "pit vipers" (rattlesnakes, copperheads, water moccasins) and coral snakes. Identifying the snake, in some cases, is not easy. Check these points:

- Fangs mean the snake is poisonous.
- Rattles identify rattlesnakes but not always; sometimes a rattler loses the rattle.
- Snakes with a sensory organ, a "pit," between nostrils and eyes, are pit vipers and are poisonous.
- Snakes striped fore-and-aft are not poisonous, except for one very rare native of the Far East.

Pit viper bites may produce the following symptoms:

- Tissue swelling at the bite gradually spreading to surrounding areas
- Swelling occurring between three to five minutes following the bite; may continue for as long as an hour and be so severe as to burst the skin
- Severe pain
- Escape of blood from the capillaries; accumulation of blood in the tissues may cause severe pain
- Severe headache, thirst
- Bleeding from internal organs into the intestines, blood in urine or stools

Bites of coral snakes, cobras, or kraits may produce some of the following symptoms:

- Irregular heartbeat, generalized weakness and exhaustion, shock
- Severe headache, dizziness, mental disturbances such as incoherent speech, stupor, mental confusion, and possible unconsciousness; extreme pain is not characteristic
- Muscular incoordination, such as inability to reach for and pick up an object or to move from place to place; sometimes muscle spasms and twitching
- Difficult or labored breathing and even respiratory paralysis
- Numbness and tingling of the skin, particularly of the lips and soles of the feet, excessive sweating
- Chills, fever
- Nausea, vomiting, diarrhea

A snakebite victim should lie down and keep calm. Remind the victim that very few people are ever disabled by snakebite. Keep the victim at rest (dry, warm, and quiet), and immobilize the affected part below the level of the heart.

For a bite on the arm or leg, place a restricting band two to four inches above the bite. The restricting band should be tight enough to stop the flow of blood in the veins but not the arteries.

Treat for shock. It may be necessary to give artificial respiration. Victims have been kept alive for two hours after becoming unconscious and have been saved by delayed injection of antivenin.

If you are alone and have been bitten by a snake, keep your emotions under control. If you can kill the snake and look for fangs, do so. Apply a restricting band two to four inches above the site of the bite, remain as quiet as possible and keep the wound lower than your heart. If pain is great, and if there is swelling, numbness, tingling, or continued oozing of blood at the bite site, you can be fairly sure that the snake is poisonous. If it is evident that no help is coming, walk slowly with a minimum of exertion.

Spiders and Scorpions

Brown recluse and black widow spiders are probably the most common of the poisonous spiders. The black widow spider bite causes pain almost at once, and the pain spreads quickly from the region of the bite to the muscles of the back, shoulders, chest, abdomen, and limbs. The pain is usually accompanied by severe spasms of the abdominal muscles.

A bite from any other poisonous spider may be felt as a sharp sting, but it is not usually accompanied by severe pain. About half an hour after the bite, the victim begins to feel painful muscle cramps near the bite, which soon spread to other muscles of the body. Shock usually develops at about the same time, though it may be delayed for as much as an hour or two. The victim becomes restless and anxious and may be very thirsty. Nausea and vomiting may occur.

Scorpions have a stinger and venom sac in the tail. Their sting causes immediate, intense pain. The area near the sting will become numb and paralyzed.

First-aid measures for spider bites and scorpion stings are limited. Clean the wound and surrounding area. Treat the victim for shock: lying down, quiet and warm. Severe muscle cramps and pain may sometimes be relieved by warm-water baths. Do not apply suction and do not incise (cut the wound in an attempt to remove the poison); it does no good. Always seek medical care.

Splinters

If a splinter is under the skin where it can be seen, it can be removed by pulling it out the same way it went in. Clean the broken skin and cover with a sterile dressing.

Transportation of the Injured

Take these precautions before transporting an injured person:

- Give necessary first aid.
- Locate all injuries to the best of your ability.
- Treat airway, breathing, and circulation problems, shock, fractures, sprains, and dislocations.
- Make the victim as comfortable as possible.

Use a regular stretcher. If you must use an improvised stretcher, be sure that it is strong enough to hold the victim. Have enough people to carry the stretcher; do not drop the victim. Fasten the victim in the stretcher so the victim cannot slip, slide, or fall off. Tie the feet together unless injuries make this impractical. Use blankets, garments, or other material to pad the stretcher and to protect the victim from exposure.

Injured persons should usually lie on their backs while being moved, but those having difficulty in breathing because of a chest wound may be more comfortable if their head and shoulders are slightly raised. Fracture cases should be moved very carefully so that the injury will not be made worse. A victim with a severe injury to the back of the head should be kept on one side. A patient should always be carried feet first unless there is some special reason for carrying him or her head first.

A *Stokes litter* is a wire basket supported by iron or aluminum rods. It is adaptable to a variety of uses and will hold a person securely in place even if it is tipped or turned. The Stokes litter is generally used on board ship, especially for transferring injured persons to and from boats. It can be used to rescue people from the water and for direct ship-to-ship transfer of injured persons.

The litter should be padded with two blankets placed lengthwise so that one will be under each of the victim's legs, and a third folded in half and placed in the upper part of the litter to protect the victim's head and shoulders. Victims should be lowered gently into the litter and made as comfortable as possible. The feet must be fastened to the end of the litter to prevent sliding up and down, and the victim must be fastened into the litter by straps over the chest, hips, and knees. The straps should go over the blanket or other covering.

Abbreviations

1MC: intercommunication voice circuit
AB: able-bodied seaman
ACE: American Council on Education
AET: avionics electrical technician
AFFF: aqueous film-forming foam
AIDS: acquired immune deficiency syndrome
AMT: aviation maintenance technician
AMVER: automated mutual assistance vessel rescue
APC: aqueous potassium carbonate
AQE: Advancement Qualification Exam
AST: aviation survival technician
ATON: aids to navigation
ATTC: Aviation Technical Training Center
AWOL: absent without leave
BI: background investigation
BM: boatswain's mate
BMOW: boatswain's mate of the watch
BW: biological warfare
C4ISR: command, control, communications, computers, intelligence, surveillance, and reconnaissance
CBRNE: chemical, biological, radiological, nuclear, and high-yield explosive (weapons or warfare)
CGA: Coast Guard Academy
CGI: Coast Guard Institute
CGMA: Coast Guard Mutual Assistance
CIC: Combat Information Center
CIWS: close-in weapon system
CLEP: College Level Examination Program
CMAA: chief master-at-arms
CMC: command master chief
CO: commanding officer
COGAP: Coast Guard Art Program
CPO: chief petty officer
CPR: cardiopulmonary resuscitation
CS: culinary specialist
CW: chemical warfare
CW: radiotelegraph (continuous wave)

DANTES: Defense Activity for Non-Traditional Education Support
DC: damage controlman
DCA: damage-control assistant
DR: dead reckoning
DV: diver
EM: electrician's mate
EMCON: emission control
EOOW: engineering officer of the watch
EPM: Enlisted Personnel Management
EPME: enlisted professional military education
EPO: engineering petty officer
EPQ: enlisted performance qualification
ESO: educational services officer
ET: electronics technician
FOUO: for official use only
FRC: fast response cutter
GE: general emergency
GM: gunner's mate
GMT: Greenwich Mean Time
GPS: global positioning system
GQ: general quarters
HIV: human immunodeficiency virus
HS: health services technician
IC: Intelligence Community
IS: intelligence specialist
IT: information system technician
LANTAREA: Coast Guard Atlantic Area
LASH: lighter aboard ship
LCVP: landing craft, vehicle, personnel
LES: leave and earnings statement
LNG/LPG: liquefied natural gas / liquefied petroleum gas
LOP: line of position
LORAN: long-range navigation
LRI: long-range interceptor
LRI-II: long-range interceptor-II
MAA: master-at-arms
MC: intercommunication voice unit ("squawk box")
MCH: multimission cutter helicopter
MCPO: master chief petty officer
ME: maritime enforcement specialist
MILSTRIP: military standard requisitioning and issuing procedures
MK: machinery technician
MLB: motor lifeboat

MMA: major maintenance availability (rehabilitation program)
MN: magnetic pole
MODU: mobile offshore drilling unit
MSDS: material safety data sheet
MSO: marine safety office
MST: marine science technician
MTS: Marine Transportation System
MU: musician
NAC: National Agency Check
NHSC: National Home Study Council
NJP: nonjudicial punishment
NP: North Pole
NSC: national security cutter
OBO: ore/bulk/oil (carrier)
OCS: Officer Candidate School
OOD: officer of the deck/officer of the day
OS: operations specialist
OSV: offshore supply vessel
OTH-IV: over the horizon-IV
PA: public affairs specialist
PACAREA: Coast Guard Pacific Area
PAL: prisoner(s) at large
PDE: Personal Data Extract
PDR: Personnel Data Record
PFD: personal flotation device
PKP: Purple-K Powder
PML: personnel marker light
PO: petty officer
POW: prisoner of war
PPE: personal protective equipment
PPEP: Pre-Commissioning Program for Enlisted Personnel
PQS: personal qualification standards
Prosign: procedure sign
Proword: procedure word
PSU: port security unit
PWCS: ports, waterways, and coastal security
RAD: roentgen
RB-HS: response boat, homeland security
RB-M: response boat, medium
RB-S: response boat, small
RO/RO: roll-on/roll-off (vessel)
RT: radiotelephone
RW: radiological warfare

SA: seaman apprentice
SAR: search and rescue
SCBA: self-contained breathing apparatus
SCPO: senior chief petty officer
SCRA: Servicemembers Civil Relief Act
SK: storekeeper
SN: seaman
SOC: Servicemembers Opportunity Colleges
SOPA: senior officer present afloat
SPAR: Semper Paratus, Always Ready (acronym used for members of
the Coast Guard Women's Reserve during World War II)
SPC-AIR: special-purpose craft, airboat
SPC-LE: special-purpose craft, law enforcement
SPO: Servicing Personnel Office
SR: seaman recruit
STD: sexually transmitted disease
SWE: service-wide examination
TIG: time in grade
TIS: time in service
TPSB: transferable port security boat
TSP: Thrift Savings Plan
UCMJ: Uniform Code of Military Justice
ULCC: ultra large crude carrier
USCG: United States Coast Guard
UTC: Universal Coordinated Time
VA: Veterans Administration
VLCC: very large crude carrier
VTS: vessel traffic service
WAGB: icebreaker
WHEC: high endurance cutter
WIX: Coast Guard barque *Eagle*
WLB: seagoing buoy tender
WLBB: Great Lakes icebreaker
WLM: coastal buoy tender
WMEC: medium endurance cutter
WMSL: national security cutter
WPB: patrol boat
WQSB: Watch, Quarter, and Station Bill
WT: watertight
WTGB: icebreaking tugs
XO: executive officer
YN: yeoman

APPENDIX A

Code of Conduct

The 12 January 2016 capture of ten U.S. Navy sailors by Iran served as an unwanted reminder for everyone wearing the uniform to review their obligations under the Code of Conduct. The Code of Conduct is an ethical guide. Its six articles deal with your chief concerns as a prisoner of war (POW); these concerns become critical when you must evade capture, resist while a prisoner, or escape from the enemy.

ARTICLE I

I am an American, fighting in the forces which guard my country and our way of life. I am prepared to give my life in their defense.

All men and women in the armed forces have the duty, at all times and under all circumstances, to oppose the enemies of the United States and support its national interests. In training or in combat, alone or with others, while evading capture or enduring captivity, this duty belongs to each American defending our nation regardless of circumstances.

ARTICLE II

I will never surrender of my own free will. If in command, I will never surrender the members of my command while they still have the means to resist.

As an individual, a member of the armed forces may never voluntarily surrender. When isolated and no longer able to inflict casualties on the enemy, members of the armed forces have an obligation to evade capture and rejoin friendly forces.

Only when evasion by an individual is impossible and further fighting would lead only to death with no significant loss to the enemy should one consider surrender. With all reasonable means of resistance exhausted and with certain death the only alternative, capture does not imply dishonor.

The responsibility and authority of a commander never extends to the surrender of a command to the enemy while the command has the power to fight and evade. When isolated, cut off, or surrounded, a unit must continue to fight until relieved or able to rejoin friendly forces through continued efforts to break out or evade the enemy.

ARTICLE III

If I am captured I will continue to resist by all means available. I will make every effort to escape and aid others to escape. I will accept neither parole nor special favors from the enemy.

The duty of a member of the armed forces to use all means available to resist the enemy is not lessened by the misfortune of captivity. A POW is still legally bound by the Uniform Code of Military Justice and ethically guided by the Code of Conduct. Under provisions of the Geneva Convention, a POW is also subject to certain rules imposed by the captor nation. When repatriated, a POW will not be condemned for having obeyed reasonable captor rules, such as sanitation regulations. The duty of a member of the armed forces to continue to resist does not mean a prisoner should engage in unreasonable harassment as a form of resistance. Retaliation by captors to the detriment of that prisoner and other prisoners is frequently the primary result of such harassment.

The Geneva Convention recognizes that a POW may have a duty to attempt escape. In fact, the Geneva Convention prohibits a captor nation from executing a POW simply for attempting to escape. Under the authority of the senior official (often called the senior-ranking officer), a POW must be prepared to escape whenever the opportunity presents itself. In a POW compound, the senior POW must consider the welfare of those remaining behind after an escape. However, as a matter of conscious determination, a POW must plan to escape, try to escape, and assist others to escape.

Contrary to the spirit of the Geneva Convention, enemies engaged by U.S. forces since 1950 have regarded the POW compound as an extension of the battlefield. In so doing, they have used a variety of tactics and pressures, including physical and mental mistreatment, torture, and medical neglect to exploit POWs for propaganda purposes, to obtain military information, or to undermine POW organization, communication, and resistance.

Such enemies have attempted to lure American POWs into accepting special favors or privileges in exchange for statements, acts, or information. Unless it is essential to the life or welfare of that person or another prisoner of war or to the success of efforts to resist or escape, a POW must neither seek nor accept special favors or privileges.

One such privilege is called *parole*. Parole is a promise by a POW to a captor to fulfill certain conditions—such as agreeing not to escape or to fight again once released—in return for such favors as relief from physical bondage, improved food and living conditions, or repatriation ahead of the sick, injured, or longer-held prisoners. An American POW will never sign or otherwise accept parole.

ARTICLE IV

If I become a prisoner of war, I will keep faith with my fellow prisoners. I will give no information nor take part in any action which might be harmful to my comrades. If I am senior, I will take command. If not, I will obey the lawful orders of those appointed over me and will back them up in every way.

Informing or any other action to the detriment of a fellow prisoner is despicable and is expressly forbidden. POWs must avoid helping the enemy identify fellow prisoners who may have knowledge of particular value to the enemy and who may, therefore, be made to suffer coercive interrogation.

Strong leadership and communication are essential to discipline. Discipline is the key to camp organization, resistance, and even survival. Personal hygiene, camp sanitation, and care of the sick and wounded are imperative. Officers and noncommissioned officers of the United States must continue to carry out their responsibilities and exercise their authority in captivity. The senior, regardless of service, must accept command. This responsibility and accountability may not be evaded.

If the senior is incapacitated or is otherwise unable to act, the next senior person will assume command. Camp leaders should make every effort to inform all POWs of the chain of command and try to represent them in dealing with enemy authorities. The responsibility of subordinates to obey the lawful orders of ranking American military personnel remains unchanged in captivity.

The Geneva Convention Relative to Treatment of Prisoners of War provides for election of a "prisoners' representative" in POW camps containing enlisted personnel but no commissioned officers. American POWs should understand that such a representative is only a spokesman for the actual senior-ranking person. Should the enemy appoint a POW chain of command for its own purposes, American POWs should make all possible efforts to adhere to the principles of Article IV.

As with other provisions of this code, common sense and the conditions of captivity will affect the way in which the senior person and the other POWs organize to carry out their responsibilities. What is important is that everyone support and work within the POW organization.

ARTICLE V

When questioned, should I become a prisoner of war, I am required to give name, rank, service number, and date of birth. I will evade answering further questions to the utmost of my ability. I will make no oral or written statements disloyal to my country and its allies or harmful to their cause.

When questioned, a POW is required by the Geneva Convention and this code to give name, rank, service number (Social Security number), and date of birth. The prisoner should make every effort to avoid giving the captor any additional information. The prisoner may communicate with captors on matters of health and welfare and additionally may write letters home and fill out a Geneva Convention "capture card."

It is a violation of the Geneva Convention to place a prisoner under physical or mental duress, torture, or any other form of coercion in an effort to secure information. If under such intense coercion, a POW discloses unauthorized information, makes an unauthorized statement or performs an unauthorized act, that prisoner's peace of mind and survival require a quick recovery of courage, dedication, and motivation to resist anew each subsequent coercion.

Actions every POW should resist include making oral or written confessions and apologies, answering questionnaires, providing personal histories, creating propaganda recordings, broadcasting appeals to other prisoners of war, providing any other material readily usable for propaganda purposes, appealing for surrender or parole, furnishing self-criticisms, and communicating on behalf of the enemy to the detriment of the United States, its allies, its armed forces, or other POWs.

Every POW should also recognize that any confession signed or any statement made may be used by the enemy as false evidence that the person is a "war criminal" rather than a POW. Several countries have made reservations to the Geneva Convention in which they assert that a "war criminal" conviction deprives the convicted individual of POW status, removes that person from protection under the Geneva Convention, and revokes all rights to repatriation until a prison sentence is served.

Recent experiences of American POWs have proved that, although enemy interrogation sessions may be harsh and cruel, one can resist brutal mistreatment when the will to resist remains intact.

The best way for a prisoner to keep faith with country, fellow prisoners, and self is to provide the enemy with as little information as possible.

ARTICLE VI

I will never forget that I am American, fighting for freedom, responsible for my actions, and dedicated to the principles which made my country free. I will trust in my God and in the United States of America.

A member of the armed forces remains responsible for personal actions at all times. A member of the armed forces who is captured has a continuing obligation to resist and to remain loyal to country, service, unit, and fellow prisoners.

Upon repatriation, POWs can expect their actions to be reviewed, as to both the circumstances of capture and the conduct during detention. The purpose of such reviews is to recognize meritorious performance as well as to investigate possible misconduct. Each review will be conducted with due regard for the rights of the individual and consideration for the conditions of captivity; captivity of itself is not a condition of culpability.

Members of the armed forces should remember that they and their dependents will be taken care of by the appropriate service and that pay and allowances, eligibility, and procedures for promotions and benefits for dependents will continue while the service member is detained. Service members should make sure that their personal affairs and family matters (such as pay, powers of attorney, current will, and provisions for family maintenance and education) are properly and currently arranged. Failure to do so can create a serious sense of guilt for a POW and place unnecessary hardship on family members.

The life of a POW is hard. Each person in this stressful situation must always sustain hope and resist enemy indoctrination. POWs standing firm and united against the enemy will support and inspire one another in surviving their ordeal and in prevailing over misfortune with honor.

APPENDIX B

Nautical Terminology

This nautical glossary contains most of the terminology you will need for effective and professional communication on board your boat or cutter.

abaft: Behind or farther aft; astern or toward the stern.

abeam: At right angles to the centerline of and outside a ship.

aboard: The position of objects in relation to a ship. "Close aboard" means near a ship.

absentee pennant: Special pennant flown to indicate absence of commanding officer, admiral, chief of staff, or officer whose flag is flying (division, squadron, or flotilla commander). Also called absence indicator.

accommodation ladder: A portable flight of steps down a ship's side.

adrift: Loose from moorings, or out of place.

aft: In, near, or toward the stern of a vessel.

aground: Resting on or touching the ground or bottom.

ahead: Forward of the bow.

ahoy: Term used to hail a boat or a ship, as "Boat ahoy!"

all hands: Entire ship's company, both officers and enlisted personnel.

allowance: Numbers, ranks, and ratings of officers and crew assigned to a ship when it is impracticable to assign a full complement. See also *complement*.

aloft: Above the ship's uppermost solid structure; overhead or high above.

alongside: By the side of the pier or ship.

amidships (or midships): In middle portion of ship, along the line of the keel. "Amidships" is used as an adverb ("He went amidships"), and "midships" is used as an adjective ("midships passageway").

anchorage: Suitable place for a ship to anchor. Area of a port or harbor.

anchor ball: Black shape hoisted in forepart of a ship to show that the ship is anchored in a fairway.

anchor cable: Chain, wire, or line running between an anchor and a ship.

anchor detail: Group of ship's personnel who handle ground tackle when the ship is anchoring or getting under way.

anchor light: White light displayed by ship at anchor. Two such lights are displayed by a ship more than 150 feet in length.

anchors aweigh: Said of an anchor when just clear of the bottom.

anchor watch: Detail of ship's personnel standing by at night as a readiness precaution while ship is at anchor.

anemometer: Instrument to measure wind velocity.

astern: Toward the stern; an object or vessel that is abaft another vessel or object.

athwart, athwartships: At right angles to the fore and aft or centerline of a ship.

auxiliary: An assisting machine or vessel, such as an air-conditioning machine or fuel ship. Not to be confused with the U.S. Coast Guard Auxiliary.

avast: A command to cease or desist from whatever is being done.

awash: So low in the water that the water is constantly washing across the surface.

aweigh: Position of an anchor just clear of the bottom.

aye, aye: Reply to an order or command to indicate that it is understood and will be carried out. ("Aye, aye, sir/ma'am," to officers.)

backstay: A stay supporting a mast from aft.

backwash: Water thrown aft by turning a ship's propeller.

bail: To dip water out of a boat with a bucket.

ballast: Heavy weight in the hold of a vessel to maintain proper stability, trim, or draft. A ship is said to be "in ballast" when it carries no cargo, only ballast.

barge: Craft used to haul material, such as a coal barge; a power boat used by flag officers, such as an admiral's barge.

barnacle: Small marine animal that attaches itself to sides and bottoms of hulls and piers.

barometer: Instrument that registers atmospheric pressure; used in forecasting weather.

batten: Long strip of steel or wood that wedges the edge of a tarpaulin against the hatch.

batten down: To cover and fasten down; to close off a hatch or watertight door.

battle lantern: A battery-powered lantern used for emergency lighting.

battle lights: Dim red lights that furnish sufficient light for personnel during darken-ship period.

beam: Width; breadth; the most extreme width (or breadth) of a nautical vessel.

bear: To be in a certain direction from the observer.

bear a hand: Speed up the action; lend a helping hand.

bearing: Direction of an object, expressed in degrees either as relative or true bearing.

Beaufort scale: A table of scales indicating various velocities of winds.

becket: Circular metal fitting on a block; a rope eye or grommet.

belay: To cancel an order; to stop; to firmly secure a line.

below: Short for "below decks"; below the main deck.

bend: A general class of knots used to join two lines together.

bend on: To secure one thing to another, such as to bend a flag onto a halyard.

berth, berthing: Space assigned a vessel for an anchoring or mooring.

bight: Middle part of a line as distinguished from the end and the standing part; a single complete turn of line; bend in a river or coastline.

bilge: Lower part of vessel where waste water and seepage collect.

billet: A crew member's assigned duties within the ship's organization.

binnacle: Large stand used to house a magnetic compass and its fittings.

bitt: Strong iron post on ship's deck for working or fastening lines; almost invariably in pairs.

bitter end: The utmost end of line.

blinker: Lamp or set of lamps, triggered to a telegraph key, used for sending flashing-light messages.

block: An item of deck gear made of one or more grooved sheaves, a frame (casing or shell), supporting hooks, eyes, or straps; may be metal or wood.

boat chock: A strong deck fitting that supports one end of a boat that is resting on deck.

boat fall: Rigging used to hoist or lower ship's boats.

boat gripe: Lashing used at sea to secure a boat hanging from the davits against the strong back and away from the ship's side.

boat hook: Wooden staff with metal hook and prod at one end; used to fend off or hold on.

boat painter: Rope attached to the stern ringbolt of a boat; used for securing it. Also a short piece of rope secured to the bow of a boat; used for towing or making fast. Not to be confused with the *sea painter*, which is a much longer rope.

boat plug: Metal or wood plug used to stop up boat's drain hole.

boats: Small open or decked-over craft propelled by oars, sails, or some type of engine. This term also applies to larger vessels built to navigate rivers and inland waters.

boat skid: Heavy wood and metal frame on ship's boat deck used to support a boat's keel.

boat sling: Rope or chain sling used for hoisting or lowering large-size boats with a single davit or crane.

boat station: Allotted place for each person when boat is being lowered.

boatswain: Warrant officer in charge of deck work. Pronounced "bo'sun."

boatswain's call: See *boatswain's pipe*.

boatswain's chair: Line-secured board on which a worker sits while aloft or over the side.

boatswain's locker: Compartment where deck gear is stowed.

boatswain's mate of the watch: Petty officer responsible for ensuring that all deck watch stations are properly manned and in order.

boatswain's pipe: Small, shrill silver whistle used by boatswain's mate to pass a call or pipe the side.

bollard: Wooden or iron post on a pier or wharf to which mooring lines are secured.

boom: Projecting spar or pole that provides an outreach for extending the foot of sails, or for mooring boats, handling cargo, and so on. Rigged horizontally or nearly so.

boot camp: Slang for recruit training center.

boots: Slang for recruits or new hands on board.

boot topping: Surface of the outside plating of ship or boat's side between light and load lines.

bow: Forward section of a vessel.

bower anchor: Either of the two anchors usually carried at the ship's bow. Most ships anchor by using one of the bowers.

bowline: One of the most used knots; used to make a temporary eye in the end of a line. Pronounced "bo'lin."

bow painter: See *boat painter*.

boxing the compass: Naming all compass points and quarter points in their proper order, from north, east, south, through west.

break: To unfurl a flag with a quick motion. In ship construction, an abrupt change in the fore-and-aft contour of a ship's main deck.

breaker: A small container for stowing drinking water carried by boats or rafts; a wave that breaks into foam against a shore.

break out: To unstow or prepare for use.

breakwater: A structure used to break the force of waves.

breast line: A mooring line running at right angles from the ship's fore-and-aft line.

bridge: Raised platform from which ship is steered, navigated, and conned; usually located in forward part of the ship.

bridle: Span of rope or chain with both ends secured.

brig: Prison on a ship or shore station.

brightwork: Metal work that is kept polished rather than painted.

broach to: Turning suddenly into the wind; to be thrown broadside into surf.

broad command pennant: Personal command pennant of an officer not a flag officer, commanding any of the following units: a division of battleships, aircraft carriers, or cruisers; a force, flotilla, or squadron of ships or craft of any type; air wing.

broad on the starboard or port beam: Bearing 090° or 315° relative to the bow of the ship.

broad on the starboard or port quarter: Bearing 135° or 225° relative to the bow of the ship.

broadside to: At right angles to the fore-and-aft line of a ship.

broken water: An area of small waves and eddies in otherwise calm water.

brow: Large gangplank leading from a ship to a pier, wharf, or float; usually equipped with rollers on the bottom and hand rails on the side.

buckler: Plating fitted over the hawsehole to prevent entry of water.

bulkhead: One of the vertical wall-like structures enclosing a compartment.

bull nose: A closed chock at the head of the bow on the forecastle deck.

bulwark: Raised plating or woodwork running along the side of a vessel above the weather deck that helps keep decks dry and prevents crew and gear from being swept overboard.

bunker: Storage space for fuel.

buoy: Floating marker anchored by a line to the bottom, which by shape and color conveys navigational information; may be lighted or unlighted, and may be equipped with whistles, bells, or gongs.

burdened vessel: See *give-way vessel.*

burgee: Swallow-tailed flag.

burgee command pennant: Personal command pennant of an officer not a flag officer, commanding any of the following units: a division of ships or craft other than aircraft carriers or cruisers.

by the head: Ship's appearance with a greater draft forward than aft. For example, "down by the head."

by the stern: Opposite of *by the head.*

cabin: On board ship, the captain's living quarters; covered compartment of a boat.

cable: See *anchor cable.*

cable jack: Crowbar-like tool used on forecastle for working the anchor cable.

cable-laid rope: Three or four plain-laid, three-stranded ropes twisted in the opposite direction to the twists of each rope; used for ropes much exposed to water.

cadet: A student at the U.S. Coast Guard Academy.

camber: The arch in a ship's deck that makes the centerline section higher than the extremities.

camel: Large fender float used for keeping vessel off wharf, pier, or quay; usually of one or more logs.

can buoy: Cylindrical metal buoy used to mark the left side of a channel (entering from seaward), painted green.

capstan or capstan head: That part of a vertical-shaft windlass around which a working line is passed.

cardinal point: One of the four principal points of the compass—north, east, south, and west.

cargo whip: Rope or chain used with a boom for handling cargo. One end has a heavy hook; the other end is rove through a block and taken to the winch. Also called cargo hoist or cargo rope.

carrick bend: Usually seen as a double carrick bend, a much-used knot for bending two lines or hawsers together.

carry away: To break loose, tear loose, or wash overboard.

carry on: An order to resume work or duties.

cast: Act of heaving the lead into the sea to determine depth of water; to throw the ship's bow in one direction or another when getting under way.

cast loose: To let go a line or lines.

cast off: To throw off; to let go; to unfurl.

cat's paw: A quickly formed twist in the bight of a line by which two eyes are formed.

centerline: Imaginary line running from ship's bow to stern.

chafe: Wearing away the surface of line, spar, or chock by rubbing.

chafing gear: Guard of canvas or rope around spars, hawsers, chocks, or rigging to prevent chafing.

chain locker: Compartment in which chain cable is stowed.

chain of command: Succession of commanding officers through which command is exercised from superior to subordinate.

chain or cable markings: A series of turns of wire or stripes of paint on certain links of each anchor chain, indicating the scope or amount of chain that has run out.

chain pipe: Heavy steel pipe that leads the chain cable through the deck to the chain locker.

chains: Platform or a general area on either side of forward part of a ship where leadsman stands as he or she takes soundings.

chain stopper: Short length of chain fitted with a pelican hook and secured to an eyebolt on the forecastle; used for quickly letting go the anchor or for securing the anchor in stowed position.

chart: Nautical map used as an aid to navigation.

charthouse or chartroom: Compartment on or near the bridge for handling and stowage of navigation equipment.

check: To slack off slowly; to stop a vessel's way gradually by a line fastened to some fixed object or to an anchor on the bottom; to ease off a rope a little, especially with a view to reducing the tension; to stop or regulate the motion, as of a cable when it is running out too fast.

chipping hammer: Small hammer with a sharp peen and face set at right angles to each other; used for chipping and scaling metal surfaces. Also called scaling hammer or boiler pick.

chock: Steel deck member, either oval or U-shaped, through which mooring lines are passed. Usually paired off with bitts.

chronometer: An especially accurate timepiece used for navigation.

clamp down: To go over a deck with damp swabs; a lesser form of swabbing down.

clap on: To clap on a rope means to catch hold in order to haul on it; to clap on a stopper or tackle means to put on a stopper or tackle; to clap on canvas means to put on more sail.

clear: To straighten out a tangled line or chain; make ready for running.

cleat: A small metal deck fitting with horns used for securing lines; also called belaying cleat. Short piece of wood nailed to brow or gangplank to give surer footing.

clinometer: Bridge and engine-room instrument that indicates amount of a ship's roll or degree of list.

close aboard: Nearby.

clove hitch: A knot much used for fastening a line to a spar or stanchion.

coaming: Raised framework around deck or bulkhead openings and cockpits of open boats to prevent entry of water.

cofferdam: Empty space between two bulkheads separating two adjacent compartments. Its purpose is to isolate one compartment from another, preventing liquid contents of one from entering the other in the event one bulkhead loses its integrity.

coil: Laying down line in circular turns, usually one turn atop the other.

collision bulkhead: Watertight bulkhead a few feet abaft the stem; used to isolate the damage due to a head-on collision.

colors: National ensign; distinguishing flag flown by a vessel to indicate her nationality. Also, the ceremonies performed at a naval activity when colors are hoisted at 0800 and hauled down at sunset.

command: A term applied to a naval unit or group of units under one officer; a definite and direct form of order.

commission: To activate a ship or aircraft; a written order giving an officer rank and authority; the rank and authority itself.

commission pennant: Long, thin, pennant flown by a ship to indicate that the ship is commissioned for service in the Coast Guard or Navy.

companionway: Set of steps or ladders leading from one deck level to another.

compartment: Space enclosed by bulkheads, deck, and overhead that corresponds to a room in a building.

compass: Instrument to indicate geographic direction.

compass rose: Diagram of a compass card on a chart; assists the navigator in laying out courses and directions on chart.

complement: Numbers, ranks, and ratings of officers and enlisted personnel as are determined necessary to fight the ship most effectively or to perform such other duties as may be required.

conn: To direct helmsman as to movement of helm, especially when navigating in narrow channels or heavy traffic. For example, to conn the vessel.

country: Living quarters, such as officer's country, wardroom country, CPO country.

coupling: Metal fitting at the ends of a length of fire hose.

course: Direction steered by a ship.

court-martial: Military court for the trial of more serious offenses.

cowl: Bell- or hood-shaped opening of a ventilator; it increases the amount of air forced into or drawn out of the ventilator.

cow's tail: Frayed end of a rope.

coxswain: Enlisted man or woman in charge of a boat; acts as helmsman. Pronounced "koksun."

cradle: A stowage rest for ship's boat.

crossing the line: Crossing the earth's equator or timeline.

crosstree: Superstructure member at top of a low mast or between two such masts; runs athwartships.

crow's nest: Lookout's stand high on a mast or crosstree.

cumshaw: A gift or something procured by trade or without payment.

cut of the jib: General appearance of a vessel or of a person.

cutter: A Coast Guard vessel sixty-five feet in length or longer; a type of sailing vessel.

cutwater: Forward edge of the stem at and below the waterline.

damage control: Measures necessary to keep ship afloat, fighting, and in operating condition.

davit: Shipboard crane that can be swung out over the side; used for hoisting and lowering boats and weights. Often found in pairs.

Davy Jones' locker: The bottom of the sea.

day's duty: Tour of duty on board ship lasting twenty-four hours.

dead ahead: Directly ahead of the ship's bow; bearing 000° relative.

dead in the water: Said of an underway ship that is making neither headway nor sternway.

dead reckoning: Navigator's estimate of ship's position from the course steered and the distance run.

deck: On a ship, corresponds to the floor of a building on land.

deck seamanship: Branch of seamanship embracing the practical side, from the simplest rudiments of marlinspike seamanship up to navigation; includes small-boat handling, ground tackle, steering, heaving the lead, signaling, and so on.

deep: The distance in fathoms between two successive marks on a lead line, as "By the deep, four."

deeps: In a lead line, the fathoms that are not marked on the line.

deep six: Slang meaning to dispose of by throwing over the side. Also, when referring to lead line use, a depth of six fathoms of water.

degaussing gear: Electrical gear that sets up neutralizing magnetic fields to protect the ship from magnetic-action mines. Pronounced "degow'sing."

department: An organizational subdivision of ship's crew into specialized functions, for example, Engineering Department, Operations Department, Supply Department.

deployment: Dispersal or disposition of ships for mission execution.

depth charge: Explosive charge used against submarines.

derelict: Abandoned vessel at sea, still afloat.

detachable link: A shackle used in connecting shots of anchor chain.

diesel: A type of oil-burning, internal-combustion engine used on some ships and boats of the Coast Guard.

dip: Lowering a flag partway in salute or in answer, and hoisting it again. A flag is "at the dip" when it is flown at about two-thirds the height of the halyards.

director: Electromechanical device for directing and controlling gunfire.

displacement: Weight of water displaced by a ship.

distance line: A light line spanning ships engaged in underway refueling or replenishment. Marked off in twenty-foot lengths, it helps station keeping.

ditty bag: Small bag used by sailors for stowage of personal articles, toilet articles, or laundry.

division: In shipboard organization, a subdivision of a ship's department for further specialization.

dock: Artificial basin for ships, fitted with gates to keep in or shut out water; water area between piers.

dock trials: Four- to six-hour trial of main engines while ship is moored alongside a pier.

dog: Metal fitting used to secure watertight doors, hatch covers, scuttles, and so on.

dogwatch: One of the two-hour watches between 1600 to 2000.

doldrums: Areas on both sides of the equator where light and variable breezes blow.

dolphin: Cluster of piles for mooring.

double-bottoms: Watertight subdivisions of ship, next to the keel and between outer and inner bottoms.

double up: To increase the number of ship-to-pier turns of a mooring line.

doubling: Rounding a point of land; for example, doubling Cape Horn.

draft: Depth of water from the surface to the ship's keel; a detail of crew members.

draft marks: Numeral figures on either side of the stem and sternpost, used to indicate the amount of ship's draft.

dressing ship: To display the national ensign at all mastheads and the flagstaff; full dressing further requires a rainbow of flags bow to stern over the mastheads.

drift lead: Sounding lead and line dropped over side of a ship to detect dragging of the vessel.

drogue: Floating anchor usually made of spars and canvas; used to keep the head of a vessel to the wind, to lessen leeway, or to check headway. Also called drag sheet, sea anchor.

duct: Large sheet-metal pipe that leads air from blowers to enclosed spaces.

duty: The responsibility to adequately perform military obligations; a period of specific military responsibility, such as a watch. See also *duty day.*

duty day: A period of duty wherein one section of the ship's in-port watch organization remains on board for a twenty-four-hour period to ensure safety on board, to operate the ship as may be required, and to maintain it in readiness. The section "having the duty" is called the "duty section." One or more watches of varying types may be stood by members of the duty section during the duty day; ship's work may also be assigned to duty section members when off watch. Most Coast Guard ships operate a four-section duty organization wherein each section has the duty every four days. A duty organization is also employed on board most shore units of the Coast Guard.

ease her: A command to reduce the amount of rudder or helm.

ease off: To ease a line; slacken it when taut.

easy: Carefully, gently.

ebb tide: Tide falling or flowing out.

echo counter: Fathometer; device for measuring depth of water by sending out vibrations that bounce back from the bottom. It measures time taken from the echo to return, and from that the distance is calculated.

eddy: A small whirlpool.

embark: To go on board a ship preparatory to sailing.

end for end: Reversing positions of an object or line.

end on: Head-to-head or stem-to-stem.

engineer officer: An officer who is responsible for the ship's machinery, fuel, and water.

engineer of the watch: An officer in charge of the engineering department while on watch.

engine order telegraph: Signaling gear for transmitting speed and direction orders from bridge to engine room.

ensign: Colors, national flag. Also, junior commissioned officer in the Coast Guard or Navy. Pronounced "en'sin."

escape hatch: In general, any hatch, usually small, that permits escape from a compartment when ordinary means of egress are blocked.

even keel: Floating level, no list.

executive officer: An officer directly subordinate to the commanding officer and second in command of the ship. The executive officer is in charge of the ship's administration, personnel, and routine.

executive petty officer: An enlisted person directly subordinate to the officer in charge of a unit. The executive petty officer is in charge of the ship's administration, personnel, and routine.

eyebolt: A metal bolt ending in an eye.

eyebrow: Curved metal arc mounted above a porthole, used to shed water.

eyes: Foremost part of weather deck in the bow of the ship.

fairlead: An eye, block, or fitting furnishing a clear lead for a line.

fair tide: Tidal current running in same direction as the ship.

fairway: In inland waters, an open channel or midchannel.

fair wind: A favoring wind.

fake: A single turn of line when line is coiled down.

fake down: Coiling down a line so that each fake of rope overlaps the one underneath and makes the line clear for running.

fall: Entire length of rope in a tackle; the end secured to the block is called the standing part; the opposite part, the *hauling part*. Also, the line used to lower and hoist a boat.

falling glass: Lowering atmospheric pressure as registered by the barometer; normally a sign of approaching bad weather.

false keel: Thin covering secured to lower side of main keel of ships; affords more protection.

fancy work: Intricate, symmetrical rope work used for decorative purposes.

fantail: Main deck section in the after part of a flush-deck ship.

fast: Snugly secured; said of a line when it is fastened securely.

fathom: A six-foot unit of length.

feather: A preparation technique whereby irregularities are leveled by sanding to produce a smooth surface for painting.

feel the way: To proceed cautiously, taking soundings while doing so.

fender: Canvas, wood, rope gear, or old rubber tire used over the side to protect a ship from chafing when alongside a pier or another ship.

fend off: To push away; pushing away from a pier or another ship when coming alongside, to prevent damage or chafing.

fiber rope: General term for cordage made of vegetable fibers such as hemp, manila, flax, cotton, or sisal.

fid: A wooden marlinspike.

field day: General cleaning day on board ship, usually before inspection.

field strip: To disassemble without further breakdown the major groups of a piece of ordnance for routine cleaning and oiling; as opposed to detailed stripping, which may be done only by authorized technicians.

fife rail: A wood or metal rail near the base of the mast bored with holes to take belaying pins, usually at the head of the flag bag.

figure-eight fake: Method of coiling rope in which the turns form a series of overlapping figure-eights advancing about one or two diameters of the rope for each turn; usually done over the lifelines.

figure-eight knot: Knot forming a large knob; easily tied.

fire control: Shipboard system of directing and controlling gun, missile, or torpedo fire; the compartment wherein fire-control functions are carried out.

fire main: System of pipes that furnish water to fireplugs.

first lieutenant: Officer in charge of cleanliness and general upkeep of a ship or shore station; the officer in charge of the deck department. This is a duty, not a rank.

fix: Determination of a ship's position by using one or more navigational methods.

flag bag: Container for stowage of signal flags and pennants; rigged with different slots to take the flag's snaps and rings.

flag officer: An officer of the rank of rear admiral or above; so called because of that officer's entitlement to fly a personal flag, which, by the numbers of stars upon it, indicates the officer's rank.

flagstaff: Small vertical spar at the stern on which the ensign is hoisted.

flank speed: A certain prescribed speed increase over standard speed; faster than full speed but less than emergency full speed.

flap: Hinged surface of airplane wing; used to increase the lift on takeoff and landing.

flare: Outward and upward curving sweep of a ship's bow; outward curve of the side from waterline to deck level. Also, a pyrotechnic device used to illuminate or attract attention.

flash burn: Burn received from the heat of explosion of a projectile or bomb or inflammable liquid.

flash plate: Protective metal plate over which the anchor cable rides. It is part of the forecastle.

flattop: Slang for an aircraft carrier.

fleet: A naval organization of ships and aircraft under one commander; normally includes all types of ships and aircraft necessary for major operations. Also, to draw the blocks of a tackle apart.

flemish: To coil line flat on deck in a clockwise direction, each fake outside the other, all laid snugly side by side; begins in the middle and works outward.

floating dry dock: Movable dock floating in water; ships of all sizes are floated into it and repaired.

flood tide: Tide rising or flowing toward land.

flotsam: Floating wreckage or goods thrown overboard. See also *jetsam*.

fluke: Flat end of an anchor that bites into the bottom.

flush deck: Continuous upper deck extending from side to side and from bow to stern.

flying bridge: An area aloft of the pilothouse often employed for lookout stations.

fore and aft: Running in the direction of the keel.

forecastle: Upper deck in the forward part of the ship. Pronounced "FOKE-sul"; abbreviated fo'csle.

forecastle deck: Partial deck of the main deck at the bow.

forestay: A stay supporting a mast from forward.

forward: Toward the bow; opposite of aft.

foul: Jammed; not clear for running.

fouled anchor: Anchor with its chain or debris twisted around it.

founder: To sink.

frame: Ribs of a vessel.

frap: To bind lightly by passing lines around; to draw together the parts of tackle or other combinations of ropes to increase tension.

frapping lines: Lines passed around the boat falls to steady them.

freeboard: Height of a ship's sides from waterline to main deck.

full speed: A prescribed speed that is greater than standard speed but less than flank speed.

funnel: Ship's smokestack; stack.

furl: Gathering up and securing a sail or awning; opposite of "spread."

fuselage: The body to which the wings and tail of an airplane are attached.

gaff: Small spar abaft the mainmast from which the national ensign is flown when the ship is under way.

gale: A wind between a strong breeze and a storm; characterized as moderate (28–33 knots), fresh (34–40), strong (41–47), or whole gale (48–55 knots).

galley: The ship's kitchen.

gangplank: See *brow*.

gangway: Opening in the bulwarks or the rail of the ship to give entrance at the head of the gangplank or brow; an order to stand aside and get out of the way.

gantline: A rope and a block on top of a mast, stack, and so on, used to hoist rigging, staging, boatswain's chairs, and other items.

gather way: To gain headway.

gear: General term for lines, ropes, blocks, fenders, and so on; personal effects.

general quarters: Battle stations for all hands.

gig: A ship's boat designated for the commanding officer's use.

gimbals: A pair of rings, one within the other, with axes at right angles to each other. Supports the compass and keeps it horizontal despite the ship's motion.

gipsey, gypsy: Cathead; drum on a horizontal shaft windlass or winch for working lines.

give-way vessel: The vessel that, by the *Rules of the Road*, must keep out of the way of another vessel. Formerly called the burdened vessel. See also *stand-on vessel*.

glass: Barometer or quartermaster's spyglass.

glasses: Binoculars.

go adrift: To break loose.

grab-rope: A rope secured above a boat boom or gangplank; used to steady oneself.

granny knot: A knot somewhat similar to a *square knot*; does not hold under strain.

grapnel: Small anchor with several arms; used for dragging for lost objects or for anchoring skiffs or dories.

gratings: Wooden or iron openwork covers for hatches, sunken decks, and so on.

gripes: Metal fastenings for securing a boat in its cradle; canvas bands fitted with thimbles in their ends and passed from the davit heads over and under the boat for securing for sea.

grommet: Ring of rope formed by a single strand laid three times around a metal ring set in canvas, cloth, or plastic.

ground: To run ashore; to strike the bottom through ignorance, violence, or accident.

ground tackle: Term referring to all anchor gear.

gudgeon: Support for a rudder; consists of metal braces bolted to the sternpost and having eyes to take the pintles, or pivot pins, on which the rudder swings.

guesswarp: A line at the outer end of a boat boom used for securing a boat to the boom. Also, a hauling line laid out by a boat.

gun mount: A gun structure with one to four guns; may be open or enclosed in a steel shield. Enclosed mounts are not as heavily armored as *turrets* and carry no gun larger than five inches.

gunwale: Upper edge or rail of a ship or boat's side. Pronounced "gun'nle."

guy: A line used to steady and support a spar in a horizontal or inclined position.

gyrocompass: Compass used to determine true direction by means of gyroscopes.

gyrocompass repeaters: Compass cards electrically connected to gyrocompass and repeating the same readings.

gyropilot: Automatic steering device connected to the repeater of a gyrocompass; designed to hold a ship on its course without a helmsman. Also called automatic steerer, iron mike, iron quartermaster.

hack chronometer: Spare or comparison chronometer, not quite as accurate as standard chronometer.

hail: To address a nearby boat or ship. Also, a ship is said to "hail from" a particular home port.

half deck: Partial deck below the main and above the lowest complete deck.

half hitch: Usually seen as two half hitches; a knot used for much the same purposes as a clove hitch.

half-mast: Position of the ensign when hoisted halfway; usually done for a day or more at a time in respect to a deceased person.

halyard: Line used for hoisting flags, pennants, or balls.

hand lead: A lead (pronounced "led") weighing from seven to fourteen pounds, secured to a line and used for measuring the depth of water or for obtaining a sample of the bottom.

handsomely: To ease a line gradually; to execute something deliberately and carefully but not necessarily slowly.

handy billy: Small, portable, power-driven water pump.

hangfire: Gun charge that does not fire immediately upon pulling the trigger but sometime later.

hatch: An opening in the ship's deck for communication or for handling stores and cargo.

haul: To pull.

hauling part: That part of the fall of a tackle to which power is applied.

hawsepipes and hawseholes: The steel castings in the bow through which anchor cables run are hawsepipes; the openings are hawseholes.

hawser: Heavy line five inches or more in circumference, used for heavy work such as towing or mooring.

head: Compartment of a ship having toilet facilities.

heading: The direction a ship's bow points at any given moment.

headroom: Clearance between decks.

headway: Forward motion of a ship.

heave: To throw or toss; to pull on a line.

heave away: An order to start heaving on a capstan or windlass, or to pull on a line.

heave in: An order to haul in a line or the anchor chain.

heave 'round: To revolve the drum of a winch or windlass so as to pull in a line or anchor cable.

heave short: An order to heave in on anchor chain until the ship is riding nearly over her anchor.

heave the lead: To employ the lead line.

heave to: To bring the ship's head into the wind or sea and hold her there by the use of engines and rudder.

heaving line: A small line with a weight on one end; the weighted end is thrown to another ship or to a pier so that a larger line attached to it may be passed.

heel: To list over.

helm: The helm proper is the *tiller*, but the term is often used to mean the rudder and the gear for turning it, such as the ship's wheel. See also *joystick*.

helmsman: The person at the wheel; the person who steers the ship.

highline: Line running between ships that are replenishing or transferring at sea.

hitch: General class of knots by which a line is fastened to another object, either directly or around it. Also, a slang expression for a term of enlistment.

hoist: Display of signal flags on halyard. Also, to raise a piece of cargo or gear.

hoisting pad: Metal piece bolted to boat's keel; has an eye to which hoisting rod is bolted.

hold: Space below decks for storage of ballast, cargo, and so on.

holiday: An imperfection or vacant space in an orderly arrangement; spots in painted work left unfinished.

holiday routine: Routine followed on board ship on authorized holidays and Sundays.

horns: Horizontal arms of a cleat or chock; projecting timbers of a stage to which rigging lines are secured.

horse latitudes: Latitudes on outer margins of trade winds (around 30° north and south) where prevailing winds are light and variable.

house: To stow or secure in a safe place; to run an anchor's shank up into the hawsepipe.

housing anchor: Anchor having one stock; houses itself in hawsepipe when hove in.

housing chain stopper: Chain stopper fitted with a screw turnbuckle, used for securing anchor in hawsepipe.

hove taut: Pulled tight.

hug: To keep close. A vessel might hug the shore.

hulk: A worn-out and stripped vessel unable to move under its own power.

hull: Framework of a vessel, together with all her decks, deckhouses, inside plating or planking but exclusive of masts, rigging, guns, and all superstructure items.

hull down: Said of a distant vessel when only her stack-tops and mast are visible above the horizon.

idler: Member of ship's company who does not stand night watches. Also known as a "dayworker."

inboard: Toward the ship's centerline.

Inland Rules: Rules enacted by Congress to govern the navigation of certain inland waters of the United States. Part of the *Rules of the Road*.

inner bottom: Top of the double bottom; consists of watertight plating.

inshore: Toward land.

intercardinal points: The four points midway between the cardinal points of the compass: northeast, southeast, southwest, northwest.

interior communication: Communications and announcement systems inside a ship.

International Rules: The rules established by agreement among maritime nations and governing the navigation of the high seas. Part of the *Rules of the Road*.

jack: Flag similar to the national ensign, flown at the *jackstaff* when in port; plug for connecting an electrical appliance to a power or phone line.

jack-o'-the-dust: Enlisted person serving as assistant to the ship's cooks.

jackstaff: Spar on the bow of a ship from which the Union Jack is flown when the vessel is in port.

Jacob's ladder: Light ladder made of rope or chain with metal or wooden rungs; used over the side and aloft.

jetsam: Goods that sink when thrown overboard. See also *flotsam*.

jettison: To throw goods overboard.

jetty: A landing; a small pier.

jigger: Light handy tackle for general work about the deck.

joystick: A type of helm employing a stick instead of a wheel to steer the ship.

junior officer of the deck: The officer or petty officer acting as an assistant to the officer of the deck. When the ship is moored or anchored, the term "junior officer of the day" is generally employed.

jury rig: Makeshift rig of mast and sail or of other gear, such as jury anchor, jury rudder; any makeshift device.

kapok: Water-resistant fiber stuff packed into lifejackets to make them buoyant.

kedge: Anchor used for kedging; that is, moving a ship a short distance at a time by taking one of the anchors out in a boat, letting it go, and then hauling the ship up to it. If this is done merely to change the heading of the ship, it is called "warping." See also *warp*.

keel: Backbone of a ship; running from stem to sternpost at the bottom.

keelhaul: To reprimand severely.

keelson: Timber or steel fabrications bolted on top of a keel to strengthen it.

king post: Short mast supporting a boom.

knife-edge: Smooth, polished edge of the coaming against which the rubber gaskets of watertight doors and scuttles press when closed; furnishes better watertight integrity.

knock off: To cease what is being done; to stop work.

knot: One nautical mile per hour. Also, a knob, tie, or fastening formed with rope.

lacing: Line used to secure canvas by passing through eyelets or grommets in the canvas.

ladder: In a ship, corresponds to stairs in a building.

landfall: First sighting of land at the end of a sea voyage.

landlubber: Seaman's term for one who has never been to sea.

landmark: Any conspicuous object on shore, used for piloting.

lash: To tie or secure by turns of line.

lay: A preliminary order, such as "Lay aloft," "Lay below," and so on. The direction of the twist of strands of rope.

lead line: Line secured to the lead used for soundings.

lead or sounding lead: Weight used for soundings, that is, for measuring the depth of the water. Pronounced "led."

leadsman: Crew member detailed to heave the sounding lead.

leave: Authorized absence of an individual from a place of duty, chargeable against the individual in accordance with applicable law.

lee: Direction away from the wind.

lee helmsman: Assistant or relief helmsman.

leeward: In a lee direction. Pronounced "lu'ard."

left-hand rope: Twisted from right to left. Strands and cables are usually left-handed.

let go by the run: Allowing a line to run free.

liberty: Any authorized absence granted for short periods to provide respite from the work environment or for other specific reasons. Liberty is not chargeable to the member.

lie to: Said of a vessel when under way with no way on.

life buoy or life ring: A ring or U-shaped buoy of cork or metal to support a person in the water; a type of personal flotation device.

lifejacket or life preserver: A belt or jacket of buoyant or inflatable material worn to keep a person afloat; a type of personal flotation device.

lifeline: Line secured along the deck to lay hold of in heavy weather; line thrown on board a wreck by a rescue crew; knotted line secured to the span of lifeboat davits for the use of the crew when hoisting and lowering.

life raft: Inflatable float craft for use in survival at sea.

lighter: Small vessel used for working (loading and unloading cargo) ships anchored in harbor.

lightship: Small ship equipped with a distinctive light and anchored near an obstruction to navigation or in entrances or shallow water to warn shipping. There are no longer any lightships in U.S. waters.

line: Seagoing term for rope or cable; the equator.

line-throwing gun: Small-caliber gun that projects a weighted-at-one-end line a long distance; surpasses a heaving line in gaining distance.

list: Inclination or heeling over of a ship to one side.

lizard: Line fitted with a thimble or thimbles and used as a leader for running rigging. A "traveling lizard" is fitted to the middle of a lifeboat's falls for use in taking up the slack when hoisting.

lock: Compartment in a canal for lowering or lifting vessels to different levels.

locker: Small metal or wooden stowage space; either a chest or closet.

log: Instrument for measuring a ship's speed through the water. Also, a short term for logbook.

longitudinal frame, longitudinal: The part of a ship's "skeleton" that runs fore and aft.

look alive: Admonishment to work faster or more diligently.

lookout: Seaman assigned duties involving watching and reporting to the officer of the deck any objects of interest; the lookouts are "the eyes of the ship."

LORAN (long-range aid to navigation): A navigational system that fixes the position of a ship by measuring the difference in the time of reception of two synchronized radio signals.

lubber's line: Line marked on inner surface of compass bowl to indicate direction of ship's bow.

lucky bag: Locker for stowage of personal gear found adrift.

magazine: Compartment; used for stowage of ammunition and explosives.

main battery: The largest-caliber guns carried by a warship.

main deck: Highest complete deck extending from stem to stern and from side to side.

mainmast: Second mast from the bow of a ship that has two or more masts. If a vessel is considered a ship but has one mast, that mast is considered the mainmast.

manhole: Round or oval hole cut in deck, bulkhead, or tank to provide access.

man-of-war: Fighting ship; warship.

manrope: Side rope to a ladder used as a handrail; rope used as a safety line anywhere on deck; rope hanging down on the side of a ship to assist in ascending the ship's side.

maritime: Pertaining to the sea.

mark: Call used in comparing watches, compass readings, or bearings; fathoms in a lead line that are marked. Also, a model or type of a piece of equipment, as Mark XIV torpedo.

marlinspike: Pointed iron instrument used in splicing line or wire.

marry: Placing two lines together, as in hoisting a boat; to sew together temporarily the ends of two lines for rendering through the block.

mast: Upright spar supporting signal yard and antennas in a naval ship. Also, the term applied to the hearing of cases of offense against discipline or for requests or commendations.

meet her: An order to shift the rudder in order to check the swing of a ship during change of course.

mess: To eat; group of crew members eating together; the compartment or location for the dining of a select group on board ship, such as the chief petty officer's mess.

messcook: Nonrated crew members assigned to clean the galley and mess deck and to help serve the meals.

mess deck: The compartment where enlisted personnel mess.

messenger: Light line used for hauling over a heavier rope or cable; for example, the messenger is sent over from the ship to the pier by the heaving line and then used to pull the heavy mooring lines across. Also, an enlisted crew member who runs errands for the officer of the deck.

midshipman: A student at the U.S. Naval Academy.

mind your rudder: Warning to helmsman to watch the ordered course carefully.

misfire: Powder charge that fails to fire when the trigger has been pulled.

monkey fist: A knot, with or without a weight enclosed, worked in the end of a heaving line to form a heavy ball to facilitate throwing the line.

mooring: Securing a ship to a pier, buoy, or another ship; or anchoring with two anchors.

mooring buoy: A large, well-anchored buoy to which one or more ships moor.

mooring line: One of the lines used for mooring a ship to a pier, wharf, or another ship.

Morse code: Code of dots and dashes used in radio and visual signaling.

motor launch: Large, sturdily built powerboat used for liberty parties and heavy workloads.

mousing: *Small stuff* for closing off a hook to prevent a load from slipping off.

muster: To assemble the crew; roll call.

nautical mile: 6,076 feet, or about a sixth longer than a statute mile.

naval stores: Oil, paint, turpentine, pitch, and other such items traditionally used for ships.

neap tide: Tide that twice during a lunar month rises and falls the least from the average level; that is, the tide with the least amount of change from high to low, occurring every twenty-eight days.

nest: Two or more vessels moored alongside one another; boat stowage in which one boat nests inside another.

net: A group of intercommunicating radio or landline communication stations; a barrier of steel mesh used to protect harbors and anchorages from torpedoes, submarines, or floating mines.

nothing to the right (left) of: Order to the helmsman not to let the ship go to the right (left) of the ordered course.

not under command: Said of a ship when disabled and uncontrollable.

nun buoy: Cone-shaped buoy used to mark channels; it is anchored on the right side, entering from seaward, and is painted red.

oakum: A caulking material made of old, tarred, hemp-fiber rope.

occulting light: A navigational aid light that blinks on an equal or greater amount of time than it blinks off; differs from flashing light, which blinks off more than it blinks on.

officer in charge: An enlisted person in charge of a unit where no officer is assigned. Officers in charge are not commanding officers but have many of the responsibilities of a commanding officer.

officer of the day (deck): The officer on watch in charge of the ship. When moored or anchored, the term "officer of the day" is generally employed.

officer of the watch: See *watch officer.*

oiler: A tanker; a vessel especially designed to carry and dispense fuel.

oiling: Taking fuel oil on board.

oil king: Petty officer in charge of fuel-oil storage.

Old Man: Seaman's term for the captain of a ship or other naval activity.

on board: Joined or embarked. "Onboard" is one word only when used in an adjectival sense: "an onboard computer."

on the bow: Bearing of an object somewhere within 45° to either side of the bow.

on the quarter: Bearing of an object somewhere astern of the ship, 45° to either side of the stern.

OOD: Officer of the deck (day).

order: Directive telling what to do but leaving the method to the discretion of the person ordered. See also *command.*

orlop: Partial deck below the lower deck; also, the lowest deck in a ship having four or more decks.

Oscar: Traditionally, the name given to the dummy employed for ship's "Man overboard" drills; the flag hoisted by a ship to indicate a man/woman overboard.

outboard: Toward the side of the vessel, or outside the vessel entirely.

out of trim: To carry a list or to be down by the head or stern.

overhand: Projection of ship's bow or stern beyond the stem or sternpost.

overhaul: To separate the blocks of a tackle; to overtake a vessel; to clear or repair anything for use.

overhead: On a ship, equivalent to the ceiling of a building ashore; ships have overheads rather than ceilings.

pad eye: Metal eye permanently secured to deck or bulkhead.

painter: A line in the bow of a boat for towing or making fast.

palm and needle: Sailor's thimble made of leather, and a large needle, used for sewing heavy canvas or leather.

parbuckle: Device for raising or lowering a heavy object along an inclined or vertical surface. A bight of rope is thrown around a secure fastening at the level to which the object is to be raised or lowered. The two ends of the rope are then passed under the object, brought all the way over it, and led back toward the bight. The two ends are then hauled or slackened together to raise or lower the object, the object itself acting as a nonmovable pulley.

parceling: Wrapping a rope spirally with long strips of canvas, following the lay of the rope and overlapping like the shingles on a roof as of a rope or line.

part: To break, as of a rope or line.

passageway: Corridor or hallway on a ship.

pass a line: To carry or send a line to or around an object, or to reeve through and make fast.

pass the word: To repeat an order or information to all hands.

pay: To fill the seams of a wooden vessel with pitch or other substance.

pay off: To turn the bow away from the wind.

pay out: To increase the scope of anchor cable; to ease off or slack a line.

peak: Topmost end of the gaff; from this point the ensign is flown while the ship is under way.

peak tank: Tank in the bow or stern of a ship; usually for water ballast.

pelican hook: Hinged hook held in place by a ring; when the ring is knocked off, the hook swings open.

pelorus: Navigational instrument used in taking bearings that consists of two sight vanes mounted on a hoop revolving about a dumb compass or a gyro repeater.

pendant: Length of rope with a block or thimble at the end.

pennant: Three- or four-sided flag that tapers off toward the end.

periscope: Optical instrument used to provide a raised line of vision where it may not be practical or possible.

pier: A harbor structure projecting out into the water with sufficient depth alongside to accommodate vessels.

pigstick: A small spar that projects above the top of the mainmast; commission pennants are usually flown on this.

pile: Pointed spar driven into the bottom and projecting above the surface of water; when driven at the corners of a pier or wharf, they are termed "fender piles."

pilot: An expert who comes on board ships in harbors or dangerous waters to advise the captain as to how the ship should be conned; also, the person at the controls of an aircraft.

pintles: Pivot pins on which a rudder turns.

pipe: An announcement over the ship's general announcing system, as in "What was that last pipe?" See also *boatswain's pipe.*

pipe down: An order to keep silent; also, used to dismiss the crew from an evolution.

pipe the side: Ceremony at the gangway in which side boys are drawn up and the boatswain's pipe is blown when a high-ranking officer or distinguished visitor comes on board.

pitch: The forward heaving and plunging motion of a vessel at sea.

pivot point: Point in a ship about which the ship turns.

plan of the day: Schedule of day's routine and events ordered by the executive officer; published daily on board ship.

platform deck: Partial deck below lowest complete deck; called first, second, and so on, from the top where there is more than one.

Plimsoll mark: A mark on the side of merchant ships to indicate allowed loading depths.

pointer: Member of gun crew who controls vertical elevation of a gun in aiming at a target; that is, the pointer positions the gun up and down. See also *trainer.*

pollywog: Person who has never crossed the line (equator).

poop deck: Partial deck at the stern over the main deck.

port: Left side of the ship facing forward; a harbor; an opening in the ship's side. The usual opening in the ship's side for light and air is also a port. The glass set in a brass frame that fits against it is called a port light.

preventer: Line used for additional safety and to prevent loss of gear under heavy strain or in case of accident.

privileged vessel: See *stand-on vessel.*

prolonged blast: Blast on the ship's whistle of from four to six seconds' duration.

protective deck: Deck fitted with heaviest protective plating.

punt: Rectangular, flat-bottomed boat usually used for painting and other work around waterline of a ship.

purchase: General term for any mechanical arrangement of tackle that increases the force applied by a combination of pulleys.

pyrotechnics: Chemicals, ammunition, or fireworks that produce smoke or lights of various colors and types.

quadrant: Metal fitting on rudder head to which steering ropes are attached.

quadrantal correctors or spheres: Two iron balls secured at either side of the binnacle; these help compensate for the ship's magnetic effect on compass.

quarter: That part of ship's side near the stern.

quarterdeck: That part of the main (or other) deck reserved for honors and ceremonies and as the station of the officer of the day in port.

quartermaster of the watch: Assists the officer of the deck in navigating when the ship is under way; a ship's quarterdeck watchstander in port.

quarters: Living space; assembly of the crew; all hands assembled at established stations for muster, drills, or inspections.

quay: A wharf; a landing place for receiving and discharging cargo. Pronounced "key."

rack: A sailor's bed.

radar (radio detection and ranging): The principle and method whereby objects are located by radio waves; a radio wave is transmitted, reflected by an object, received, and illuminated by an oscilloscope or cathode ray screen.

radio detection finder: Apparatus for taking bearings on the source of radio transmission.

rail: Top pipe of the lifeline pipes that extend along various outboard sections of weather decks; uppermost edge of a bulwark.

rail loading: Loading a davit- or crane-supported boat while it is swung out and even with the deck.

rake: Angle of a vessel's masts and stacks from the vertical or sloped end of a barge.

rakish: Having a rake to the masts; smart; speedy in appearance.

range: Distance in yards from ship to target; two or more objects in line to indicate direction.

rank: Grade of official standing of commissioned and warrant officers.

rate: Grade of official standing of enlisted personnel. A rate identifies an enlisted member by pay grade or level of advancement with a *rating*; a rate reflects levels of aptitude, training, experience, knowledge, skill, and responsibility. See also *rating*.

rat guard: A sheet-metal disk constructed in conical form with a hole in the center and slit from the center to the edge. It is installed over the mooring lines to prevent rats from boarding ship from the shore over the mooring lines.

rating: Name given to an enlisted occupational specialty that consists of specific aptitudes, training, experience, knowledge, and skills.

ratline: Short length of small stuff running horizontally across shrouds, used for a step.

recognition: Process of determining friendly or enemy character of a ship, plane, or other object or person.

reducer: Metal fitting between fire-main outlet and hose coupling of smaller diameter.

reef: Chain or ridge of rocks, coral, or sand in shallow water.

reeve: To pass the end of a rope through any lead, such as a sheave or fairlead.

relative bearing: Bearing or direction of an object in degrees in relation to the bow of the ship. The bow of the ship is taken as 000° and an imaginary circle is drawn clockwise around the ship; objects are then reported as being along a line of bearing through any degree division of this circle.

relieving (the watch, the duty, and so on): To take over the duty and responsibilities, such as when one sentry relieves another. Those who relieve are reliefs.

request mast: Mast held by captain or executive officer to hear special requests for leave, liberty, and so on.

rig: General description of a ship's upper works; to set up, fit out, or put together.

rigging: General term for all ropes, chains, and gear used for supporting and operating masts, yards, booms, gaffs, and sails. Rigging is of two kinds: standing rigging, or lines that support but ordinarily do not move, and running rigging, or lines that move to operate equipment.

rig ship for visitors: Word passed as a warning to all hands to have the ship and their persons in neat order for expected visitors.

riser: Vertical branch pipe; that is, a pipe going up and down between decks and having branch connections or offshoots.

roll: The side-to-side motion of a ship at sea.

rope: General term for cordage more than one inch in diameter. If smaller, it is known as cord, twine, line, or string. It is constructed by twisting fibers or metal wire. The size is designated by the diameter (for wire rope) or by the circumference (for fiber rope). The length is given in fathoms or feet.

ropeyarn Sunday: A time for repairing clothing or other personal gear.

round line: Three-stranded, right-handed small stuff; used for fine seizing.

rouse in: To haul in, especially by manpower.

rudder: A flat, vertical, mobile structure at the stern of a vessel; used to control the vessel's heading.

ruffles: Roll of the drum used in rendering honors.

Rules of the Road: Regulations enacted to prevent collisions between watercraft.

runner: Line fastened at one end to a fixed object, such as an eyebolt, on deck and rove through a single block. It has an eye on its other end to which a tackle is clapped on. The term is also loosely applied to any line rove through a block.

running bowline: Bowline made over the standing part of its own rope so that it forms a free-sliding noose.

running lights: Lights required by law to be shown by ship or plane when under way between sunset and sunrise.

salvage: To save a ship or cargo from danger; to recover a ship or cargo from disaster and wreckage.

Samson post: In small craft, a single bitt amidships.

scope: Length of anchor cable out.

scow: Large, open, flat-bottomed boat for transporting sand, gravel, mud, and so on.

screw: The propeller; the rotating, bladed device that propels a vessel through the water.

scullery: Compartment for washing and sterilizing eating utensils.

scupper: Opening in the side of a ship to carry off water.

scuttle: A small opening through hatch, deck, or bulkhead to provide access; a similar hole in side or bottom of ship; a cover for such an opening; to sink a ship intentionally by boring holes in the bottom or by opening seacocks.

scuttlebutt: Drinking fountain. Also, a rumor, usually of local importance.

sea anchor: See *drogue*.

seabag: Large canvas bag for stowing a sailor's gear and clothing.

sea chest: Intake between ship's side and sea valve or seacock.

seacock: Valve in pipe connected to the sea; a vessel may be flooded by opening the seacocks.

sea ladder: Rope ladder, usually with wooden steps, for use over the side.

sea lawyer: Enlisted member who likes to argue; usually one who thinks the regulations and standing orders can be twisted to favor his or her personal inclinations.

sea marker: Dye for brightly coloring the water to facilitate search and rescue.

sea painter: A long line running from well forward on the ship and secured by a toggle over the inboard gunwale in the bow of a boat.

seaworthy: Capable of putting to sea and meeting usual sea conditions.

second deck: Complete deck below the main deck.

section: A unit of a shipboard division.

secure: To make fast; to tie; an order given on completion of a drill, exercise, or evolution that means to withdraw from the corresponding stations and duties.

secure for sea: Extra prescribed lashings on all movable objects.

seize: To bind with small rope.

seizing stuff: Small cordage for seizing.

semaphore: Code indicated by the position of the arms; hand flags are used to increase readability.

sennet, sennit: Ornamental, braided, fancy ropework formed by plaiting (interweaving) a number of strands.

service stripes: Diagonal stripes on the lower left sleeve of an enlisted member's uniform denoting periods of enlistment. Usually referred to as "hash marks."

serving: Additional protection over parceling consisting of continuous round turns of small stuff.

serving mallet: Wooden mallet with a groove cut lengthwise in its head; used for serving large rope.

set: Direction of the leeway of a ship or of a tide or current.

set taut: An order to take in the slack and take a strain on running gear before heaving it in.

set the course: To give the helmsman the desired course to be steered.

set the watch: The order to station the first watch.

shackle: U-shaped piece of iron or steel with eyes in the ends through which a bolt passes to close the U.

shaft alley: Spaces within a ship surrounding the propeller shaft.

shakedown: Cruise of newly commissioned ship to test and adjust all machinery and equipment and to train the crew as a working unit.

sheave: Wheel of a block over which the rope reeves. Pronounced "shiv."

sheer: Longitudinal upward curve of a deck; amount by which the deck at the bow is higher than the deck at the stern. Also, a sudden change of course.

sheer off: To turn suddenly away.

shell: Casing of a block.

shellback: Person who has crossed the equator and been initiated.

shift the rudder: An order to swing the rudder an equal distance in the opposite direction.

ship: A general term for large oceangoing craft or vessels; to enlist or reenlist, in other words, to "ship over."

ship's company: All the enlisted members and officers serving in, and attached to, a ship; *all hands.*

shipshape: Neat, orderly.

shore patrol: Naval personnel detailed to maintain discipline, to aid local police in handling naval personnel on liberty or leave, and to assist naval personnel in difficulties ashore.

shore up: To prop up.

short blast: Whistle, horn, or siren blast of about one second's duration.

short stay: When anchor chain has been hauled in until amount of chain out is only slightly greater than depth of water and ship is riding almost directly over the anchor.

shot: Short length of chain, usually fifteen fathoms.

shove off: To leave; an order to a boat to leave a landing or a ship's side.

shroud: Side stay of hemp or wire running from masthead to rail to support the mast.

sick bay: Ship's hospital or dispensary.

side boys: Nonrated personnel manning the side when visiting senior officers or distinguished visitors come on board.

side lights: Red and green running lights carried on port and starboard sides, respectively.

single up: To reduce the number of mooring lines out to a pier preparatory to sailing; that is, to leave only one easily cast off line in each place where mooring lines were doubled up for greater security.

sister hooks: Twin hooks on the same swivel or ring; closed, they form an eye.

skeg: Continuation of the keel aft to protect the propeller.

skids: Beams fitted over decks for stowage of heavy boats.

skivvies: Slang for underclothing.

slack: The part of a line hanging loose; to ease off; state of the tide when there is no horizontal motion.

slings: Fittings for hoisting a boat or other heavy lift by crane or boom; they consist of a metal ring with four pendants. Two of these pendants are for athwartships steadying lines, the other two shackle to chain bridles permanently bolted to the keel of the boat.

slip: To let go by unshackling, as an anchor cable; space between two piers; waste motion of a propeller.

small craft: Generally, all vessels less than small-ship size.

small stuff: Small cordage designated by the number of threads (nine-thread, twelve-thread, and so on) or by special names, such as marline, ratline stuff.

smart: Snappy, seamanlike.

smoking lamp: A lamp on board old-time ships used by personnel to light their pipes; now used in the phrase, "The smoking lamp is lit (or out)" to indicate when smoking is allowed or forbidden throughout the ship or a specified area therein.

snipes: Slang for members of the engineering department.

sonar (sound navigation and ranging): Device for locating objects under water by emitting vibrations similar to sound and for measuring the time taken for these vibrations to bounce back from anything in their path.

soogie: To wash or scrub. In the Coast Guard, scouring powder is called soogie powder.

SOPA: Senior officer present afloat.

sound: To measure depth of water by means of a lead line. Also, to measure depth of liquids in oil tanks, voids, blisters, and other compartments or tanks.

sound-powered phone: Shipboard telephone powered by voice alone.

span: Line made fast at both ends with a purchase hooked to its bight; wire rope located between davit heads and set up by a turnbuckle. Also, to bridge or reach across.

spanner: A tool for coupling hoses.

spar: Steel or wood pole serving as a mast, boom, gaff, pile, and so on.

spar buoy: Long, thin, wooden spar used to mark channels.

speed cone: Cone-shaped, bright yellow signal used when steaming in formation to indicate engine speeds.

speed light: White or red light mounted high on a ship to indicate changes in speed at night.

spitkit: Derisive term for small, unseaworthy vessel; slang for a sailor's toilet kit.

splice: To join two lines by tucking the strands of each into the other.

splinter screen: Protective plating around a gun mount.

spring: Mooring line leading at an angle of about 45° off centerline of vessel; to turn a vessel with a line.

spur shore: Wooden spar used to hold vessel clear of a pier.

squall: A sudden storm-like gust of wind and rain.

square away: To get things settled down or in order.

square knot: An ancient and simple binding knot used to secure a rope or a line around an object.

squeegee: Drier for decks and glass made of a flat piece of wood with a rubber blade and a long wooden handle.

stack: Ship's smoke pipe. See *funnel*.

stadimeter: Instrument for measuring distance from an object.

stage: Platform rigged over ship's side for painting or repair work.

stanchion: Wood or metal upright used as a support.

stand: Condition of tide when there is no vertical motion. See also *slack*.

standard speed: Speed set as basic speed by officer in command of a unit.

stand by: Preparatory order meaning "Get ready" or "Prepare to."

stand-on vessel: The vessel with the right of way. Formerly called the privileged vessel. See also *give-way vessel*.

starboard: Right side of a ship looking forward.

starboard pennant: A trapezoid-shaped green-and-white pennant indicating that the commanding officer of the ship from which it flies is the senior officer present afloat (SOPA). Also called the SOPA pennant.

station keeping: The art of keeping a ship in its proper position in a formation of ships.

stay: Piece of rigging, either wire or fiber, used to support a mast.

steady: Order to helmsman to hold ship on course.

steerageway: Slowest speed at which a ship can be steered.

stem: Upright post or bar at most forward part of the bow of a ship or boat. It may be a casting, forging, welding, or made of wood.

stern: After part of a ship.

sternway: Backward movement of a ship.

stopper: Short length of rope or chain firmly secured at one end and used in securing or checking a running line.

stove: Broken in; crushed in.

stow: To put gear in its proper place.

strake: Continuous line of planks or plates running the length of a vessel.

strand: Part of a line or rope made up of yarns.

striker: Enlisted member in training for a particular rating.

strip ship: To prepare ship for battle action by getting rid of any unnecessary gear.

strongback: Spar lashed to a pair of boat davits; acts as a spreader for the davits and provides a brace for more secure stowage of a lifeboat at sea.

superstructure: All equipment and fittings, except armament, extending above the hull.

superstructure deck: Partial deck higher than the main, forecastle, and poop decks, and not extending to the ship's sides.

survey: Examination by authorized competent personnel to determine whether a piece of gear, equipment, stores, or supplies should be discarded or retained.

swab: A rope or yarn mop.

swamp: To sink by filling with water.

swash plates: Plates pierced with a number of holes fixed in tanks to prevent liquids from moving too violently when ship rolls or pitches.

swing ship: Moving the ship through the compass points to check the magnetic compass on different headings and make up a deviation table.

swivel: Metal link with an eye at one end fitted to revolve freely and thus keep turns out of a chain.

tachometer: Mechanical device indicating shaft revolutions.

tackle: Arrangement of ropes and blocks to give mechanical advantage; a purchase, that is, a rig of lines and pulleys to increase available hauling force. Pronounced "take-el."

tackline: Short length of line used to separate flags in a hoist.

taffrail: A rail at the stern of a ship.

taffrail log: Device that indicates the speed of the ship through the water. It is trailed on a line from the taffrail and consists of a rotator and a recording instrument.

take a turn: To pass a turn around a cleat, bitts, or bollard with a line and hold on.

taps: Lights out for the night.

tarpaulin: Heavy canvas used as protective covering.

task force: Temporary grouping of units under one commander formed for purpose of carrying out a specific operation or mission.

taut: With no slack. Also, strict as to discipline and orderliness.

telephone talker: Crew member who handles sound-powered phone for drills and evolutions.

tend: Direction the cable leads when ship is anchored. Also, to oversee (for example, to tend to your duties).

tender: An auxiliary vessel that supplies and repairs ships or aircraft.

thimble: Iron ring grooved on outside for rope grommet.

thwart: Crosspiece used as a seat for a boat.

tide: The vertical rise and fall of the sea caused by gravitational effect of sun and moon.

tier: To stow cable in chain locker.

tiller: Short piece of metal or wood fitted into the head of the rudder and used to turn a boat's rudder.

toggle: Wooden or metal pin slipped into a becket; furnishes a rapid release.

tompion, tampion: Plug placed in muzzle of gun to keep dampness and foreign objects out. Pronounced "tompkin."

top: Platform at top of mast; to "top a boom" is to lift up its end.

topping lift: Line used for topping a boom and taking its weight.

topside: Above decks.

tow: To pull through the water; vessels so towed. The usual towing vessels in naval talk are tugs, not towboats.

towing lights: Special white lights displayed by a towing vessel at night.

track: Path of a vessel.

tracking: Keeping a gun directed at a moving target.

trades: Generally steady winds of the tropics that blow toward the equator; northeast in the Northern Hemisphere and southeast in the Southern Hemisphere.

train: Auxiliary vessels; also, to move a gun horizontally onto a target.

trainer: Gun-crew member who controls horizontal movement of gun in aiming it at a target.

transom: Athwartships piece bolted to sternpost; planking across stern of square-sterned boat.

trice up: To hitch up or hook up, such as to trice up a rack.

trick: Period of time a helmsman is at the wheel, as "to take a trick at the wheel."

trim: Angle to the horizontal at which a ship rides; that is, how level the ship sits in the water; shipshape.

trimming tanks: Tanks used for water ballast. By flooding or emptying these tanks, the ship may be trimmed, that is, balanced in water at various angles.

truck: Flat, circular piece secured at top of mast or at top of flagstaff and jackstaff. Also, uppermost part of a mast.

trunk: A vertical shaft.

tumble home: Amount vessel's sides come in at the deck from the perpendicular.

turbine: High-speed rotor turned by steam or other hot gases.

turnbuckle: Metal appliance consisting of a thread and screw capable of being set taut or slacked and used for setting up standing rigging.

turn to: An order to begin work.

turret: Heavily armored housing containing a grouping of main battery guns. It extends downward through decks and includes ammunition-handling rooms and hoists. See also *gun mount*.

two-blocked: When two blocks of a tackle have been drawn as closely together as possible.

veer: To let anchor cable, line, or chain run out by its own weight. Also, when the wind changes direction clockwise or to the right, it is said to veer.

very well: Reply of a senior (or officer) to a junior (or enlisted person) to indicate that information given is understood, or that permission is granted.

void: Empty compartment below decks.

wake: The track left in the water behind a ship.

wale shores: Stout timbers of various lengths used to prevent a dry-docked vessel from toppling over. They are rigged between the vessel's sides and the sides of the dry dock.

walk back: An order to keep the gear in hand but to walk back with it toward the belaying point.

wall knot: Knot made at the end of a rope by back-splicing the ends, thus forming a knot. Used for finishing off seizing and, on the end of a rope, to prevent the rope from unreeving.

wardroom: Officer's mess and lounge on board a ship.

warp: To move a vessel by a line or laid-out anchor, "Warp the ship into the slip." See also *kedge*.

watch: A period of duty, usually of four hours' duration. Watches call for a variety of duties and are of many types: quarterdeck watch, messenger watch, signal watch, radio watch, and so on.

watchcap: Knitted wool cap worn in cool or cold weather; canvas cover placed over a stack when not in use.

watch officer: An officer regularly assigned to duty in charge of a watch or of a portion thereof; for example, the officer of the deck or the engineering officer of the watch.

Watch, Quarter, and Station Bill: A large chart showing every crew member's location and duties in a Coast Guard or Navy ship's organization of drills and evolutions.

water breaker: Drinking-water cask or container carried in boats.

waterline: Point to which the ship sinks in water; line painted on hull showing point to which ship sinks in water when properly trimmed.

waterlogged: Filled or soaked with water but still afloat.

watertight integrity: System of keeping ship afloat by maintaining watertightness.

waterway: Gutter at side of ship's deck to carry water to scuppers.

weather: Exposed to wind and rain; to the windward, for example, "to face the weather," or "to weather a storm."

weather cloth: Canvas spread for protection from wind and weather.

weather deck: Portion of main, forecastle, poop, and upper deck exposed to weather.

weather eye: To "keep a weather eye" is to be on the alert.

weigh: To lift the anchor off the bottom.

well deck: A low weather deck.

wharf: Harbor structure alongside which vessels moor. A wharf generally is built along the water's edge; a *pier* extends well out into the harbor.

wheelhouse: Pilothouse; the topside compartment where on most ships the officer of the deck, helmsman, quartermaster of the watch, and so on stand their watches.

where away: An answering call requesting the location of object sighted by lookout.

whipping: Keeping the ends of a rope from unlaying by wrapping with turns of twine and tucking the ends.

wildcat: Sprocket wheel on windlass for taking the links of the chain cable.

winch: Hoisting engine secured to the deck; used to haul lines by turns around a horizontally driven drum or gypsy.

windlass: Anchor engine used for heaving in the anchor.

wind scoop: Metal scoop fitted into a port to direct air into the ship for ventilation.

wind ship: To turn a ship end for end, usually with lines at a pier. Pronounced "wined."

windward: Into the wind; toward the direction from which wind is blowing; opposite of *leeward*.

wire rope: Rope made of wire strands, as distinguished from *fiber rope*.

with the sun: In clockwise direction; the proper direction in which to coil a line; right-handed.

work a ship: To handle ship by means of engines and other gear; for example, to work a ship into a slip using engines, rudder, and lines to dock.

worming: Filling the lays of a wire rope preparatory to parceling and serving.

yard: Spar attached at the middle of a mast and running athwartships, used as a support for signal halyards or signal lights; also, a place used for ship-building and as a repair depot, such as the Coast Guard Yard, Baltimore, Maryland.

yardarm: Either side of a yard.

yardarm blinker: Signal light mounted above the end of a yardarm and flashed on and off to send messages.

yarn: Twisted fibers used for rough seizings, which may be twisted into strands; also, a story, as to "spin a yarn," meaning to tell a story that is not necessarily true.

yaw: Zigzagging motion of a vessel as it is carried off its heading by strong overtaking seas. This motion swings the ship back and forth across the intended course.

yoke: The piece fitting across the head of a boat's rudder, to the end of which steering lines are attached.

APPENDIX C

Allied Naval Signal Pennants and Flags and International Alphabet Flags

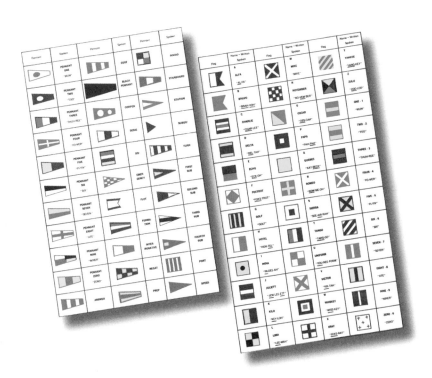

Pennant	Spoken	Pennant	Spoken	Pennant	Spoken
	PENNANT ONE "WUN"		CODE		SQUAD
	PENNANT TWO "TOO"		BLACK PENNANT		STARBOARD
	PENNANT THREE "THUH-REE"		CORPEN		STATION
	PENNANT FOUR "FO-WER"		DESIG		SUBDIV
	PENNANT FIVE "FI-YIV"		DIV		TURN
	PENNANT SIX "SIX"		EMER-GENCY		FIRST SUB
	PENNANT SEVEN "SEVEN"		FLOT		SECOND SUB
	PENNANT EIGHT "ATE"		FORMA-TION		THIRD SUB
	PENNANT NINE "NINER"		INTER-ROGATIVE		FOURTH SUB
	PENNANT ZERO "ZERO"		NEGAT		PORT
	ANSWER		PREP		SPEED

Flag	Name — Written / Spoken	Flag	Name — Written / Spoken	Flag	Name — Written / Spoken
	A ALFA "AL-FA"		M MIKE "MIKE"		Y YANKEE "YANG-KEY"
	B BRAVO "BRAH-VOH"		N NOVEMBER "NO-VEM-BER"		Z ZULU "ZOO-LOO"
	C CHARLIE "CHAR-LEE"		O OSCAR "OSS-CAH"		ONE - 1 "WUN"
	D DELTA "DEL-TAH"		P PAPA "PAH-PAH"		TWO - 2 "TOO"
	E ECHO "ECK-OH"		Q QUEBEC "KAY-BECK"		THREE - 3 "THUH-REE"
	F FOXTROT "FOKS-TROT"		R ROMEO "ROW-ME-OH"		FOUR - 4 "FO-WER"
	G GOLF "GOLF"		S SIERRA "SEE-AIR-RAH"		FIVE - 5 "FI-YIV"
	H HOTEL "HOH-TEL"		T TANGO "TANG-GO"		SIX - 6 "SIX"
	I INDIA "IN-DEE-AH"		U UNIFORM "YOU-NEE-FORM"		SEVEN - 7 "SEVEN"
	J JULIETT "JEW-LEE-ETT"		V VICTOR "VIK-TAH"		EIGHT - 8 "ATE"
	K KILO "KEY-LOH"		W WHISKEY "WISS-KEY"		NINE - 9 "NINER"
	L LIMA "LEE-MAH"		X XRAY "ECKS-RAY"		ZERO - 0 "ZERO"

APPENDIX D

IALA Maritime Buoyage System
Lateral Marks Region B

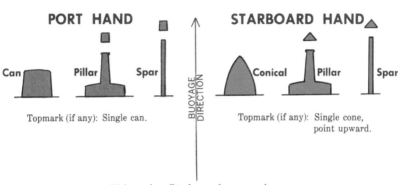

PORT HAND

Can Pillar Spar

BUOYAGE DIRECTION

STARBOARD HAND

Conical Pillar Spar

Topmark (if any): Single can.

Topmark (if any): Single cone, point upward.

Lights, when fitted, may have any phase
characteristic other than that used
for preferred channels.

Examples
Quick Flashing
Flashing
Long Flashing
Group Flashing

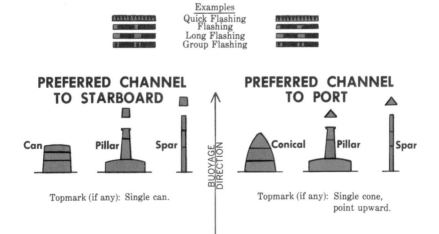

**PREFERRED CHANNEL
TO STARBOARD**

Can Pillar Spar

BUOYAGE DIRECTION

**PREFERRED CHANNEL
TO PORT**

Conical Pillar Spar

Topmark (if any): Single can.

Topmark (if any): Single cone, point upward.

Lights, when fitted, are composite
group flashing Fl (2+1).

APPENDIX E

Storm Warning Signals

DAYTIME SIGNALS

SMALL CRAFT · GALE · WHOLE GALE · HURRICANE

NIGHT SIGNALS

SMALL CRAFT · GALE · WHOLE GALE · HURRICANE

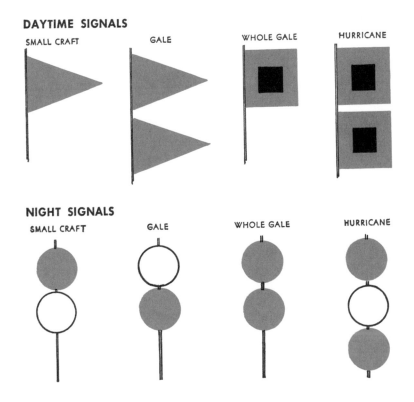

APPENDIX F

Beaufort Wind Scale

Developed in 1805 by RADM Sir Francis Beaufort, U.K. Royal Navy, the Beaufort Wind Scale is a system of estimating and reporting wind speed. Estimation of the wind's speed and a constant awareness of the state of the sea are important responsibilities of a mariner. Sea state can have a significant influence upon a ship. Course changes, leeway, dead reckoning, engine speeds, and even rudder angles are affected by wind and sea.

Force	Wind (Knots)	WMO★ Classification	Appearance of Wind Effects on Water
0	Less than 1	Calm	Sea surface smooth and mirror-like
1	1–3	Light air	Scaly ripples, no foam crests
2	4–6	Light breeze	Small wavelets, crests glassy appearance, no breaking
3	7–10	Gentle breeze	Large wavelets, crests begin to break, scattered whitecaps
4	11–16	Moderate breeze	Small waves 1–4 ft. becoming longer, numerous whitecaps
5	17–21	Fresh breeze	Moderate waves 4–8 ft. taking longer form, many whitecaps, some spray
6	22–27	Strong breeze	Larger waves 8–13 ft. form, whitecaps common, more spray
7	28–33	Near gale	Sea heaps up, waves 13–19 ft., white foam streaks off breakers
8	34–40	Gale	Moderately high (18–25 ft.) waves of greater length, edges of crests begin to break into spindrift, foam blown in streaks
9	41–47	Strong gale	High waves (23–32 ft.), sea begins to roll, dense streaks of foam, spray may reduce visibility
10	48–55	Storm	Very high waves (29–41 ft.) with overhanging crests, sea white with densely blown foam, heavy rolling, lowered visibility

Force	Wind (Knots)	WMO* Classification	Appearance of Wind Effects on Water
11	56–63	Violent storm	Exceptionally high (37–52 ft.) waves, foam patches cover sea, visibility more reduced
12	64+	Hurricane	Air filled with foam, waves more than 45 ft., sea completely white with driving spray, visibility greatly reduced

Source: http://www.spc.noaa.gov/faq/tornado/beaufort.html.
* WMO stands for the World Meteorological Organization.

Index

About the Editor

Jim Dolbow received his commission in the U.S. Coast Guard Reserve on 31 July 2002. Recalled to active duty in 2003 to support Operation Noble Eagle, Dolbow served as a watchstander in the Joint Operations Command Center in Washington, D.C., coordinator of a round of port security grant applications, and leader for a harbor patrol team. Activated again in 2005, he took part in post–Hurricane Katrina operations and served as a legislative affairs analyst on the staff of ADM Thad Allen, USCG, who led those efforts.

A lifelong student of national security affairs, Dolbow has studied and/or trained at the Institute of World Politics (a graduate school in national security in Washington, D.C.), the U.S. Naval War College, Joint Special Operations University, and the U.S. Air Force Special Operations School. A frequent writer, he has been published in *Proceedings*, *National Defense*, *Washington Post*, *Washington Times*, *Sea Power*, *Navy Times*, and *Armed Forces Journal International*. In 2013 he was awarded First Prize along with CAPT Jim Howe, USCG (Ret.) as *Proceedings* Authors of the Year.

Dolbow is a native of Penns Grove, New Jersey, and is a life member of the U.S. Naval Institute and a former congressional defense staffer. He served on the editorial board of the U.S. Naval Institute from 2005 to 2010.

He is a senior acquisitions editor for professional development books at the Naval Institute Press and Fleet Professor of Theater Security Decision Making for the U.S. Naval War College's Fleet Seminar Program in Washington, D.C. He is the proud owner of Conan, a rescued Black and Tan Coonhound.

Suggestions and feedback regarding this book are welcomed. Email him at jdolbow@usni.org.